Springer Series in
Surface Sciences

34

Editor: Robert Gomer

Springer Series in **Surface Sciences**

Editors: G. Ertl, R. Gomer, and D.L. Mills — Managing Editor: H.K.V. Lotsch

1 **Physisorption Kinetics**
By H.J. Kreuzer, Z.W. Gortel

2 **The Structure of Surfaces**
Editors: M.A. Van Hove, S.Y. Tong

3 **Dynamical Phenomena at Surfaces, Interfaces and Superlattices**
Editors: F. Nizzoli, K.-H. Rieder, R.F. Willis

4 **Desorption Induced by Electronic Transitions, DIET II**
Editors: W. Brenig, D. Menzel

5 **Chemistry and Physics of Solid Surfaces VI**
Editors: R. Vanselow, R. Howe

6 **Low-Energy Electron Diffraction**
Experiment, Theory and Surface Structure Determination
By M.A. Van Hove, W.H. Weinberg, C.-M. Chan

7 **Electronic Phenomena in Adsorption and Catalysis**
By V.F. Kiselev, O.V. Krylov

8 **Kinetics of Interface Reactions**
Editors: M. Grunze, H.J. Kreuzer

9 **Adsorption and Catalysis on Transition Metals and Their Oxides**
By V.F. Kiselev, O.V. Krylov

10 **Chemistry and Physics of Solid Surfaces VII**
Editors: R. Vanselow, R. Howe

11 **The Structure of Surfaces II**
Editors: J.F. van der Veen, M.A. Van Hove

12 **Diffusion at Interfaces: Microscopic Concepts**
Editors: M. Grunze, H.J. Kreuzer, J.J. Weimer

13 **Desorption Induced by Electronic Transitions, DIET III**
Editors: R.H. Stulen, M.L. Knotek

14 **Solvay Conference on Surface Science**
Editor: F.W. de Wette

15 **Surfaces and Interfaces of Solids**
By H. Lüth

16 **Atomic and Electronic Structure of Surfaces**
Theoretical Foundations
By M. Lannoo, P. Friedel

17 **Adhesion and Friction**
Editors: M. Grunze, H.J. Kreuzer

18 **Auger Spectroscopy and Electronic Structure**
Editors: G. Cubiotti, G. Mondio, K. Wandelt

19 **Desorption Induced by Electronic Transitions DIET IV**
Editors: G. Betz, P. Varga

20 **Scanning Tunneling Microscopy I**
General Principles and Applications to Clean and Adsorbate-Covered Surfaces
Editors: H.-J. Güntherodt, R. Wiesendanger

21 **Surface Phonons**
Editors: W. Kress, F.W. de Wette

22 **Chemistry and Physics of Solid Surfaces VIII**
Editors: R. Vanselow, R. Howe

23 **Surface Analysis Methods in Materials Science**
Editors: D.J. O'Connor, B.A. Sexton, R. St. C. Smart

24 **The Structure of Surfaces III**
Editors: S.Y. Tong, M.A. Van Hove, K. Takayanagi, X.D. Xie

25 **NEXAFS Spectroscopy**
By J. Stöhr

26 **Semiconductor Surfaces and Interfaces**
By W. Mönch

27 **Helium Atom Scattering from Surfaces**
Editor: E. Hulpke

28 **Scanning Tunneling Microscopy II**
Further Applications and Related Scanning Techniques
Editors: R. Wiesendanger, H.-J. Güntherodt

29 **Scanning Tunneling Microscopy III**
Theory of STM and Related Scanning Probe Methods
Editors: R. Wiesendanger, H.-J. Güntherodt

30 **Concepts in Surface Physics**
By M.C. Desjonquères, D. Spanjaard

31 **Desorption Induced by Electronic Transitions DIET V**
Editors: A.R. Burns, E.B. Stechel, D.R. Jennison

32 **Scanning Tunneling Microscopy and Related Techniques**
By Ch. Bai

33 **Adsorption on Ordered Surfaces of Ionic Solids**
Editors: H.J. Freund, E. Umbach

34 **Surface Reactions**
Editor: R.J. Madix

R.J. Madix (Ed.)

Surface Reactions

With Contributions by
D.J. Auerbach B.E. Bent L.H. Dubois
C.M. Friend R.M. Lambert R.J. Madix
H.A. Michelsen C.B. Mullins R.G. Nuzzo
R.M. Ormerod C.T. Rettner J.T. Roberts
W.H. Weinberg

With 150 Figures

Springer-Verlag
Berlin Heidelberg New York London Paris
Tokyo Hong Kong Barcelona Budapest

Sep/AE
PHYS

Professor R.J. Madix
Department of Chemical Engineering
Stanford University
Stanford, CA 94305-5025, USA

Series Editors

Professor Dr. Gerhard Ertl
Fritz-Haber-Institut der Max-Planck-Gesellschaft, Faradayweg 4–6,
D-14195 Berlin, Germany

Professor Robert Gomer, Ph.D.
The James Franck Institute, The University of Chicago, 5640 Ellis Avenue,
Chicago, IL 60637, USA

Professor Douglas L. Mills, Ph.D.
Department of Physics, University of California,
Irvine, CA 92717, USA

Managing Editor: Dr. Helmut K.V. Lotsch
Springer-Verlag, Tiergartenstrasse 17,
D-69121 Heidelberg, Germany

ISBN 3-540-57605-3 Springer-Verlag Berlin Heidelberg New York
ISBN 0-387-57605-3 Springer-Verlag New York Berlin Heidelberg

Library of Congress Cataloging-in-Publication Data. Surface reactions/Robert J. Madix (ed.) p. cm. — (Springer series in surface sciences; 34) Includes bibliographical references and index. ISBN 0-387-57605-3 1. Surfaces (Physics) 2. Surface chemistry. I. Madix, Robert J., 1938– . II. Series. QC173. 4. S94S9638 1994 546'. 6—dc20 93-49867

Typesetting: Macmillan India Ltd., Bangalore-25

SPIN: 10068806   54/3140/SPS – 5 4 3 2 1 0 – Printed on acid-free paper

Preface

In the past ten years the study of the mechanisms of chemical transformations on metal surfaces has advanced appreciably. Today complex reaction networks can be unraveled by combining several spectroscopies, derived principally from the practice of ultrahigh-vacuum surface physics. Of paramount importance in this field is the combination of mass spectrometric methods for the identification of reaction products with spectroscopies which help identify surface-bound reactive intermediates. This quasi-monograph highlights the progress in this field with studies which clearly exemplify such research and at the same time provide more general understanding of chemical reactivity at surfaces.

This book was constructed to be a resource to all scientists interested in the chemical reactivity of metals, including those whose primary interest may lie in fields outside surface reactivity. The book is intended to be an advanced case-study text, not a "review" in the standard sense. Each chapter develops principles and illustrates the use of experimental methods. Consequently, more attention is given to experimentation than normally found in journal articles or review articles. My intent in organizing these chapters was to make this field accessible to professionals and graduate students in the fields of chemistry, material science, and physics. Even so, we hope that experts in the field of surface reactivity will also find these chapters informative.

After the introduction (Chap. 1) the book consists of chapters on the mechanism of selective oxidation by silver (Chap. 2 by R.J. Madix and J.T. Roberts), desulfurization (Chap. 3 by C.M. Friend), alkyne cyclization (Chap. 4 by R.M. Lambert and R.M. Ormerod), and organic rearrangements on aluminum (Chap. 5 by L.H. Dubois, B.E. Bent and R.G. Nuzzo). In each of these chapters illustrations of the use of isotopic labelling, temperature programmed reaction spectroscopy, photoelectron spectroscopies and vibrational spectroscopy in dissecting complex surface reaction processes are presented. Where possible, the relationships between the heterogeneous reactions studied and the counterparts in homogeneous chemistry are discussed.

Whereas the first four chapters address the important issue of molecular rearrangements and the elementary bond-breaking and bond-forming reactions that occur, Chapters 6 and 7 are concerned with the dynamics and kinetics of activated adsorption. Chapter 7 (by H.A. Michelson, C.T. Rettner and D.J. Auerbach) concerns the activated adsorption of dihydrogen on copper – the surface analogue of $H + H_2$ in the gas phase – from both the state integrated and state-to-state viewpoints. The book concludes with a discussion of the activated adsorption of alkanes (Chapter 7 by C.B. Mullins and W.H. Weinberg).

It is my pleasure to thank all the contributors to this volume. I am also grateful to Springer-Verlag for their interest in publishing this book and to Professor Robert Gomer for inviting it. Finally, all of us owe thanks to the many scientists who have contributed significantly to the rapid growth of this field to those who laid the foundation for modern surface chemical physics and chemistry through the development of the many methods in use today.

Stanford, April 1994

R.J. Madix

Contents

1. Introduction
R.J. Madix 1
References 3

2. The Problem of Heterogeneously Catalyzed Partial Oxidation: Model Studies on Single Crystal Surfaces
R.J. Madix and *J.T. Roberts* (With 27 Figures) 5
2.1 Modes of Oxygen Chemisorption on Metal Surfaces 6
2.2 Reactions of Molecularly Chemisorbed Oxygen 13
2.3 Reactions of Atomically Chemisorbed Oxygen 16
2.3.1 Atomic Oxygen as a Nucleophile on Silver 17
2.3.2 Atomic Oxygen as a Brønsted Base on Silver 19
2.3.3 Addition of Atomic Oxygen to Carbon-Carbon Double Bonds on Silver 38
2.3.4 Generalization to Other Metals 43
2.4 Conclusion 49
References 50

3. Desulfurization Reactions Induced by Transition Metal Surfaces
C.M. Friend (With 16 Figures) 55
3.1 Background 55
3.2 The Reactions of Thiols on Transition-Metal Surfaces 58
3.2.1 Spectroscopic Identification and Characterization 58
3.2.2 Structural Studies of Adsorbed Intermediates 66
3.2.3 Chemical Probes of the Mechanism 72
3.2.4 Coverage Dependence of Reactivity 76
3.3 Desulfurization of Cyclic Sulfur-Containing Molecules 76
3.4 Conclusions 85
References and Notes 86

4. Tricyclisation and Heterocyclisation Reactions of Ethyne over Well-Defined Palladium Surfaces
R. M. Lambert and *R.M. Ormerod* (With 26 Figures) 89
4.1 Background 89

4.2 Mechanistic Studies of Ethyne Tricyclisation 90
4.2.1 Molecular Beam Results, Temperature-Programmed Reaction and Isotope Labelling: Molecular Formula of the Reaction Intermediate 90
4.2.2 Characterisation of the C_4H_4 Intermediate 95
4.2.3 The Reactively Formed Benzene is Tilted: Effect of Surface Packing Density on the Conformation, Yield and Desorption Kinetics of Benzene Formation 100
4.3 Studies at High Pressures 105
4.4 The Effects of Promoters, Poisons and Other Coadsorbed Species 110
4.5 The Structure and Bonding of Ethyne Chemisorbed on Transition Metal Surfaces 113
4.6 Why is Tricyclisation so Specific to Palladium and Why is the (111) Plane so Strongly Favoured? 122
4.7 Other Cyclisation Reactions 127
4.8 Conclusions 131
References 131

5. Model Organic Rearrangements on Aluminum Surfaces
L.H. Dubois, *B.E. Bent*, and *R.G. Nuzzo* (With 21 Figures) 135
5.1 Background 135
5.2 Carbon-Halogen Bond Cleavage 138
5.2.1 Reactive Sticking Probability 138
5.2.2 High-Resolution EELS and TPRS Observations of C-X Bond Cleavage 141
5.3 Integrated Desorption Mass Spectrometry 147
5.4 Alkyl Surface Chemistry 149
5.4.1 Iodoalkanes with β-Hydrogens 152
5.4.2 Dihaloalkanes 160
5.4.3 Radical Participation in Aluminum Alkyl Chemistry 172
5.5 Etching of Aluminum Surfaces with Alkyl Halides 174
5.6 Model and Real Systems: A Comparison 175
5.7 Aluminum Surfaces vs. Aluminum Compounds: A Summary 179
5.8 Conclusion 180
References and Notes 181

6. The Adsorption of Hydrogen at Copper Surfaces: A Model System for the Study of Activated Adsorption
H.A. Michelsen, *C.T. Rettner*, and *D.J. Auerbach* (With 34 Figures) 185
6.1 Introductory Remarks 185

6.2 A One-Dimensional Description: The Translational Degree of Freedom 190
6.2.1 The Activation Barrier 190
6.2.2 Early Adsorption Measurements with Molecular Beams. . . . 193
6.2.3 Early Desorption Measurements and Detailed Balance. . . 194
6.3 A Two-Dimensional Description: The Translational and Vibrational Degrees of Freedom 198
6.3.1 The 2-D Potential Energy Surface 198
6.3.2 Recent Adsorption Measurements with Molecular Beams. . . . 201
6.3.3 Quantitative Treatment of Adsorption and Desorption Data 208
a) Sticking Probability Models 209
b) Quantitative Comparison of Adsorption Data 212
c) Desorption and the Role of Surface Motion 213
6.3.4 State-Resolved Scattering Measurements 220
a) Reflection Probability Measurements 220
b) Inelastic Scattering Measurements: Vibrational Excitation 221
6.3.5 State-Resolved Desorption Measurements: $S_0(v,E_i)$ via Detailed Balance 224
6.4 A Multidimensional Description: The Degrees of Freedom Including Translation, Vibration, Rotation, Molecular Orientation, and Impact Parameter. . . . 228
6.4.1 Theoretical Descriptions for More than Two Dimensions 228
6.4.2 Activation Energy Measurements: The Effect of Rotation on Adsorption Rate. . . . 229
6.4.3 State-Resolved Desorption Measurements: The Role of Rotation in Adsorption and Desorption 229
6.4.4 Inelastic Scattering Measurements: The Influence of Rotation on Vibrational Excitation 233
6.5 Summary 234
References 235

7. Kinetics and Dynamics of Alkane Activation on Transition Metal Surfaces
C.B. Mullins and *W.H. Weinberg* (With 26 Figures). . . . 239
7.1 Background. . . . 239
7.2 Trapping-Mediated Dissociative Chemisorption 249
7.2.1 Trapping Dynamics. . . . 250
7.2.2 Kinetics. . . . 255
7.2.3 Microscopic Reaction Mechanism 265

7.3 Direct Dissociative Chemisorption 267
7.3.1 Activation via Translational Energy 268
7.3.2 Activation via Vibrational Energy 271
7.3.3 Collision-Induced Activation 273
References 275

Subject Index 279

Contributors

D.J. Auerbach

IBM Research Division, Almaden Research Center, 650 Harry Road, San Jose, CA 95120-6099, USA

B.E. Bent

Department of Chemistry, Columbia University, New York, NY 10027, USA

L.H. Dubois

AT&T Bell Laboratories, Murray Hill, NJ 07974, USA

C.M. Friend

Department of Chemistry, Harvard University, Cambridge, MA 02138, USA

R.M. Lambert

Department of Chemistry, University of Cambridge, Lensfield Road, Cambridge CB2 1EW, UK

R.J. Madix

Department of Chemical Engineering, Stanford University, Stanford, CA 94305-5025, USA

H.A. Michelsen

Department of Chemistry, Harvard University, 12 Oxford Street, Cambridge, MA 02138, USA

C.B. Mullins

Department of Chemical Engineering, University of Texas at Austin, Austin, TX 78712-1062, USA

R.G. Nuzzo

Departments of Materials Science and Engineering and Chemistry, University of Illinois Urbana, IL 61801, USA

R.M. Ormerod

Department of Chemistry, University of Cambridge, Lensfield Road, Cambridge CB2 1EW, UK

C.T. Rettner

IBM Research Division, Almaden Research Center, 650 Harry Road, San Jose, CA 95120-6099, USA

J.T. Roberts

Department of Chemistry, University of Minnesota, Minneapolis, MN 55455-0431, USA

W.H. Weinberg

Department of Chemical Engineering, University of California, Santa Barbara, CA 93106, USA

1. Introduction

R.J. Madix

Historical Background of the Development of Mechanistic Studies of Complex Surface Reactions

In 1956 in a classic experiment *Ehrlich* and *Hickmott* determined the evolution of the pressure of N_2 evolved from a tungsten filament with time, measured by an ionization gauge, revealing the presence of several kinetic routes for desorption with different energetics [1.1]. Shortly thereafter *Redhead* extended this method to a continuously pumped volume, yielding pressure bursts which were the derivatives of the pressure buildup curves of the method of *Ehrlich* and *Hickmott* and from which activation energies and preexponential factors could be easily estimated [1.2].

At this time there was little known about the mechanism of reactions on metal surfaces or about rate constants for elementary steps on surfaces. There were serious difficulties with the determination of kinetics and mechanism on a molecular level because reactions were normally studied under steady-state conditions on metallic catalysts supported on high surface area materials. Mechanistic arguments were based solely on the compliance of overall rate expressions with the proposed network of elementary steps. Rate expressions often were the result of the combination of many steps, sometimes including transport effects, and isolation and identification of *the* rate determining step was extremely difficult [1.3]. Furthermore, surface concentrations of reactive intermediates were not measurable. Without these concentrations it was impossible to determine preexponential factors.

Temperature Programmed Reaction Spectroscopy (TPRS) was developed in order to circumvent these problems [1.4]. In favorable cases it enables one to determine the identity of rate-determining reactive intermediates and the rate-constant parameters for the slowest (rate-determining) step in complex reaction sequences on surfaces. It was an outgrowth of the "flash desorption" methods, but it embodies additional logic necessary to identify the rate-determining reactive intermediates and thus to understand complex surface chemical reactions. Quantitative mass spectrometry is used to discriminate the rate of evolution of different products from the surface. Examples of the use of this method are given in Chaps. 2–5. In favorable cases TPRS provides a complete map of the reaction network. During the linear temperature program reactions reach their peak rate when the temperature is approximately equal to 16 E,

Springer Series in Surface Sciences, Vol. 34
Surface Reactions Ed.: R.J. Madix

where E is the activation energy [kcal/mol] for the reaction in question. Thus, for complex reactions in which a multiplicity of reactions may be involved, the different reaction channels are separated in temperature (time) by approximately 16 ΔE, where ΔE [kcal/mol] is the difference in activation energies of any two reaction channels. Since temperature differences of a few degrees Kelvin are easily measured in such an experiment, activation energy differences of a fraction of a kcal/mol can be measured (assuming, for the sake of example, that the preexponential factors are equal). Furthermore, an important corollary follows; namely, two products showing the same peak shape and peak temperature in a temperature programmed reaction spectrum are the products of the same rate-limiting reaction step, and therefore result from the reaction of a common intermediate(s). Consequently, the composition of the rate-limiting reaction intermediate can be determined. In many cases this composition leads straightforwardly to the identification of the rate-determining reactive intermediate.

Vibrational spectroscopy is the most useful compliment to TPRS. For cases in which the evolution of product into the gas phase is rate-limited by surface reaction, not desorption of the products, both high-resolution Electron Energy Loss Spectroscopy (EELS) and Fourier Transform Infrared Spectroscopy (FTIR) may reveal structural information about the reactive intermediate as well as confirm its identity. For cases in which product evolution into the gas phase is *desorption-limited*, TPRS does not reveal the stoichiometry of the reactive intermediate, and other spectroscopies *must* be used to identify reactive intermediates.

The combined use of TPRS and EELS is nicely illustrated by the study of the oxidation of ethylene glycol ($HOCH_2CH_2OH$) on Ag (1 1 0) in Chap. 2 [1.5]. The oxidation proceeds via a dialkoxide intermediate, ethylenedioxy silver, formed by the selective removal of the -OH protons by adsorbed oxygen atoms below 170 K. The dialkoxide can be isolated by annealing the surface to 325 K to desorb the water in order to study its vibrational modes. Further reaction of the dialkoxide begins above 350 K. At 375 K there is indication of cleavage of the first C–D bond in the dialkoxide, since D_2 appears in the TPRS. Indeed, formation of the aldehydic function is evident in EELS following annealing the surface to 360 K. Other intriguing examples of the interplay between these two methods are discussed throughout the book.

The understanding of mechanistic surface chemistry is far from complete. On the contrary, the field has just begun. If we are to understand the origin of the enormous differences in reactivity of different classes of molecules on a surface, or the great differences of reactivity of different metals with one class of molecules, we must know the elementary steps involved and the identity of the rate-limiting reactive intermediate. With this knowledge we can proceed to determine the structure of this species and relate its reactivity to the electronic nature of the complex formed by the metal and this intermediate through theory.

References

1.1 T. Hickmott, G. Ehrlich: J. Chem. Phys. **24**, 1263 (1956)
G. Ehrlich, T. Hickmott: Nature **177**, 1045 (1956)

1.2 P.A. Redhead: Vacuum **12**, 203 (1962); Trans. Far. Soc. **57**, 641 (1961)

1.3 See for example, the review of the formic acid decomposition by P. Mars, J.J.F. Scholten, P. Zwietering: Adv. Catalysis **14**, 35 (1963)

1.4 J. Falconer, J. McCarty, R.J. Madix: Surf. Sci. **42**, 329 (1974)
I.E. Wachs, R.J. Madix: Surf. Sci. 531 (1978)
I.E. Wachs, R.J. Madix: J. Catalysis **53**, 208 (1978)

1.5 A.J. Capote, R.J. Madix: J. Am. Soc. **111,** 276 (1989); Surf. Sci. **214**, 276 (1989)

2. The Problem of Heterogeneously Catalyzed Partial Oxidation: Model Studies on Single Crystal Surfaces

R.J. Madix and *J.T. Roberts*

The oxidation of organic molecules mediated by metal surfaces is a subject of immense importance. Either complete oxidation to water and carbon dioxide or partial and selective oxidation to valuable chemical intermediates may be desired. These two extremes are obviously mutually exclusive, and identification of the surface-bound intermediates involved and their rates of reaction is needed to intelligently design reaction systems favoring partial or complete oxidation. The current understanding of metal mediated oxidation processes is the subject of this chapter.

An intriguing and difficult set of problems is found in the area of heterogeneously catalyzed partial oxidation. Particularly for the partial oxidation of organic molecules, selectivity is difficult to control because the products of total oxidation, carbon dioxide and water, are so thermodynamically favorable. Still, impressive advances have been made in the selectivity for partial oxidation. Silver, for instance, catalyzes the conversion of ethylene to ethylene epoxide [2.1]. For some silver-based catalysts, the selectivity for epoxide formation reaches into the high 80% range, with the remainder of the ethylene forming CO_2 and water. A second example of a selective partial oxidation catalyst is vanadyl pyrophosphate (VPO), which converts butane to maleic anhydride [2.2]. This fourteen electron oxidation is remarkable both for its selectivity ($\geq 75\%$) and because it involves the reaction of unactivated sp^3-hybridized C–H bonds. Perhaps the most difficult current problem is that of methane activation, with the object of converting methane to higher molecular weight hydrocarbons and/or oxygenated species like methanol [2.3]. Not surprisingly, selectivity is difficult to control, but substantial gains have been made [2.4].

One approach for the determination of the elementary reaction steps in a metal mediated oxidation process involves the use of single-crystal surfaces, prepared and treated in a controlled fashion under ultrahigh vacuum, as idealized catalysts. Such studies are invaluable for elucidating mechanistic aspects of surface reactions for the following reasons. First, the study of surface reactions at extremely low ambient pressures insures that reactions occurring in the adsorbed phase can be isolated and investigated without the interference of competing gas-phase processes. Second, under ultrahigh vacuum, adsorption of background gases is generally of negligible importance, and reactions can be conducted on clean surfaces or in the presence of controlled amounts of model poisons and promoters. Third, structures of chemisorbed reaction intermediates

Springer Series in Surface Sciences, Vol. 34
Surface Reactions Ed.: R.J. Madix

can often be spectroscopically determined. This feature is of immeasurable importance for the definition of reaction pathways. Finally, the issue of "structure-sensitivity" may be addressed by examining the same reaction on different crystallographic planes of a single metal. Many problems related to understanding heterogeneous reactivity are therefore greatly simplified in these highly idealized systems.

The purpose of this chapter is to review the current state of knowledge of oxidation of organic molecules on single-crystal surfaces, generally under ultrahigh vacuum. The work described draws heavily on work conducted in our laboratory and largely concerns reactions on silver, since silver has unique characteristics as a partial oxidation catalyst. The chapter is divided into four parts. First is a brief description of the modes of oxygen chemisorption on metal surfaces, with discussion of both experimental and theoretical investigations. Then follows a discussion of the reactions of molecularly chemisorbed oxygen on metal surfaces. The third and largest section concerns reactions of atomically chemisorbed oxygen, with an emphasis on the remarkable degree to which reactions of atomic oxygen have been systematized into three general reaction types. Particularly on silver, the reactivity of atomic oxygen is so well understood that the reaction pathway(s) of a previously unstudied molecule can be predicted *a priori*. The chapter concludes with a brief assessment of our present understanding of oxidation on single crystal surfaces, and touches on the likely avenues of future progress in this important field.

2.1 Modes of Oxygen Chemisorption on Metal Surfaces

Oxygen adsorption has been studied on many different single-crystal and polycrystalline transition-metal surfaces. In general, oxygen chemisorption can be characterized as either associative (the O–O bond does not break) or dissociative. On reactive metals like iron, tungsten, and molybdenum, O_2 adsorption is almost invariably dissociative and irreversible. On these metals the surface and/or bulk oxide is extremely stable [2.5], so that O_2 desorption via recombination of atomically chemisorbed oxygen ($O_{(a)}$) does not occur. Metals on the right side of the transition-metal series are generally much less active for dissociative oxygen chemisorption. Moreover, on metals such as silver [2.6], gold [2.7], palladium, [2.8], and platinum [2.9] $O_{(a)}$ chemisorption is reversible and $O_{(a)}$ recombination with concomitant $O_{2(g)}$ evolution occurs readily. Also, on these less reactive surfaces, associative oxygen chemisorption sometimes occurs. Under the right conditions, silver [2.6], platinum [2.9], and $O_{(a)}$-passivated palladium [2.8] all adsorb $O_{2(g)}$ into a molecular state, $O_{2(a)}$. The stability of dissociatively adsorbed oxygen correlates well with the stability of the corresponding bulk oxide, as shown in Fig. 2.1. The following discussion concerns oxygen chemisorption on these latter surfaces, where both $O_{2(a)}$ and $O_{(a)}$ are stable surface species.

Co	Ni	Cu
stable oxide	stable oxide	stable oxide

Rh	Pd	Ag
(RhO) 1300 K (RhO_2) 1320 K (O_a) 1000–1100 K BE = 87 kcal/mol	(PdO) 1148 K (O_a) 830–900 K BE = 86 kcal/mol	(Ag_2O) 450 K (O_a) 500–600 K BE=80 kcal/mol

Ir	Pt	Au
(IrO_2) 1373 K (O_a) 900–1100 K BE = 89 kcal/mol	(PtO_2) 923 K (O_a) 690–800 K BE = 83 kcal/mol	(Au_2O_2) 423 K (O_a) 650 K BE < 79 kcal/mol

Fig. 2.1. The stability of dissociatively adsorbed oxygen compared to that of the corresponding bulk oxide for various transition metals. Listed are the temperatures for decomposition of the bulk oxides, the temperatures at which $O_{(a)}$ recombines and desorbs in vacuum, and the metal-oxygen binding energies determined from temperature programmed desorption

Oxygen chemisorption has extensively been examined on two single-crystal silver surfaces, Ag (1 1 1) [2.10] and Ag (1 1 0) [2.6]. Both $O_{(a)}$ and $O_{2(a)}$ have been identified on these surfaces [2.11], but the ease of dissociation differs strongly on these two surfaces [2.10a, 12–15]. On Ag (1 1 0) the dissociation probability is near 10^{-3}, whereas on Ag (1 1 1) it is approximately 10^{-6}. At low surface temperatures (< 150 K) O_2 adsorption is wholly molecular. On Ag (1 1 1) molecularly chemisorbed oxygen ($O_{2(a)}$) desorbs quantitatively without dissociation at 220 K upon heating, but on Ag (1 1 0) $O_{2(a)}$ desorption competes with dissociation into $2O_{(a)}$ near 180 K. Atomically chemisorbed oxygen on Ag (1 1 0) undergoes recombination and desorption at approximately 600 K, with the exact temperature somewhat dependent on coverage and heating rate. Oxygen chemisorption on Ag (1 1 0) is dissociative for surface temperatures above ≈ 180 K. $O_{(a)}$ on Ag (1 1 1) can be most easily formed via the adsorption of oxygen atoms generated in the gas-phase from $O_{2(g)}$ dissociation at a hot filament or by $O_{2(g)}$ exposure at higher pressures. Recombination of $O_{(a)}$ with O_2 and subsequent desorption on Ag (1 1 1) occurs at ≈ 580 K, very close to the

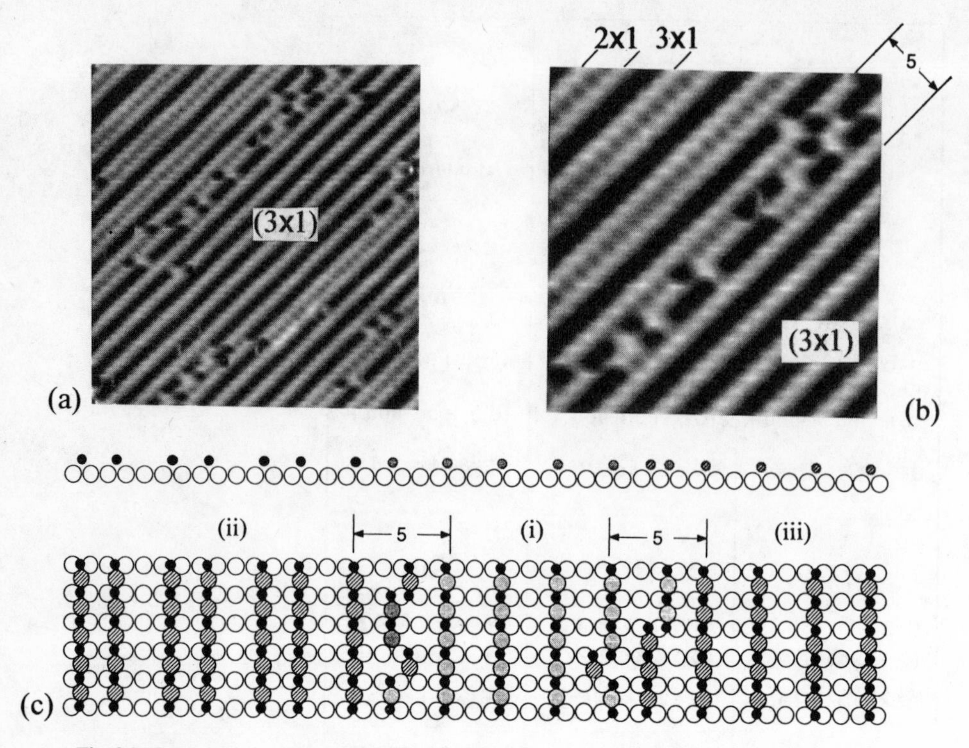

Fig. 2.2. Scanning tunnelling micrograph image of $O_{(a)}$ on Ag(1 1 0) [2.19]

$O_{(a)}$ recombination temperature on Ag (1 1 0). The surface structures formed by $O_{(a)}$ on Ag (1 1 0) are coverage dependent [2.16–18]. At a coverage of 0.5 oxygen atoms per silver atom in the outermost surface layer, a (2×1) structure is formed on Ag (1 1 0), with the oxygen atoms residing in the troughs. These structures are beautifully shown in recent scanning tunnelling micrographs (Fig. 2.2) [2.19]. At an oxygen pressure of 5 torr and a surface temperature of 490 K, chemisorbed oxygen atoms form $c(6 \times 2)$ and $p(4 \times 4)$ structures on the (1 1 0) and (1 1 1) surfaces, respectively.

Evolution of $O_{2(g)}$ from Ag (1 1 0), via $O_{2(a)}$ desorption at 180 K or $O_{(a)}$ recombination at 600 K, can be readily detected and distinguished using temperature programmed desorption (TPD), as shown in Fig. 2.3. From the temperature for desorption of $O_{2(a)}$, the activation energy for molecular oxygen desorption may be estimated as 11 kcal/mol, assuming first-order kinetics and a frequency factor of $10^{13}\ s^{-1}$ [2.20]. The activation energy (E_{act}) and frequency factor for $O_{(a)}$ recombination on Ag (1 1 0), a process which is effectively first order due to adsorbate-adsorbate interactions, are 41 kcal/mol and $4 \times 10^{14}\ s^{-1}$, respectively [2.20b]. Similar values are obtained for Ag (1 1 1) [2.10a]. Using

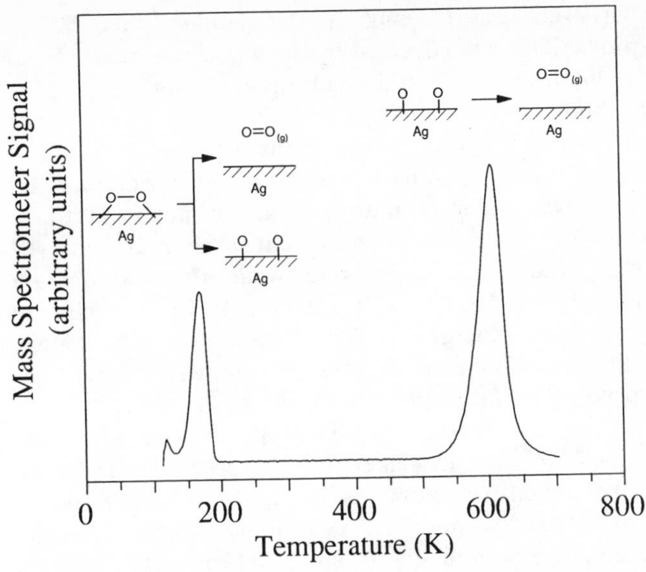

Fig. 2.3. Temperature programmed desorption of oxygen from Ag(1 1 0). Oxygen adsorption occurred at 100 K, and the heating rate was approximately 6 K/s. The low temperature state is due to the desorption of molecularly adsorbed oxygen, and the high temperature state to the recombination of two adsorbed oxygen atoms followed by immediate oxygen desorption

well established methods, accurate values of both E_{act} and the frequency factor can be determined [2.21].

Surface-$O_{2(a)}$ interactions are quite strong on Ag (1 1 0). Vibrational spectroscopy shows that O–O stretch [ν(O–O)] is reduced from 1580 cm^{-1} in the gas-phase to 640 cm^{-1} upon chemisorption [2.22], consistent with the formation of a silver-peroxide type species. The large reduction in ν(O–O) implies significant charge transfer from the Ag surface into the unoccupied $O_2 \pi^*$ levels upon chemisorption. Generalized valence-band calculations on the interaction of O_2 with a 24 atom Ag cluster also find that charge transfer from Ag to the unoccupied $O_2 \pi^*$ level is extensive (1–1.5 electrons) in the equilibrium chemisorption geometry [2.23]. This charge transfer results in the projection of considerable $O_2 \pi$-electron density away from the Ag (1 1 0) surface. These calculations also successfully predict ν(O–O) (627 cm^{-1} vs. the experimental value of 640 cm^{-1}) and ν(Ag–O_2) (278 vs. 240 cm^{-1}). The most stable O_2 geometry is one in which the dioxygen species resides in the trough, aligned along the $[1\bar{1}0]$ azimuth (the closest-packed direction), 0.94 Å above the surface plane. The O–O bond length is calculated to be 1.55 Å. As shown below, the calculated geometry is in excellent agreement with experimental results.

The bonding geometry of $O_{2(a)}$ on Ag(1 1 0) was deduced using *N*ear *E*dge *X*-ray *A*bsorption *F*ine *S*tructure (NEXAFS) spectroscopy [2.24]. NEXAFS is an extremely powerful technique which exploits the intense, tuneable, and highly polarized soft X-ray synchrotron radiation to probe electronic transitions from adsorbate core levels (C(1s), O(1s), etc.) to unoccupied molecular orbitals [2.25]. NEXAFS transition probabilities are governed by the dipole selection rule, so the intensity of a transition from a 1s level depends upon the alignment of the electric field vector of the incident X-radiation with the orbital into which a transition is occurring. Thus, by varying the angle of the surface with respect to the incident electric field vector (***E***), one can deduce the *directionality* of chemisorbate bonds. For example, in chemisorbed carbon monoxide, transitions from the C(1s) and O(1s) levels into π^* antibonding orbitals are observed only when ***E*** is perpendicular to the C–O bond, while σ^* transitions are observed when ***E*** is along the bond direction. In practice, NEXAFS spectra are recorded by scanning the photon energy through a transition for a fixed angle between crystal and the photon source. Spectra are typically recorded for radiation incident perpendicular to the surface (***E*** parallel to the surface) and for radiation incident $\approx 80°$ from the surface normal (***E*** nearly perpendicular to the surface). For highly anisotropic surfaces like Ag(1 1 0), spectra can also be recorded with ***E*** aligned along different azimuths. In Fig. 2.4, NEXAFS spectra at the O(1s) edge were obtained at normal ($\theta = 90°$) and glancing ($\theta = 10°$) incidence, with ***E*** aligned along both the $[1\bar{1}0]$ and $[001]$ azimuths. A single resonance is observed in the $O_{2(a)}$ NEXAFS spectra, due to transitions from the O(1s) level into the O–O σ^* antibonding level. In contrast to gas phase O_2, a transition into the π^* level was not observed, implying that it is substantially occupied via charge transfer from the surface. The polarization dependence of

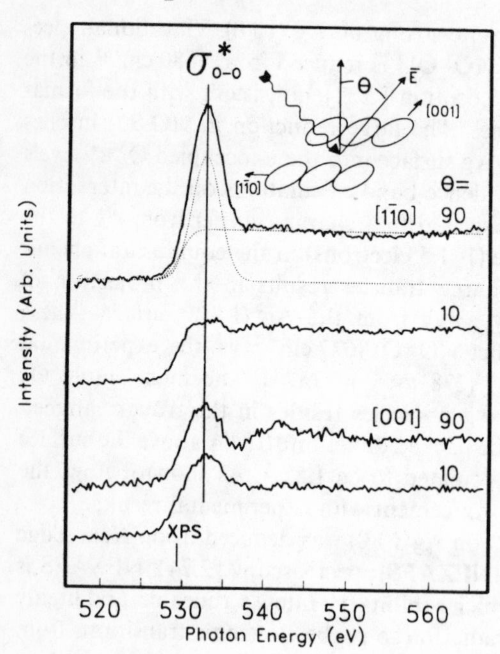

Fig. 2.4. NEXAFS spectra from excitation of the O(1 s) level for $O_{2(a)}$ on Ag(1 1 0) as a function of polar and azimuthal ***E*** orientations. The O–O σ^* resonance is strongest when ***E*** is oriented along the O–O bond, and this occurs when ***E*** is aligned along both the $[1\bar{1}0]$ azimuth and parallel to the surface ($\theta = 90°$). The line at 529.3 eV marks the O(1 s) binding energy relative to the Fermi level for $O_{2(a)}$ on Ag(1 1 0)

the σ^* resonance demonstrates that $O_{2(a)}$ is bound parallel to the Ag(1 1 0) surface, with O–O molecular axis aligned along the $[1\bar{1}0]$ azimuth. This geometry is clearly indicated because the O(1s) $\rightarrow \sigma^*$ transition is most intense when $\boldsymbol{E}$ is parallel to surface plane, aligned along the $[1\bar{1}0]$ direction [2.26]. More recent NEXAFS results with higher resolution show a weak π^* resonance, in accord with theoretical predictions of the π^* occupancy.

The O–O bond length in $O_{2(a)}$ is estimated to be 1.47 ± 0.05 Å from the energy of the NEXAFS σ^* transition [2.24]. The estimate is derived from an empirical correlation between the energies of σ^* transition and O–O bond lengths in a number of gas-phase molecules [2.27]. The O–O bond length of 1.47 Å in $O_{2(a)}$ compares with 1.21 Å in gaseous O_2 and 1.45–1.48 Å in gaseous hydrogen peroxide [2.28], indicating a reduction in bond order from 2 to nearly 1 upon chemisorption. Molecularly chemisorbed oxygen may therefore be thought of as a surface peroxide.

The NEXAFS observations, $O_{2(a)}$ vibrational spectra, and generalized valence bond calculations all indicate substantial charge transfer from Ag(1 1 0) to O_2. Additionally, the geometry deduced from NEXAFS agrees with the lowest energy calculated geometry, with O–O parallel to the surface along the $[1\bar{1}0]$ azimuth. Finally, the calculated and experimentally-derived bond lengths (1.55 and 1.47 Å, respectively) are nearly identical, within experimental error. Taken together, these data provide a very clear picture of molecular oxygen chemisorption on Ag(1 1 0). The O–O bond in $O_{2(a)}$ is oriented, and the bond order is reduced from 2 to 1. The very large reduction in bond order and the orientation of the adsorbed molecule rationalize the relative ease with which O_2 dissociates on Ag(1 1 0) [2.6] compared to Ag(1 1 1) [2.10]. Spectroscopic data for $O_{2(a)}$/Ag(1 1 1) are not nearly as complete, but X-ray photoemission data suggest that charge transfer from Ag(1 1 1) is not as extensive as from Ag(1 1 0) (≈ 1 vs. ≈ 2 electrons) [2.10b], implying a much smaller decrease in bond order upon chemisorption. However, recent results cite an O–O stretching frequency of 670 cm^{-1} for $O_{2(a)}$ on Ag(1 1 1) [2.29]. The chemisorption of O_2 on Ag(1 1 1) requires further investigation for complete understanding, particularly using vibrational spectroscopy and NEXAFS.

Molecularly chemisorbed oxygen has also been identified on Pt(1 1 1) [2.9] as well as $O_{2(a)}$-precovered Pd(1 0 0) [2.8a], Pd(1 1 0] [2.8b], and Rh(1 1 1) [2.30]. On Pt(1 1 1), $O_{2(a)}$ desorption competes with dissociation at ≈ 140 K [2.9a], in a process similar to that which occurs on Ag(1 1 0). Two dioxygen species have been detected on Pt(1 1 1) by surface vibrational spectroscopy, the two species exhibiting O–O stretching frequencies of 690 and 860 cm^{-1}. The relative populations of the two $O_{2(a)}$ species are dependent upon surface temperature and $O_{2(a)}$ coverage, but in general the species with ν(O–O) at 860 cm^{-1} is more thermally stable. NEXAFS studies of $O_{2(a)}$ on Pt(1 1 1) indicate that the O–O bond is tilted slightly with respect to the surface plane [2.24]. Additionally, transitions from O(1s) to π^* levels on O_2 are detected. Thus, charge transfer from the surface to O_2 appears not to be as extensive as on Ag(1 1 0). O_2 on Pt(1 1 1) is more properly considered a superoxide (bond order

≈ 1.5), rather than a peroxide. On Pd(1 0 0) and Pd(1 1 0) stable $O_{2(a)}$ cannot be formed until the surface is passivated by the adsorption of approximately 0.25 monolayers of $O_{(a)}$ [2.8]. The binding of O_2 to those two Pd surfaces is similar, with desorption occurring at approximately 120 K. On Pd(1 0 0), ν(O–O) is 730 cm^{-1}, [2.8a] indicating charge transfer from Pd to oxygen, as on Pt and Ag. The data do not allow unambiguous discrimination between peroxidic and superoxidic O_2 on Pd surfaces, however. On Rh(1 1 1), O_2 initially adsorbs dissociatively at 100 K [2.30, 31]. At higher coverages, however, a molecularly chemisorbed state is populated which desorbs at 140 K [2.30]. No vibrational data are currently available concerning $O_{2(a)}$ on Rh(1 1 1).

Atomically chemisorbed oxygen on Ag(1 1 0) and Ag(1 1 1) has been studied with a number of different experimental tools. The maximum coverage of $O_{(a)}$ that can be attained on Ag(1 1 0) under ultrahigh vacuum is 0.50 monolayers, as deduced from $O_{2(g)}$ thermal desorption yields, Low-Energy Electron Diffraction (LEED), and X-ray Photoelectron Spectroscopy (XPS) [2.6]. If adsorption occurs at high O_2 pressures (≈ 50 torr) and elevated silver temperatures (> 475 K), the maximum $O_{(a)}$ coverage attained is 0.67 monolayers [2.12]. The $O_{(a)}$ recombination temperature of this high coverage state is 565 K, approximately 35 K lower than the recombination temperature for $O_{(a)}$ coverages ≤ 0.50 monolayers. The bonding of $O_{(a)}$ to Ag(1 1 0) has been studied using SEXAFS, a technique related to NEXAFS [2.32]. The results were interpreted in terms of a structure in which oxygen atoms chemisorb in the Ag(1 1 0) troughs, bridging across two Ag atoms, 0.2 Å above the Ag plane. The vibrational spectrum of $O_{(a)}$ on Ag(1 1 0) shows a single mode at 315 cm^{-1}, assigned to the vibration of $O_{(a)}$ against the Ag surface [2.22]. LEED demonstrates that for $O_{(a)}$ coverages as low as ≈ 0.1 monolayers, $O_{(a)}$–$O_{(a)}$ interactions are repulsive along the $[1\bar{1}0]$ azimuth and attractive along the [0 0 1] azimuth [2.16]. $O_{(a)}$ therefore chemisorbs on Ag(1 1 0) such that the distances along the troughs are maximized, but across the troughs (in the [0 0 1] direction) they are minimized. Strong lateral interactions between chemisorbed oxygen atoms lead to $O_{(a)}$ recombination kinetics that are impossible to interpret using simple second-order desorption kinetics.

Although the temperatures of $O_{(a)}$ recombination to $O_{2(g)}$ on Ag(1 1 0) and Ag(1 1 1) are similar, there are differences in the way $O_{(a)}$ chemisorbs on the two surfaces. The Ag(1 1 1) surface is more isotropic than Ag(1 1 0), so that competing attractive and repulsive interactions should not be as pronounced. In fact, attractive interactions among $O_{(a)}$ clearly dominate on Ag(1 1 1). Diffraction measurements show that even for small coverages, $O_{(a)}$ "islands" form on Ag(1 1 1) at 490 K with a p(4 × 4) diffraction pattern and a local $O_{(a)}$ concentration of ≈ 0.41 monolayers [2.10a]. The vibrational spectrum for $O_{(a)}$ on Ag(1 1 1) exhibits a Ag–$O_{(a)}$ stretch at 220 cm^{-1}, nearly 100 cm^{-1} lower than on Ag(1 1 0) [2.14]. The value of ν(Ag–O) on Ag(1 1 1) should be regarded with some reservation, however, because experiments were conducted in vacuum chamber with a high water partial pressure. The spectrum exhibited some unassigned modes, including one in the region expected for ν(O–H). Additionally, the Ag–H_2O stretch on Ag(1 1 0) is at 240 cm^{-1}, very close to 220 cm^{-1} [2.33].

The bonding of $O_{(a)}$ to silver surfaces has been studied theoretically, using Generalized Valence Bond (GVB) methods, on a three atom silver cluster [2.34]. The most provocative finding of these calculations is that two nearly degenerate forms of $O_{(a)}$ may exist on a silver surface. The first of these is a closed-shell, electron-rich species which bonds to Ag via two σ bonds. The second stable $O_{(a)}$ species is an oxyradical with an unpaired electron on O projecting into space. Somewhat paradoxically, although the species are calculated to be isoenergetic, the closed-shell species is proposed to preferentially form on Ag(1 1 0) at low $O_{(a)}$ coverages. However, below 0.5 monolayers coverage, XPS [2.35] and HREELS [2.22] data point to a single $O_{(a)}$ chemical state, and NEXAFS measurements provide no evidence for an unfilled π-symmetry orbital localized on $O_{(a)}$, as evidenced by the lack of a π^* resonance [2.36]. It is very unlikely that such differences could be distinguished by NEXAFS, however. Interestingly, similar GVB calculations on a larger Ag cluster also suggest that two stable $O_{(a)}$ states can be formed, but the oxyradical is less stable than the closed-shell di-σ species by ≈ 15 kcal/mol [2.37]. The implications of the possible formation of an oxyradical on the mechanism for ethylene epoxidation over silver will be discussed later in this chapter.

The chemisorption of $O_{2(a)}$ and $O_{(a)}$ on Ag, Pt, Pd, and Rh is a complex subject, but some generalizations can be made. First, molecular adsorption occurs on these surfaces at temperatures below ≈ 120 K. Desorption of associatively adsorbed oxygen, often accompanied by competitive dissociation into 2 $O_{(a)}$, is usually complete by 200 K. Associative chemisorption on these surfaces invariably involves significant weakening of the O–O bond, as evidenced by a reduction in ν(O–O), suggesting that chemisorption may activate O_2 toward reactions which might not otherwise be observed. In the next section reactions of $O_{2(a)}$ will be discussed. Atomically chemisorbed oxygen on these surfaces generally undergoes recombination and desorption as $O_{2(g)}$ between 600 and 800 K. The fact that $O_{(a)}$ chemisorption is reversible on these surfaces implies that $O_{(a)}$ is relatively mobile and weakly bound compared to $O_{(a)}$ on earlier transition metals. Thus, $O_{(a)}$ can actually participate in chemical reactions with coadsorbed molecules, rather than acting principally to block sites or modify the surface electronic structure of the metal.

2.2 Reactions of Molecularly Chemisorbed Oxygen

As the discussion above makes clear, chemisorption of molecular oxygen on silver, platinum, and palladium surfaces invariably leads to reduction of the oxygen-oxygen bond order. The possibility therefore exists that chemisorption activates molecular oxygen toward reaction with coadsorbed molecules. This expectation of enhanced reactivity led us to investigate the chemistry of molecularly chemisorbed oxygen on Ag(1 1 0) [2.38]. On this surface, molecular oxygen is indeed quite reactive, but only toward molecules which are good π-electron acceptors. For instance, sulfur dioxide is oxidized by molecular oxygen to form chemisorbed sulfite ($SO_{3(a)}$). Surface sulfite was unambiguously identified by *H*igh-*R*esolution *E*lectron *E*nergy *L*oss vibrational *S*pectroscopy (HREELS) [2.38]. Vibrational spectra were recorded after (a) adsorption of sulfur dioxide multilayers on Ag(1 1 0), and (b) adsorption of SO_2 onto Ag(1 1 0) precovered with molecularly chemisorbed oxygen, as shown in Fig. 2.5. In each case adsorption occurred at 90 K. The symmetric S–O stretch, at 1160 cm^{-1} in SO_2 multilayers and 1010 cm^{-1} in chemisorbed SO_2 [2.39], is absent in Fig. 2.5b, clearly indicating that reaction has occurred. Additional confirmation that $O_{2(a)}$ reacts is provided by the absence of the O–O stretch at 660 cm^{-1} [2.22]. The frequencies of the species formed from reaction of SO_2 with $O_{2(a)}$ match those previously assigned to $SO_{3(a)}$/Ag(1 1 0) (Table 2.1) [2.39]. Notably, only $SO_{3(a)}$ was formed; a stretch at 1285 cm^{-1} that is prominent in the

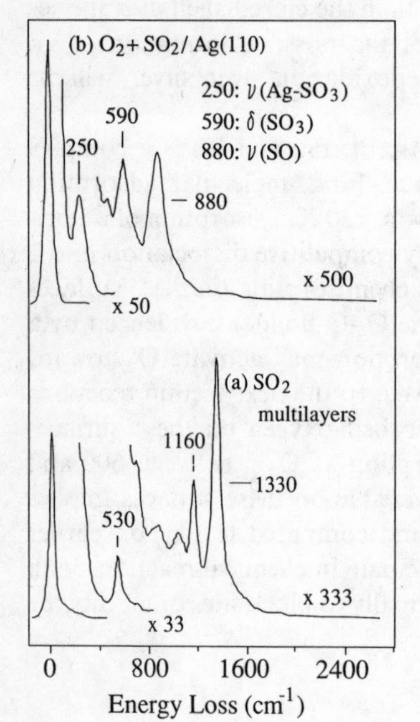

Fig. 2.5. Electron energy loss vibrational spectra of (a) SO_2 multilayers on Ag(1 1 0), and (b) 0.30 monolayers $O_{2(a)} + SO_2$ on Ag(1 1 0). In each case, adsorption occurred at 95 K. Multiplication factors in each spectrum are referenced to their respective elastic peaks

Table 2.1. Summary of the vibrational frequencies (in cm^{-1}) of SO_3/Ag(1 1 0) formed from the reaction of $SO_{2(a)}$ with $O_{2(a)}$, compared to those previously measured for SO_3/Ag(1 1 0)

Assignment	$O_{2(a)} + SO_{2(a)}$/Ag(1 1 0)	SO_3/Ag(1 1 0)
ν(Ag–SO_3)	250	220
δ(SO_3)	590	585
ν(SO)	880	835, 885

HREELS vibrational spectrum of $SO_{4(a)}$ is absent in Fig. 2.5b. The fact that SO_2 oxidation by $O_{2(a)}$ occurs at 90 K indicates that the activation energy for reaction is very low. Furthermore, reaction occurs 90 K below the temperature for $O_{2(a)}$ dissociation, excluding the possibility that atomically chemisorbed oxygen initiates the reaction.

Carbon monoxide is another π-acid that reacts with $O_{2(a)}$ [2.38]. The reaction product is chemisorbed carbonate ($CO_{3(a)}$), also identified on the basis of its HREELS vibrational spectrum [2.40]. Again, reaction occurs at 90 K, implying a nearly unactivated oxidation pathway. In the presence of excess coadsorbed CO, the integrated intensity of the molecular O_2 desorption state at 180 K is approximately 10% of that expected in the absence of CO for an initial $O_{2(a)}$ coverage of 0.3 monolayers, indicating that the reaction between $CO_{(a)}$ and $O_{2(a)}$ is nearly quantitative [2.38]. As expected on the basis of previous work, $CO_{3(a)}$ decomposes at ≈ 480 K to $CO_{2(g)}$ and $O_{(a)}$ [2.6a, 41].

Notably, molecularly adsorbed oxygen is unreactive toward molecules which are not good π-electron acceptors. Among the molecules which do not react with $O_{2(a)}$ on Ag(1 1 0) are ethylene, cyclohexene, and acetonitrile [2.42], all of which react with atomically chemisorbed oxygen. Even water, ammonia, and formic acid, which readily transfer protons to $O_{(a)}$, do not react with $O_{2(a)}$. The fact that poor π-acids are unreactive towards $O_{2(a)}$ is important because the nature of the oxygen species which is active for ethylene epoxidation on a silver catalyst is somewhat controversial. Several groups have proposed that $O_{2(a)}$ is the oxygen species active for epoxidation [2.43], but these results show that $O_{2(a)}$ does not react readily with weak π-acids, at least on Ag(1 1 0) under ultrahigh vacuum. As we will see below, atomically adsorbed oxygen on Ag(1 1 0) does epoxidize C–C double bonds on Ag(1 1 0) under ultrahigh vacuum [2.44]. For this and other reasons discussed below, it is likely, therefore, that $O_{2(a)}$ is not an important intermediate in silver catalyzed partial oxidation reactions.

The reactivity patterns observed for $O_{2(a)}$ on Ag(1 1 0) are entirely in accord with the calculated electronic structure of $O_{2(a)}$ on a 24 atom Ag cluster discussed above. Generalized valence bond calculations show that Ag transfers 1–1.5 electrons into the $O_2\,\pi^*$ orbitals, resulting in the projection of considerable π-electron density away from the Ag surface [2.23]. Qualitatively, then, the best interaction is likely to be between $O_{2(a)}$ and molecules which are good π-electron acceptors such as CO and SO_2.

It has also been suggested that HCN [2.45] and CO [2.46] react with molecularly adsorbed oxygen on Pt(1 1 1). Both of these reactions were studied using temperature programmed reaction mass spectrometry (TPRS). TPRS is a tool for measuring the reaction kinetics and product distributions of surface reactions that result in the formation of gas-phase products. In a TPRS experiment, the molecule of interest is adsorbed on the surface with the surface temperature held fixed at a temperature low enough to preclude reaction upon adsorption, usually ≈ 100 K. The crystal temperature is subsequently increased at a constant heating rate β, and the gaseous products are detected with a mass spectrometer. If the pumping speed of the reaction chamber is fast, relative to the gas evolution rate, then the mass spectrometer signal at temperature T is proportional to the reaction rate at T. Temperature-programmed reaction mass spectrometry is an enormously useful tool for measuring the activation energies (E_{act}) and frequency factors (A) of surface reactions [2.20, 21].

The reactivity of $O_{2(a)}$ toward HCN on Pt(1 1 1) was inferred from a comparison of the TPRS for HCN oxidation by $O_{2(a)}$ to those for oxidation by atomically chemisorbed oxygen. The HCN oxidation products from reaction with $O_{2(a)}$ (CO, CO_2, and H_2O) are all evolved into the gas phase at temperatures well below the temperatures at which they evolve during reaction with $O_{2(a)}$ [2.45]. Carbon monoxide was proposed to react with $O_{2(a)}$ because temperature programmed reaction of a coadsorbed CO/O_2 mixture resulted in the coincident desorption of CO_2 and O_2 at 160 K. When CO was reacted on a mixed $^{18}O_{2(a)}/^{16}O_{(a)}$ adlayer, ^{18}O was preferentially incorporated into the low temperature CO_2 product, further suggesting that $O_{2(a)}$ is reactive toward CO on Pt(1 1 1) [2.46]. Unfortunately, it cannot be concluded from these results that direct reaction between $O_{2(a)}$ and HCN or CO occurs. First, for neither the CO nor HCN reactions were the products different from those formed from reaction with $O_{(a)}$. Moreover, in these cases temperature programmed reaction spectroscopy alone cannot distinguish between the reactions of molecular and atomic oxygen on Pt(1 1 1) because O_2 dissociates below the temperature at which products actually evolve from the surface [2.9b]. Indeed, unpublished EELS vibrational spectroscopic studies indicate that coadsorbed CO and O_2 on Pt(1 1 1) do not react below the onset of O_2 dissociation [2.47]. Atomic oxygen formed during $O_{2(a)}$ dissociation may be more reactive than $O_{(a)}$ which is thermally equilibrated with the surface. Hot-atom effects of this nature have been proposed for oxidation of carbon monoxide on aluminium [2.48] and ammonia on zinc and magnesium [2.49]. These reactions must be investigated further using in situ surface spectroscopies like HREELS to determine if reaction actually occurs below the temperature of $O_{2(a)}$ dissociation.

The question of how molecularly adsorbed oxygen reacts on transition-metal surfaces remains to be investigated in greater detail. On Ag(1 1 0) there is clear evidence that $O_{2(a)}$ reacts with coadsorbed π-acids, but the reactivity of $O_{2(a)}$ is not understood nearly as well as that of $O_{(a)}$. There are tantalizing suggestions that $O_{2(a)}$ reacts with both CO and HCN on Pt(1 1 1) [2.45, 46], but unequivocal proof is lacking. Given the possible importance of $O_{2(a)}$ as an intermediate in catalytic oxidation chemistry, these reaction systems all merit further study.

2.3 Reactions of Atomically Chemisorbed Oxygen

Reactions of atomically chemisorbed oxygen ($O_{(a)}$) have been examined on a number of different surfaces, but by far the preponderance of investigations have concerned Ag(1 1 0). The number of reactions in which $O_{(a)}$ participates on this surface is truly noteworthy, but more remarkable is the degree to which these reactions can be systematized and grouped into but three general types. Atomically adsorbed oxygen can act as (i) a nucleophile [2.50], (ii) a Brønsted base [2.51], and (iii) an oxygenation agent for carbon-carbon double bonds [2.44]. It also leads to increased binding energy of electron donors such as nitriles, alcohols, and alkenes, presumably by creation of electron-deficient silver atoms or Lewis acid sites. Both the structural and physical properties of a molecule determine how it reacts with $O_{(a)}$. Of key importance are (a) the presence of electron-deficient centers which are subject to nucleophilic attack (e.g. aldehydic carbon atoms), and (b) the gas phase acidity of the reacting molecule, which governs the propensity of proton transfer to $O_{(a)}$. The following subsections describe in detail these aspects of the surface reactivity of $O_{(a)}$, with an emphasis on the relationships between adsorbate structure and reactivity.

2.3.1 Atomic Oxygen as a Nucleophile on Silver

That oxygen acts as a nucleophile on Ag(1 1 0) is perhaps not surprising given that quantum-mechanical calculations suggest $O_{(a)}$ can exist as a closed-shell, electron-rich species [2.34]. The range of adsorbates which react with $O_{(a)}$ via attack at an electron deficient center is quite broad, ranging from inorganic oxides to organic aldehydes and ketones. Nitrogen dioxide, for instance, reacts with $O_{(a)}$ via oxygen addition at the electron deficient nitrogen center [2.52]:

$$O_{(a)} + NO_{2(a)} \rightarrow NO_{3(a)}.$$

Surface nitrate ($NO_{3(a)}$) was identified by its vibrational spectrum. Sulfur dioxide reacts similarly, initially undergoing oxidation at 240 K to a surface sulfite ($SO_{3(a)}$) and then disproportionating to surface sulfate ($SO_{4(a)}$) at 500 K [2.39]. Both sulfite and sulfate intermediates initially form metastable monodentate structures which convert to bidentate species on further heating.

Among organic molecules that react via nucleophilic addition of $O_{(a)}$ are formaldehyde (H_2CO) [2.53] and acetaldehyde (CH_3COH) [2.54]. Nucleophilic attack by $O_{(a)}$ at the electropositive carbonyl center in formaldehyde leads to the reversible formation of a η^2-methylenedioxy ($H_2CO_{2(a)}$) intermediate [2.53, 55], as shown in Scheme 2.1a. Oxygen addition to form η^2-methylenedioxy is quite facile, occurring by 225 K as evidenced by vibrational spectroscopy. The η^2-methylenedioxy is thermally unstable, however, and decomposes by 250 K to form gaseous dihydrogen and surface formate ($HCO_{2(a)}$) [2.53] (Scheme 2.1b). The analogy to classical organic nucleophilic addition/displacement reactions is perhaps most precise in the addition of $O_{(a)}$ to methyl formate ($HCOOCH_3$) [2.55]. In this reaction, $O_{(a)}$ actually acts as a *displacing agent* and the methyl formate methoxy group as a *leaving group* to form adsorbed formate and methoxy ($CH_3O_{(a)}$). According to the vibrational spectra, this reaction occurs upon adsorption at 160 K. Experiments with isotopically labeled oxygen confirm the nucleophilic nature of $O_{(a)}$:

$$^{18}O_{(a)} + HC^{16}O^{16}OCH_{3(a)} \rightarrow HC^{18}O^{16}O_{(a)} + CH_3{}^{16}O_{(a)}. \tag{2.1}$$

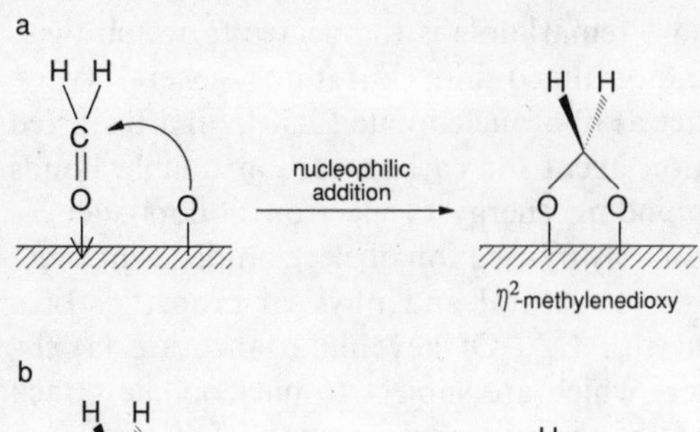

Scheme 2.1. (a) Oxidation of formaldehyde to chemisorbed η^2-methylenedioxy by $O_{(a)}$ on Ag(1 1 0). (b) Decomposition of η^2-methylenedioxy to surface formate

Displacement reactions are also observed which produce only isotopic substitution. For example, oxygen exchange between $O_{(a)}$ and acetone occurs readily below 200 K [2.56]:

$$^{18}O_{(a)} + (CH_3)_2C^{16}O_{(a)} \rightarrow {}^{16}O_{(a)} + (CH_3)_2C^{18}O_{(a)}. \tag{2.2}$$

As a final example of nucleophilic addition on Ag(1 1 0), we will consider the oxidation of chemisorbed carbon dioxide to surface carbonate ($CO_{3(a)}$). Due to its unsaturation and electron deficiency, carbon dioxide is highly susceptible to nucleophilic attack at carbon. Chemisorbed carbon monoxide does indeed react with $O_{(a)}$ to form $CO_{3(a)}$ by 200 K [2.6a, 40, 41]. Adsorbed carbonate is stable on Ag(1 1 0) until 480 K, where it decomposes to $CO_{2(g)}$ and $O_{(a)}$. Carbonate formed from the reaction of $^{18}O_{(a)}$ with CO_2 leads to a distribution of ^{18}O in the carbonate decomposition products that is entirely statistical, demonstrating that the three C–O bonds are *kinetically* equivalent [2.6a]. The structure of $CO_{3(a)}$ has been investigated using several different spectroscopic probes. X-ray Photoelectron Spectroscopy (XPS) reveals that the three carbonate oxygens are chemically equivalent [2.57]. Furthermore, both the C(1s) and O(1s) XPS binding energies are consistent with an anionic species. Near Edge X-ray Absorption Fine Structure (NEXAFS) measurements recorded from excitation at both the C(1s) and O(1s) edges show that the $CO_{(3)}$ plane is parallel to the Ag(1 1 0) surface ($\pm 10°$) with no azimuthal ordering even on the highly anisotropic Ag(1 1 0) plane [2.58]. The energy of σ^* NEXAFS is highly sensitive to the length of the bond being excited. As shown in Fig. 2.6, the energy of the σ^*_{C-O} transition is identical to that of $CdCO_3$, suggesting that the electronic structure of $CO_{3(a)}$ is indistinguishable from that of a bulk ionic carbonate. Quantum mechanical calculations also suggest that carbonate on Ag(1 1 0) is ionic in nature [2.59] in agreement with both the XPS and NEXAFS results. HREELS vibrational spectra are entirely consistent with all other data if the symmetry of CO_3 is assigned as C_s [2.58].

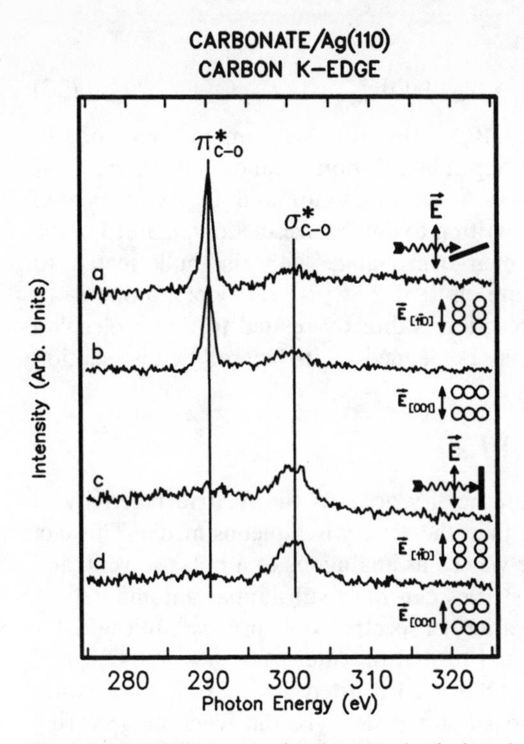

Fig. 2.6. NEXAFS spectra of carbonate adsorbed on Ag(1 1 0) ($CO_{3(a)}$) compared to those of bulk $CdCO_3$. Spectra were recorded for excitation at both the carbon and oxygen K-edges (C(1 s) and O(1 s), respectively). The close agreement of the $CO_{3(a)}$ and $CdCO_3$ σ^* positions indicates that the C–O bond length in $CO_{3(a)}$ is nearly equal to that in bulk $CdCO_3$, 1.29Å

In summary, atomically chemisorbed oxygen on Ag(1 1 0) acts as a strong nucleophile. Inorganic oxides (SO_2, NO_2, CO_2, etc.) and electron deficient organics (e.g., formaldehyde, acetaldehyde, acetone, and methyl formate) all react with $O_{(a)}$ to form nucleophilic addition and/or displacement products. These reactions all point to an atomically chemisorbed oxygen species that is very electron rich (i.e. nucleophilic), and therefore reactive towards electron deficient (i.e., electrophilic) centers in neighboring chemisorbed molecules. In the absence of other reactive sites (especially acidic bonds, as explained below) any chemisorbed species with an electrophilic center should be expected to react with $O_{(a)}$ via nucleophilic attack.

2.3.2 Atomic Oxygen as a Brønsted Base on Silver

Atomically chemisorbed oxygen on Ag(1 1 0) also reacts readily with chemisorbed molecules that possess acidic C–H and O–H bonds (that is, bonds capable of transferring protons to suitable acceptors). A generalized Brønsted acid–base reaction may be written:

$$(\gamma + \alpha - \delta)\,e^- + HB_{(a)} + O^{\delta-}_{(a)} \rightarrow B^{\gamma-}_{(a)} + OH^{\alpha-}_{(a)}\,. \tag{2.3}$$

where $HB_{(a)}$ is the chemisorbed acid, $B^{\gamma-}_{(a)}$ is the adsorbed conjugate base of HB, and $OH^{\alpha-}_{(a)}$ is a surface hydroxyl group. The reaction is called a Brønsted acid-base reaction because $HB_{(a)}$ acts as a proton donor and $O_{(a)}$ as a proton acceptor. The reaction equation is written to emphasize the ionic nature of the adsorbed species and the facile charge exchange with the bulk metal. In subsequent equations the charges are omitted. The propensity of a molecule to react with $O_{(a)}$ via an acid–base reaction is directly related to that molecule's gas-phase acidity, ΔH_{acid}, which may be defined by its heterolytic dissociation energy [2.60]:

$$HB_{(g)} \rightarrow H^+ + B^-, \qquad \Delta H = \Delta H_{acid}\,. \tag{2.4}$$

It is noteworthy that the gas-phase acidity governs the relative reactivity of organic molecules with $O_{(a)}$ rather than the acidity in aqueous media. This fact suggests that the surface *cannot* be viewed as analogous to a polar solvent, and supports the view that adsorbed species can have substantial anionic nature, a property directly observed by a variety of spectroscopic probes. Molecules for which ΔH_{acid} is greater than or equal to that of water ($\Delta H_{acid} = 391$ kcal/mol [2.61]) are observed to react with $O_{(a)}$ as a Brønsted acid [2.62]. Thus, water itself reacts with $O_{(a)}$ to form two surface hydroxyls, the reaction occurring between 200 and 250 K [2.63]:

$$H_2O_{(a)} + O_{(a)} \rightarrow 2\,OH_{(a)}\,. \tag{2.5}$$

Surface hydroxyls have been characterized by X-ray photoelectron [2.35] and HREELS vibrational [2.33] spectroscopies. For coadsorbed OH and H_2O, O–H stretches are observed at 3380 and 3640 cm^{-1}. Isolated hydroxyl groups exhibit a single O–H stretch at 3380 cm^{-1}. The hydroxyl groups are stable until ≈ 320 K, at which they disproportionate to yield water, which desorbs, and $O_{(a)}$ with an activation energy of 22.2 ± 0.3 kcal/mol and a second-order frequency factor of 0.7 cm^2/s [2.64]. Excess $O_{(a)}$ increases the barrier for $OH_{(a)}$ disproportionation, possibly due to hydrogen bonding between $OH_{(a)}$ and $O_{(a)}$.

It is important to stress that the reactivity of C–H and O–H bonds toward $O_{(a)}$ on Ag(1 1 0) is related to their heterolytic, not homolytic, dissociation energy. This has been demonstrated for many different reaction systems, but we mention two here. First, methylacetylene (CH_3CCH, $\Delta H_{acid} = 381$ kcal/mol [2.65]) initially reacts with $O_{(a)}$ via activation of the acetylenic C–H bond [2.66]. If homolytic bond energy governed reactivity, the weaker methyl carbon–hydrogen bonds should react preferentially with $O_{(a)}$. Instead the stronger [2.67] (in a homolytic dissociation sense) but more acidic [2.65] acetylenic C–H bond reacts first. The importance of gas-phase acidity in determining adsorbate stability on Ag(1 1 0) is further established by the displacement of an adsorbed conjugate base $B_{(a)}$ upon adsorption of a second acid HB′. If ΔH_{acid} of HB′ is greater than that of HB, displacement readily occurs

[2.60]:

$$B_{(a)} + HB'_{(a)} \rightarrow HB_{(a)} + B'_{(a)} \qquad \Delta H_{acid}(HB') > \Delta H_{acid}(HB). \tag{2.6}$$

For example, formic acid (ΔH_{acid} 345 kcal/mol [2.68]) displaces chemisorbed ethoxy ($C_2H_5O_{(a)}$) to form chemisorbed ethanol (ΔH_{acid} 378 kcal/mol [2.65]) via proton transfer from formic acid [2.60]:

$$HCOOH_{(a)} + C_2H_5O_{(a)} \rightarrow HCOO_{(a)} + C_2H_5OH_{(a)}. \tag{2.7}$$

Displacement occurs even though the O–H bond in formic acid is stronger than in ethanol (112 vs. 104 kcal/mol, respectively [2.69]). C_2D_2 (ΔH_{acid} 375.4 kcal/mol) can also be used to displace the conjugate base of a weaker acid containing only C–H bonds, thereby facilitating the identification of intermediates by mass spectrometry.

Reactions of alcohols with $O_{(a)}$. The fact that good gas-phase acids react so readily with $O_{(a)}$ allows for the preparation of a virtually limitless number of stable surface compounds whose structure and reactivity can then be studied in depth. As we shall see, the selective oxidation of organic molecules on silver generally follows a two step process involving initial low–temperature activation of any acidic bonds by $O_{(a)}$, followed by scission ($O_{(a)}$-assisted or not) of the C–H bond(s). Two types of C–H bond scission can occur: (i) that of activated C–H bonds near 300 K and (ii) that of unactivated C–H bonds at much higher temperatures. Unactivated C–H bond scission often occurs via a disproportionation process.

Surface alkoxy species ($RO_{(a)}$) can be prepared via reaction of the parent alcohol with $O_{(a)}$. Ethanol forms an ethoxy ($C_2H_5O_{(a)}$) species that is stable until $\approx$275 K, at which temperature one of the C–H bonds located at the carbon adjacent to the ethoxy oxygen (β- to the metal) is activated by silver, leading to the formation of acetaldehyde (CH_3CHO) which immediately desorbs [2.70]. Some of the resulting adsorbed hydrogen atoms combine to form $H_{2(g)}$, while the rest react with some of the remaining $C_2H_5O_{(a)}$ to form gaseous ethanol. Similar β-C–H bond activation reactions are observed for other molecules with C–H bonds located adjacent to oxygen. Thus, surface methoxy ($CH_3O_{(a)}$) decomposes near 300 K to gas-phase formaldehyde [2.71] and the alkoxy species from cyclohexanol forms cyclohexanone at 300 K [2.72]. The activation energies for all of these processes are similar, near 18–20 kcal/mol. More precise values are given in the original papers.

Another alkoxy intermediate, somewhat more complicated than those derived from ethanol and methanol, is the surface allyloxy species **1**:

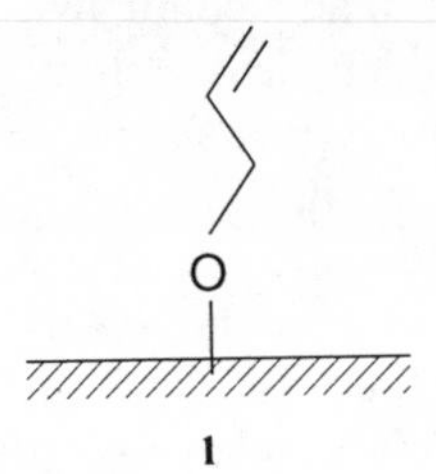

1

As was the case for the simpler alkoxys, **1** was prepared on Ag(1 1 0) from reaction of the parent alcohol with $O_{(a)}$ [2.73]. Surface allyloxy was investigated to determine the effect of unsaturation in the carbon skeleton on reactivity. On this surface the reactivity of allyloxy is not substantially different from that of the saturated alkoxy species. Allyloxy decomposes at 310 K during the temperature-programmed reaction to form H_2, acrolein (CH_2CHCHO), and regenerated allyl alcohol. Interaction of the carbon–carbon double bond with the surface apparently does not stabilize the surface alkoxy, nor is it hydrogenated to form adsorbed propyloxy ($C_3H_7O_{(a)}$), despite the fact NEXAFS measurements show evidence for π-donor bonding between the carbon–carbon double bond and the surface [2.74]. At 95 K, the allyloxy double bond is oriented nearly parallel to the surface and directed nearly perpendicular to the closest-packed direction (the [1 $\bar{1}$ 0] azimuth). In contrast to Ag(1 1 0), on Cu(1 1 0), allyloxy is more stable than the corresponding saturated alkoxys, decomposing at 370 K to form acrolein and regenerated allyl alcohol [2.75]. For comparison, decomposition of surface ethoxy on Cu(1 1 0) occurs at 320 K [2.76]. Furthermore, on Cu(1 1 0) some allyloxy carbon–carbon double-bond hydrogenation by surface hydrogen atoms occurs, leading to the formation of a small amount of surface propyloxy ($C_3H_7O_{(a)}$) [2.75], which decomposes as expected to gaseous H_2, propionaldehyde (C_2H_5CHO), and propanol at 320 K [2.76]. Finally, some surface allyloxy reacts to form an intermediate that decomposes at 435 K to gaseous acrolein, propionaldehyde, dihydrogen, and water. The key intermediate leading to the formation of the products at 435 K is proposed to be a surface metallacycle:

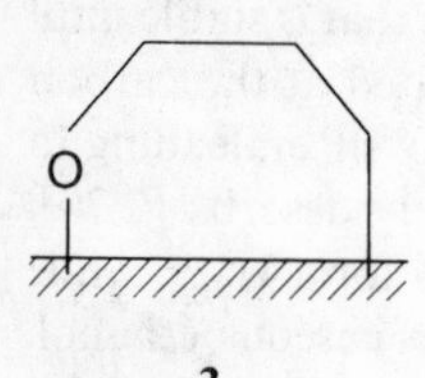

2

The reason that allyloxy forms a metallacycle on Cu(1 1 0) but not on Ag(1 1 0) is not yet clear, but stronger interactions between the C–C double bond and the Cu(1 1 0) surface may facilitate the C–H bond formation step(s) that lead to metallacycle formation.

As exemplified by allyl alcohol, even unsaturated alkoxy species react in a highly selective manner on Ag(1 1 0) via activation of the C–H bond located adjacent to the alkoxy oxygen. The highly selective oxidation of an alcohol, first to a surface alkoxy and then to the corresponding gaseous aldehyde or ketone, can be envisioned to occur on Ag(1 1 0) via a simple catalytic cycle, as shown in Fig. 2.7. Dissociative adsorption of oxygen to produce $O_{(a)}$ is the initial reaction step, followed by conversion of the alcohol to the surface alkoxy. At this step in the cycle, water, produced from the reaction of $O_{(a)}$ with two alcoholic OH bonds, evolves into the gas phase. The slow step in the reaction is dehydrogenation of the alkoxy intermediate to an aldehyde or ketone, with nearly

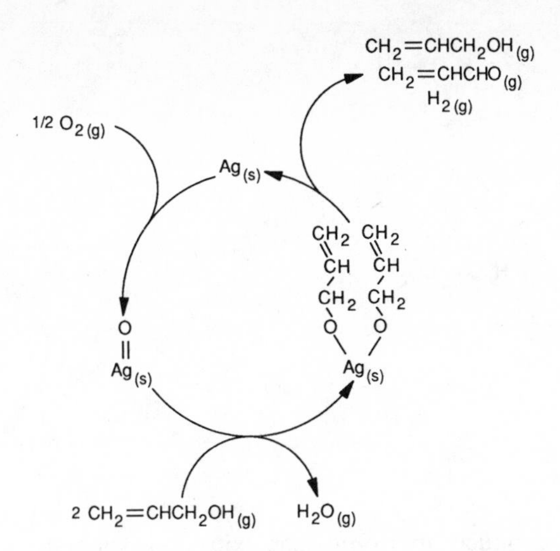

Fig. 2.7. Catalytic cycle for the oxidation of allyl alcohol on silver. The extended silver surface is schematically represented by $Ag_{(s)}$; the figure is not intended to imply that reaction occurs on a single silver atom

simultaneous production of H_2 and the regenerated alcohol (formed via a recombination reaction between the alkoxide and $H_{(a)}$). At this stage, the silver surface is clean and can complete another catalytic cycle. The oxidation of allyl alcohol over a silver sponge catalyst at 490 K at steady state has been studied [2.77], and the results are in complete agreement with those obtained for allyl alcohol oxidation on Ag(1 1 0) under ultrahigh vacuum. Acrolein is the principal oxidation product, formed with a selectivity of 99.8% at a conversion of 96.6%. A small amount of total oxidation to CO_2 is observed over the silver-sponge catalyst, in accord with the observation that some allyl alcohol decomposes during reaction $O_{(a)}$ on Ag(1 1 0) under ultrahigh vacuum. Unfortunately, the kinetics for allyl alcohol oxidation over the silver sponge were not determined, so a more quantitative comparison is not possible. This general mechanism also governs the selective oxidation of methanol to formaldehyde over a silver catalyst [2.78].

So far we have examined alcohols for which the rate-limiting step in oxidation over silver is the scission of an *activated* C–H bond, that is a bond adjacent to an oxygen center. Carbon-hydrogen bonds located at carbon atoms adjacent to electronegative (O, N, etc.) or unsaturated C = C, C = O, etc.) centers are generally 4–7 kcal/mol weaker than C–H bonds not adjacent to such centers [2.79], and are therefore *activated* toward many reactions. The reaction of *tert*-butoxy, an alkoxy without active C–H bonds therefore provides an interesting contrast to the reaction of activated alkoxys like those described above (Scheme 2.2). The *tert*-butoxy intermediate is easily prepared via the reaction of $O_{(a)}$ with *tert*-butanol [2.80]. Because there are no activated C–H bonds (i.e., those at positions located adjacent to the alkoxy oxygen), there is no pathway analogous to that which leads to aldehydes or ketones from activated alkoxys. Indeed, *tert*-butoxy is much more stable on Ag(1 1 0) compared to any unactivated alkoxy. While alkoxys with *activated* C–H bonds decompose at ≈ 300 K [2.70–72], tert-butoxy does not begin to decompose until 440 K

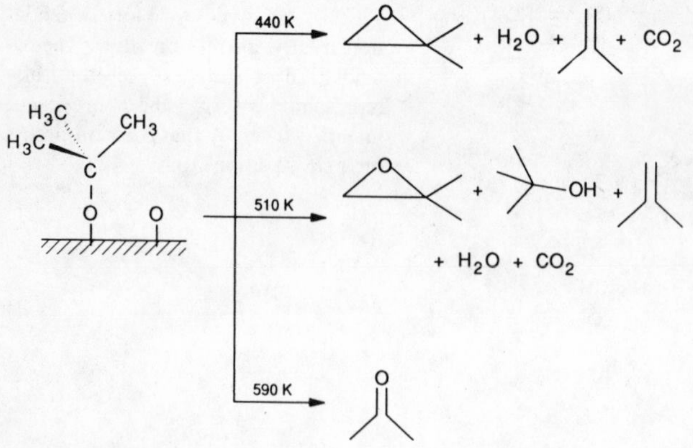

Scheme 2.2. Decomposition of *tert*-butoxy on Ag(1 1 0)

[2.80]. The major decomposition products are isobutylene oxide (**3a**), isobutylene (**3b**), and carbon dioxide.

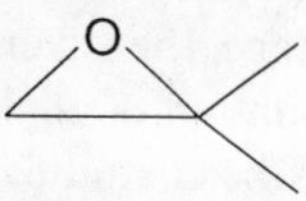

3a 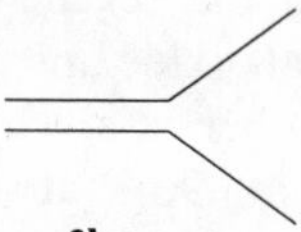 **3b**

The competing reaction pathways of *tert*-butoxy are shown schematically in Scheme 2.2. Formation of an oxametallacycle as a transient intermediate is clearly implied by the simultaneous evolution of isobutylene and isobutylene oxide, which form via two competing reactions of this intermediate. γ-Hydride elimination (i.e. elimination of the H atom located γ relative to the metal surface) of *tert*-butoxy to isobutylene has some precedent in the reaction of alkyl thiolates on Mo(1 1 0), which also undergo γ-elimination to alkenes [2.81]. The mechanism for *tert*-butoxy decomposition on Ag(1 1 0) is not yet understood in detail, but it is clear from kinetic isotope effect measurements for isobutylene oxide and acetone formation that the rate-limiting step for decomposition is cleavage of a C–H bond [2.80b]. The stability of these C–H bonds compared to those of other alkoxys, such as ethoxy and allyloxy, is at least partly ascribed to the fact that the *tert*-butoxy bonds are not activated, as described above.

The stability of *tert*-butoxy on Cu(1 1 0) is also enhanced relative to unactivated alkoxys on the same surface [2.82]. Decomposition occurs at $\approx$ 600 K to form isobutylene, the regenerated alcohol, and water only. The absence of isobutylene oxide, in contrast to what is observed on silver, is attributed to the fact that the Cu–O bond is stronger than the Ag–O bond, rendering formation of the epoxide less favorable. Interestingly, in contrast to primary alcohols like methanol [2.83] and ethanol [2.76], the O–H bond of *tert*-butanol is not activated by the clean Cu(1 1 0) surface [2.82]; *tert*-butoxy is formed via the reaction of the parent alcohol with $O_{(a)}$, in a process analogous to that on Ag(1 1 0). The lack of reaction between clean Cu(1 1 0) and *tert*-butanol may be due to steric congestion around the tertiary carbon, leading to unfavorable kinetics for O–H bond activation compared to reversible desorption.

As a final example of alkoxy oxidation on Ag(1 1 0), we will consider the reaction of dialkoxy, derived from the reaction of ethylene glycol [$(CH_2OH)_2$] with $O_{(a)}$. This example nicely illustrates the interplay between temperature programmed reaction spectroscopy and vibrational spectroscopy in following the course of surface reactions. Furthermore, the reactions between ethylene glycol and $O_{(a)}$ on Ag(1 1 0) are particularly interesting because the ethylene glycol molecule presents two identical reactive functional groups to $O_{(a)}$, introducing several possible complexities to the reaction sequence. It is well known, for instance, that discrete dioxametallacyclic complexes are orders of magnitude more stable than the analogous alkoxide complexes [2.84], presumably due to the strain involved in bringing a C–H bond of a dioxametallacycle into proximity with a metal center. As expected from the reactions of monofunctional alcohols, ethylene glycol reacts with $O_{(a)}$ via activation of both O–H bonds, forming a chemisorbed ethylenedioxy intermediate (**4a**), a dioxametallacycle [2.85]:

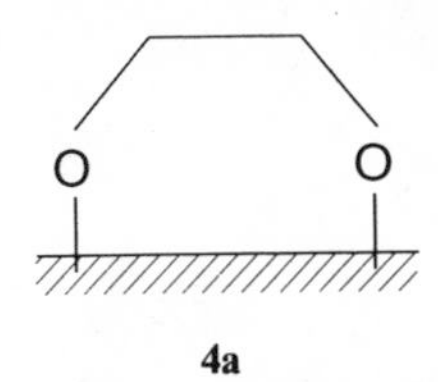

4a

Removal of the hydroxyl hydrogens is evident from the temperature programmed reaction spectrum shown in Fig. 2.8, which is the result of reaction of [$(CD_2OH)_2$] with 0.25 monolayers of $O_{(a)}$ on Ag(1 1 0). Since only H_2O and no HDO or D_2O is evolved, it is clear that only the O–H bonds are activated by $O_{(a)}$. Formation of the dialkoxide is complete by 300 K, as evidenced by the absence of O–H stretches or bending modes in the vibrational spectra obtained after annealing to that temperature (Fig. 2.9). All modes in the vibrational spectra obtained at 125 and 300 K are readily assigned to the diol and the dialkoxide, respectively (Table 2.2). If the reaction between ethylene glycol and $O_{(a)}$ is conducted under "lean oxygen" conditions, such that all initial $O_{(a)}$ is consumed during alkoxy formation, the initial step in the decomposition of ethylenedioxy is cleavage of an activated C–H bond to form the monoalkoxy intermediate **4b** at 365 K [2.85a].

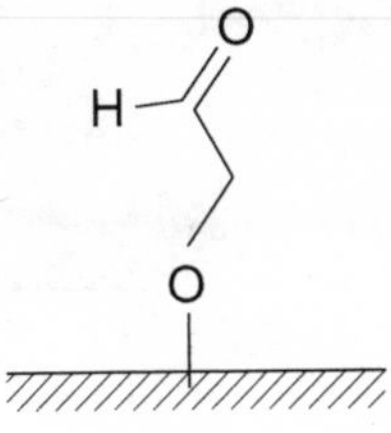

O
H
O

4b

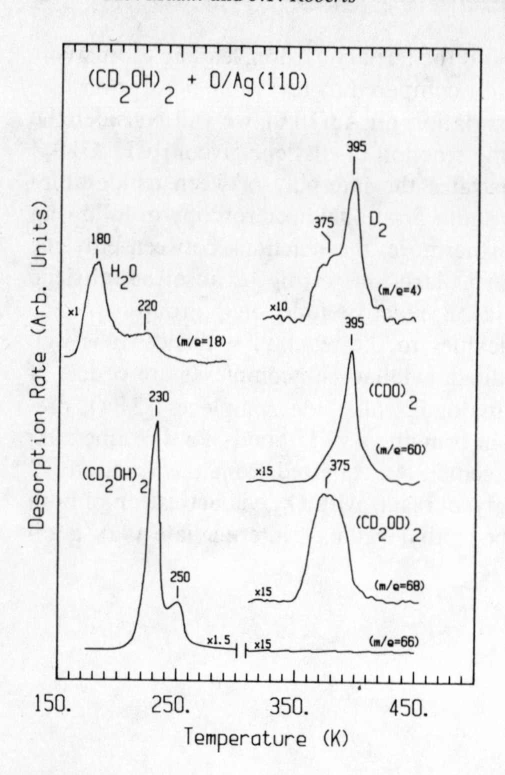

Fig. 2.8. Temperature programmed reaction spectra of ethylene glycol multilayers on Ag(1 1 0) precovered with 0.25 monolayers of atomically adsorbed oxygen

Table 2.2. Summary of the vibrational frequencies [in cm^{-1}] for $(CH_2OH)_{2(a)}$, $(CH_2O)_{2(a)}$, and $O=CHCH_2O_{(a)}$ on Ag(1 1 0)

vibrational mode	liquid	$K_2[OCH_2CH_2O]$	$(CH_2OH_2)_2$/Ag(1 1 0)–0		
			125 K	300 K	360 K
ν(OH)	3275 s (b)		3350 s (b)		
ν_{as}(CH)	2935 s	2899 vs			2730 w[a]
ν_s(CH)	2875 s	2833 s	2900 m (b)	2860 m	2860 m
$\delta(CH_2)$	1459 m	1445 s	1450 m	1450 w	nr
δ(COH)	1405 m (vb)		1410 sh		
$\rho_w(CH_2)$	1332 w	1370 s	nr	1340 w	1330 w
$\rho_t(CH_2)$	1212 w	1290 m	nr	nr	nr
ν_s(CO)	1087 vs	1093 s	1090 s	1090 vs	1090 s
					1730[b]
ν_{as}(CO)	1038 vs	1042 s	nr	nr	
$\rho_r(CH_2)$	887 s	883 m			
ν(CC)	864 m	858 sh	850 m	890 m	890 m
τ(OH)	700 s (b)		710 s (b)		
δ(CCO)	478 w		nr		
τ(CC)	360 w	nr	nr	nr	nr
ν(Ag–O)			nr	240 vs	240 vs

[a] 2730 cm^{-1} mode due to ν(CH) of aldehyde group in $O=CHCH_2O_{(a)}$.
[b] 1730 cm^{-1} mode due to ν(CO) in $O=CHCH_2O_{(a)}$.

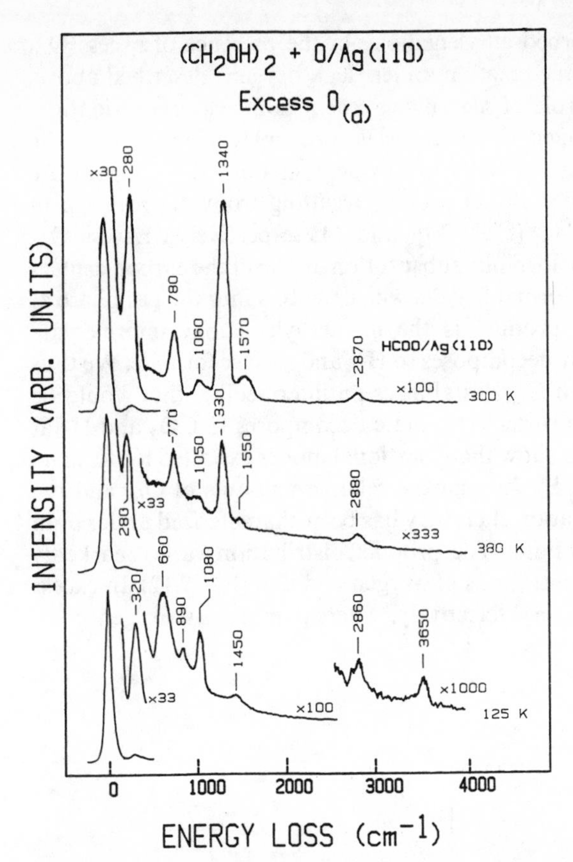

Fig. 2.9. Electron energy loss vibrational spectra of (a) chemisorbed ethylene glycol on Ag(1 1 0), (b) chemisorbed $(CH_2O)_2$ produced from annealing coadsorbed ethylene glycol and $O_{(a)}$ to 300 K, and (c) intermediate **4a** produced from annealing coadsorbed ethylene glycol and $O_{(a)}$ to 360 K

Hydrogen resulting from C–H bond cleavage in formation of **4b** evolves into the gas phase at 365 K via recominative processes, both as H_2 and as regenerated ethylene glycol (Fig. 2.8). Evidence for the formation of **4b** is compelling: HREELS vibrational spectra (Fig. 2.9) clearly demonstrate the presence of both a carbonyl C–O stretch ($\nu(CO) = 1730\ cm^{-1}$) and an aldehydic C–H stretch ($\nu(CH) = 2730\ cm^{-1}$). Intermediate **4b** itself undergoes C–H bond activation at 380 K, producing the dialdehyde glyoxal [$(CHO)_2$] in the gas phase. These reactions are clearly shown in the temperature-programmed reaction spectra shown in Fig. 2.8. It is important to note that cyclic ethylenedioxy intermediate **4a** (which decomposes at 365 K) is considerably more stable than the analogous acyclic methoxy and ethoxy species (decomposition temperature approximately 300 K). Kinetic measurements show that the extra stability of the dialkoxide is due to a higher *entropy* of activation for C–H bond activation in **4a**, the reaction step leading to formation of **4b**. The stability of chemisorbed ethylenedioxy over methoxy thus mirrors the greater stability of platinum ethylenedioxy transition metal complexes compared to their dimethoxy analogs [2.84], though due to the presence of neighboring coordinatively unsaturated metal centers on a transition metal surface the stabilization is much less dramatic.

The reaction of chemisorbed ethylenedioxy in the presence of excess $O_{(a)}$ differs in some respects from the reaction under "lean oxygen" described above, and illustrates the potential role of *scavenging reactions* in selective oxidation. Such reactions can be predicted using the basic principles which govern the reactivity of $O_{(a)}$. Scavenging reactions are evident in both the temperature programmed reaction and vibrational spectra resulting from the reaction of ethylene glycol with excess $O_{(a)}$ (Figs. 2.10 and 11, respectively). Excess $O_{(a)}$ reacts with **4a** at 300 K via nucleophilic substitution at one of the carbon centers [2.85b]. The leaving group is formaldehyde, which evolves into the gas phase at 320 K, and the substitution product is the η^2-methylenedioxy intermediate described above, which rapidly decomposes to $H_{(a)}$ and surface formate. We thus find here an example in which $O_{(a)}$ destabilizes an intermediate that would be stable on the other-wise clean surface. Formate decomposes to CO_2 and H_2 at 410 K. The vibrational spectra show the vibrational modes expected for formate from previous studies [2.33]. By varying the relative amounts of $O_{(a)}$ and the adsorbate of interest, the oxidation chemistry has been characterized under both lean and rich oxidizing conditions. The product distribution varies markedly with the relative surface concentrations of oxygen and diol (Fig. 2.12). In excess diol, the aldehyde is formed very selectively, whereas in excess oxygen combustion products predominate.

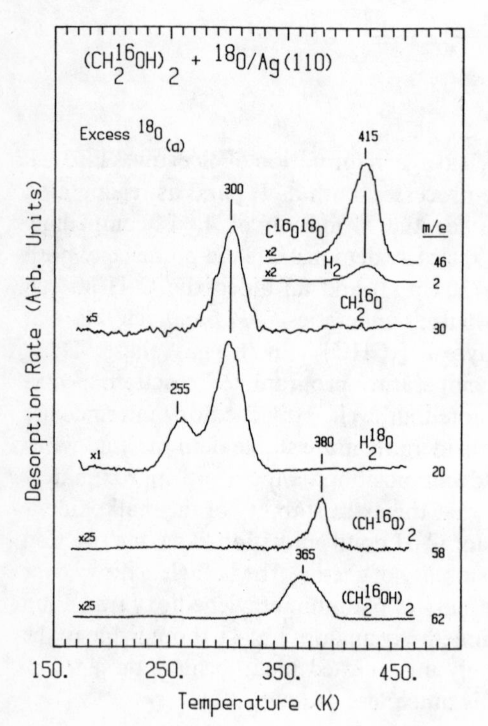

Fig. 2.10. Temperature programmed reaction spectra resulting from the reaction of ethylene glycol with excess $O_{(a)}$ on Ag(1 1 0)

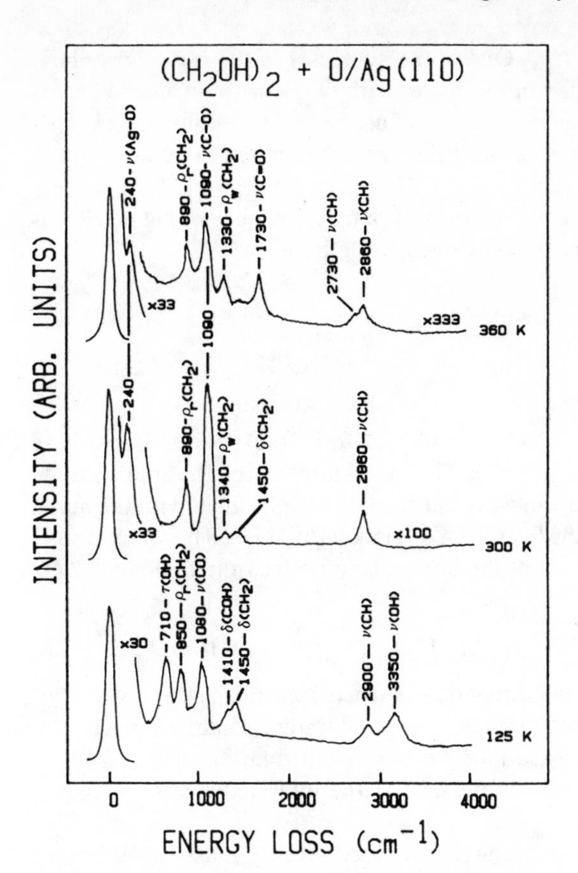

Fig. 2.11. Electron energy loss vibrational spectra resulting from the reaction of ethylene glycol with excess $O_{(a)}$ on Ag(1 1 0)

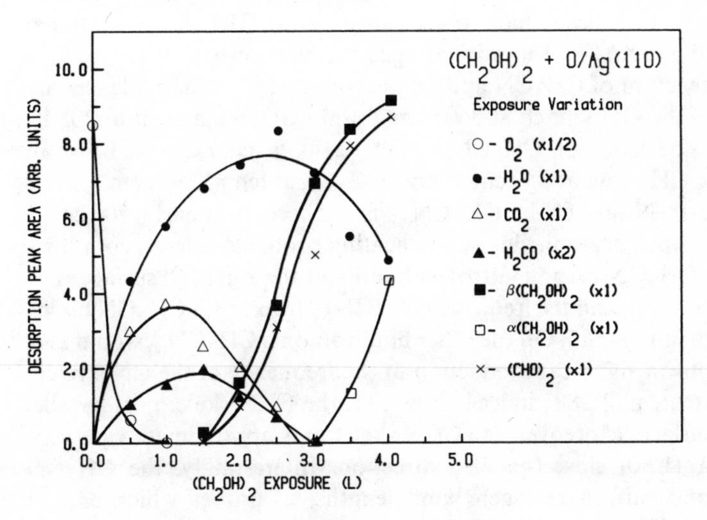

Fig. 2.12. Relative product distributions resulting from the reaction of ethylene glycol with $O_{(a)}$ on Ag(1 1 0) as a function of the oxygen atom precoverage

Reactions of nitriles with $O_{(a)}$. One of the successes of the use of gas phase acidities to predict the reactivity of molecules with $O_{(a)}$ is with nitriles. Whereas ethane is completely unreactive with $O_{(a)}$, acetonitrile readily reacts. sp^3-Hybridized C–H bonds adjacent to a nitrile group are much more acidic than those in a molecule like ethane ($\Delta H_{acid} \approx 373$[55] vs. 421 kcal/mol [2.86]) because the p-orbital which contains the lone-electron pair of a nitrile conjugate base is resonance stabilized by the carbon–nitrogen triple bond:

$$R_2C^{\ominus}{-}C{\equiv}N \longleftrightarrow R_2C{=}C{=}N^{\ominus}$$

Thus, acetonitrile (CH_3CN) reacts with $O_{(a)}$ to form the Brønsted acid-base products $CH_2CN_{(a)}$ and $OH_{(a)}$ [2.87]. The stoichiometry of the intermediate produced by C–H bond activation was confirmed by displacing it with the much stronger gas-phase acid HCOO*D* ($\Delta H_{acid} = 345$ kcal/mol [2.68]), which results in the quantitative displacement of the acetonitrile-derived intermediate [2.60, 87]:

$$CH_xCN_{(a)} + (3\text{-}x)\,HCOOD_{(a)} \rightarrow CH_2D_{3-x}CN_{(a)} + (3\text{-}x)\,HCOO_{(a)}\,. \quad (2.8)$$

Mass spectral analysis of the subsequently desorbed acetonitrile showed that at most one deuterium was incorporated into the intermediate displaced by deuterated formic acid, thus excluding the possible formation of $CHCN_{(a)}$ or $CCN_{(a)}$, leading to the incisive identification of the intermediate as $CH_2CN_{(a)}$ (Fig. 2.13).

Positive identification of these novel intermediates by mass spectrometry thus allows one to study their structure and bonding to the surface. Chemisorbed CH_2CN has also been characterized using both HREELS vibrational spectroscopy and NEXAFS. Vibrational spectroscopy reveals that CH_2CN formation from reaction of CH_3CN and $O_{(a)}$ occurs by 205 K, as evidenced by the appearance of the OH stretch at 3605 cm^{-1} following the anneal to 205 K. The vibrational spectrum of $CH_2CN_{(a)}$ (Fig. 2.14) is characterized by an unusually intense CH_2 wag at 850 cm^{-1} and a C–N stretch at 2075 cm^{-1}. The frequency of the C–N stretch in $CH_2CN_{(a)}$ is reduced by nearly 200 cm^{-1} relative to chemisorbed acetonitrile itself indicating probable side-on coordination between the CH_2CN carbon–nitrogen bond and the Ag(1 1 0) surface. The intensity of the CH_2 wag and the frequency of ν(CN) suggest that CH_2CN bonds to silver, as shown in Fig. 2.15. On the other hand, anionic $CH_2CN_{(a)}$ would also be expected to show a significant reduction in the frequency of the CN stretch. NEXAFS measurements [2.88] indeed show that the CCN skeleton is parallel to the Ag(1 1 0) surface. Moreover, the CCN skeleton is approximately perpendicular to the Ag(1 1 0) closest-packed direction. Interestingly, the CH_2CN carbon–carbon and carbon–nitrogen bond lengths, quantities which can be extracted from the NEXAFS spectra [2.27] are not significantly perturbed from those in chemisorbed acetonitrile; no rehybridization of the CN bond occurs.

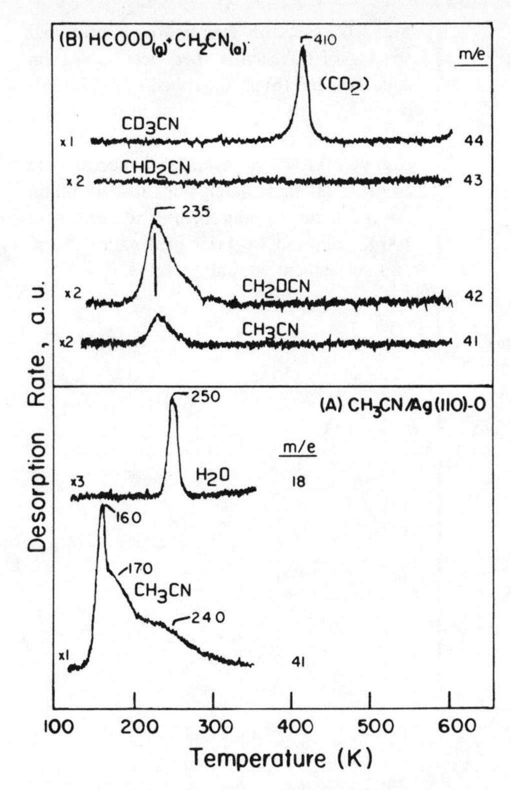

Fig. 2.13. Displacement of $CH_2CN_{(a)}$ by HCOOD on Ag(1 1 0). (a) Temperature programmed reaction spectra recorded after the adsorption of CH_3CN multilayers on Ag(1 1 0) precovered with 0.10 monolayers $O_{(a)}$. The products were acetonitrile (*m/e* 41) and water (*m/e* 18). Acetonitrile multilayers desorb at 160 K, followed by two broad peaks at 170 and 240 K due to the desorption of reversibly chemisorbed acetonitrile. Water, produced via the activation of acidic C–H bonds by $O_{(a)}$ and $OH_{(a)}$, evolves in a single peak at 250 K. Reaction was stopped at 350 K to prevent decomposition of the $CH_2CN_{(a)}$ intermediate produced from the reaction of $CH_3CN_{(a)}$ with $O_{(a)}$. (b) Temperature programmed reaction spectra recorded after subsequently cooling the surface from Fig. 2.7a to 140 K and exposing it to HCOOD. Adsorption of the strong acid HCOOD resulted in displacement of the adsorbed acetonitrile conjugate base back to adsorbed molecular acetonitrile. Spectra of CH_3CN (*m/e* 41), CH_2DCN (*m/e* 42), CHD_2CN (*m/e* 43), and CD_3CN (*m/e* 44) were recorded. No CD_3CN or CHD_2CN was detected, indicating that the stoichiometry of the intermediate from reaction of CH_3CN with $O_{(a)}$ is CH_2CN

Furthermore, the π-orbital perpendicular to the surface plane is significantly broadened by inter-action with the surface, whereas the π-orbital parallel to the surface is unperturbed, demonstrating that the bonding interaction occurs between π_{perp} and the surface.

The thermal stability of CH_2CN on Ag(1 1 0) has been examined using temperature programmed reaction mass spectrometry [2.87]. The intermediate

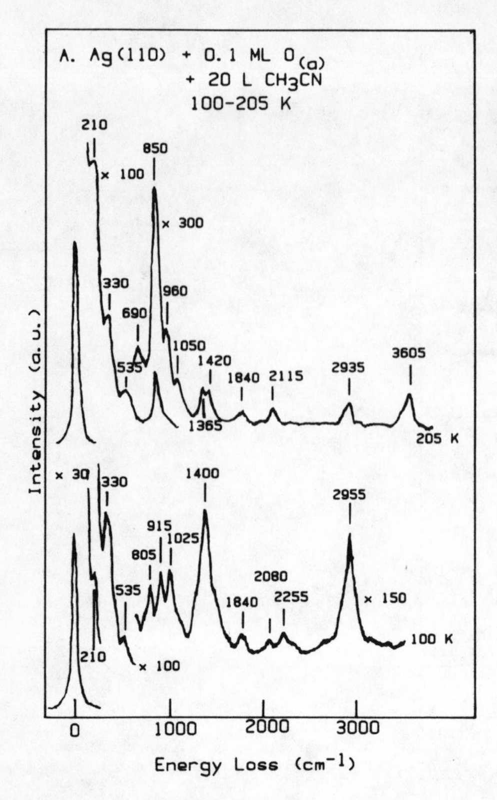

Fig. 2.14. Electron energy loss vibrational spectra of (a) chemisorbed acetonitrile on Ag(1 1 0), and (b) chemisorbed CH_2CN produced from the reaction of acetonitrile with $O_{(a)}$ on Ag(1 1 0). The $O_{(a)}$ coverage in each case was 0.10 monolayers, and spectra were recorded (a) immediately after adsorption on the 100 K surface, and (b) after adsorption at 100 K followed by brief heating to 205 K and subsequent cooling to 100 K

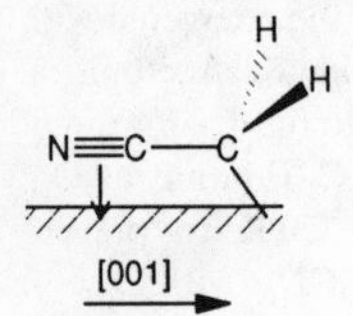

Fig 2.15. Geometry proposed for $CH_2CN_{(a)}$ on Ag(1 1 0)

is stable until ≈ 475 K, when it begins to react via an unusual disproportionation reaction:

$$2\,CH_2CN_{(a)} \rightarrow CH_3CN_{(g)} + HCN_{(g)} + C_{(a)}\,. \tag{2.9}$$

Temperature-programmed reaction mass spectrometry reveals that acetonitrile and hydrogen cyanide evolve into the gas-phase nearly simultaneously during disproportionation. The activation energy for disproportionation is estimated to be 29.6 kcal/mol, assuming that the reaction is first-order with a frequency factor of 10 [2.13, 12]. The mechanism for CH_2CN disproportionation is unknown, but the rate-limiting step likely involves C–H bond scission as judged by the magnitude of the kinetic isotope effect. Surface atomic carbon generated from CH_2CN decomposition can be removed by reaction with $O_{(a)}$ to form gaseous CO_2.

The presence of excess $O_{(a)}$ lowers the barrier for CH_2CN disproportionation on Ag(1 1 0) from 29.6 to 25.8 kcal/mol [2.76]. An analogous effect is also seen for ethylenedioxy and *tert*-butoxy decomposition. Moreover, the *mechanism* of $CH_2CN_{(a)}$ decomposition changes; CH_2CN reacts with $O_{(a)}$ to produce water, chemisorbed nitrile ($CN_{(a)}$) groups, and surface formate ($HCO_{2(a)}$) [2.89]. Experiments conducted with $CH_3^{13}CN$ show that the methyl carbon is oxidized to CO_2 with the kinetics expected from $HCO_{2(a)}$ decomposition. The oxidation of acetonitrile in the presence of excess $O_{(a)}$ may be written:

$$CH_3\,^{13}CN_{(a)} + 3\,O_{(a)} \rightarrow H_2O_{(g)} + HCO_{2(a)} + {}^{13}CN_{(a)}\,. \tag{2.10}$$

As expected, surface formate decomposes to $CO_{2(g)}$ and $H_{2(g)}$ at 410 K. $CN_{(a)}$ is further oxidized by $O_{(a)}$ to $CO_{2(g)}$, $N_{2(g)}$, and $NO_{(g)}$. The mechanism for conversion of $CH_2CN_{(a)}$ to $HCO_{2(a)}$ and $CN_{(a)}$ is not known, but the pathway is believed to involve either (a) insertion of $O_{(a)}$ into the $CH_2CN_{(a)}$ surface-carbon bond, or (b) further deprotonation of $CH_2CN_{(a)}$ by $O_{(a)}$:

$$\text{a)}\; CH_2CN_{(a)} + O_{(a)} \rightarrow OCH_2CN_{(a)}\,,$$

$$\text{b)}\; CH_2CN_{(a)} + O_{(a)} \rightarrow CHCN_{(a)}\,. \tag{2.11}$$

The CHCN intermediate can be thought of as a Fischer-type carbene, a type of ligand well known in organometallic chemistry [2.90]. Fischer-type carbenes are susceptible to nucleophilic attack at carbon [2.91]. To the best of our knowledge OCH_2CN has no equivalent in organometallic chemistry. Either of these intermediates could oxidize further to $HCOCN_{(a)}$, either via deprotonation of $OCH_2CN_{(a)}$ or nucleophilic attack by $O_{(a)}$ on $CHCN_{(a)}$ (Scheme 2.3a).

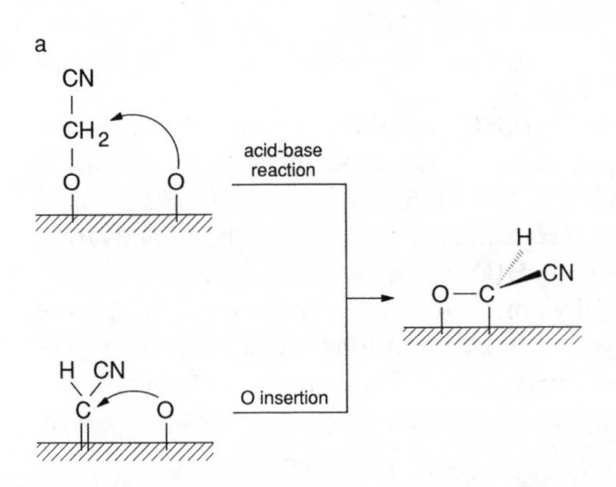

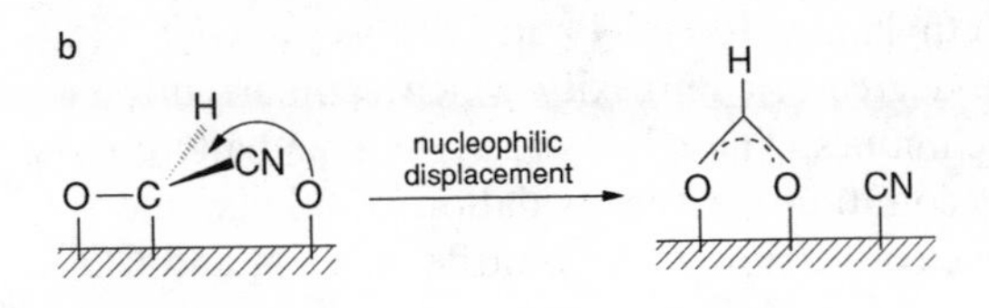

Scheme 2.3. (a) Pathways for the conversion of $OCH_2CN_{(a)}$ and $CHCN_{(a)}$ to $HCOCN_{(a)}$ on Ag(1 1 0). (b) Nucleophilic displacement of $HCOCN_{(a)}$ by atomically chemisorbed oxygen to give adsorbed formate ($HCO_{2(a)}$)

$HCOCN_{(a)}$ could subsequently be oxidized to formate via nucleophilic addition by $O_{(a)}$ at carbon, followed by C–CN bond cleavage (Scheme 2.3b). Note that the pathway proposed for $HCOCN_{(a)}$ oxidation is similar to that for the conversion of $H_2CO_{(a)}$ to formate [2.53].

Reactions of amines with $O_{(a)}$. Surprisingly, N–H bonds in both ammonia [2.92] and ethylamine [2.93] are activated by $O_{(a)}$, despite the fact that their gas phase acidities ($\Delta H_{\mathrm{acid}} = 404$ and 399 kcal/mol, respectively [2.94]) are lower than that of water ($\Delta H_{\mathrm{acid}} = 391$ kcal/mol [2.61]). Ammonia is dehydrogenated by $O_{(a)}$ to $NH_{2(a)}$, $NH_{(a)}$, and then $N_{(a)}$ [2.92]; the exact temperature(s) of N–H bond activation are not yet precisely defined [2.92]. $N_{(a)}$ undergoes recombination to $N_{2(g)}$ at 530 K, or, in the presence of excess $O_{(a)}$, forms $NO_{(g)}$. The temperatures at which $O_{(a)}$ activates the N–H bonds of ethylamine are better defined. Cleavage of the first N–H bond occurs by 155 K, as evidenced by the appearance of a hydroxyl O–H stretch in the HREELS vibrational spectrum [2.93]. The resulting $C_2H_5NH_{(a)}$ intermediate is stable until 360 K, when it reacts in a fashion analogous to hydroxyl disproportionation:

$$2\,C_2H_5NH_{(a)} \rightarrow C_2H_5NH_{2(g)} + C_2H_5N_{(a)}\,. \tag{2.12}$$

This reaction proceeds via the *direct* transfer of H between two $C_2H_5NH_{(a)}$ species and underscores the inability of silver (not $O_{(a)}$) to activate N–H bonds. The $C_2H_5N_{(a)}$ product from $C_2H_5NH_{(a)}$ disproportionation is unstable with respect to C–H metallation by the silver surface at 370 K, decomposing very selectively by cleavage of the activated C–H bond to acetonitrile and hydrogen:

$$C_2H_5N_{(a)} \rightarrow CH_3CN_{(g)} + 2\,H_{(a)}\,,$$
$$2\,H_{(a)} \rightarrow H_{2(g)}\,. \tag{2.13}$$

The rate-limiting step in ethylamine oxidation is therefore metallation of the C–H bonds in $C_2H_5N_{(a)}$ by silver, a reaction that proceeds with an activation energy of 15.0 ± 2.4 kcal/mol and a first-order frequency factor of $10^{9.7 \pm 1.6}\,\mathrm{s}^{-1}$. Experiments conducted with $C_2D_5NH_2$ confirm that all N–H bond activation occurs via reaction with $O_{(a)}$, while C–H(D) bonds are metallated directly by silver [2.93]. The oxidation of ethylamine to acetonitrile is similar to the oxidation of ethanol to acetaldehyde with the exception that two activated C–H bonds are cleaved during ethylamine oxidation. This difference probably reflects the greater thermodynamic stability of the nitrile (CN) group over the C = NH group.

The reactions of ethylamine and acetonitrile with $O_{(a)}$ under "oxygen-lean" conditions are now sufficiently well understood that a catalytic cycle can be constructed for the oxidation of ethylamine to HCN and CO_2, as shown in Fig. 2.16. This cycle is typical of the degree of complexity which can currently be studied using surface science techniques. The complete cycle is actually composed of two cycles, the first accounting for the oxidation of ethylamine to acetonitrile and the second for the conversion of acetonitrile to HCN and CO_2

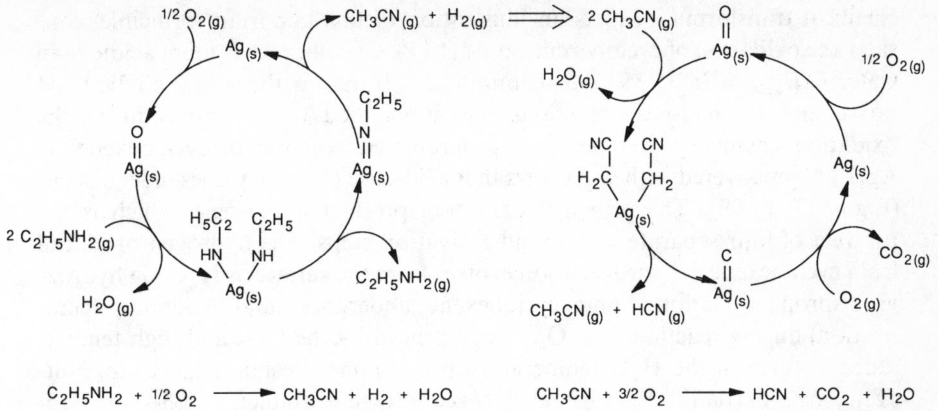
1/2 O2(g)
CH3CN(g) + H2(g)
2 CH3CN(g)
Ag(s)
C2H5
N
Ag(s)
O
2 C2H5NH2(g)
H5C2 C2H5
HN NH
H2O(g)
C2H5NH2(g)
NC CN
H2C CH2
CO2(g)
O2(g)
C
CH3CN(g) + HCN(g)
C2H5NH2 + 1/2 O2 → CH3CN + H2 + H2O
CH3CN + 3/2 O2 → HCN + CO2 + H2O

Fig. 2.16. Catalytic cycle for the oxidation of ethylamine to HCN, CO_2, and H_2O on Ag(1 1 0). The complete cycle is broken into two smaller cycles, one for the oxidation of ethylamine to acetonitrile, H_2O, and H_2, and the second for the oxidation of acetonitrile to HCN and H_2O. The extended silver surface is schematically represented by $Ag_{(s)}$; the figure is not intended to imply that reaction occurs on a single silver atom

in excess O_2. The oxidation cycle is remarkable for its selectivity: no products other than HCN, H_2, H_2O, and CO_2 are formed. Every rate-limiting intermediate on the oxidation pathway has been characterized spectroscopically, and the kinetics for all rate determining steps are well understood.

Reactions of hydrocarbon acids with $O_{(a)}$. As a final illustration of acid–base chemistry on Ag(1 1 0), we will consider the reaction of hydrocarbon acids with $O_{(a)}$. Although $O_{(a)}$ is completely unreactive toward weak hydrocarbon acids like ethane and ethylene on Ag(1 1 0), it readily reacts with good acids; i.e, molecules with acetylenic or allylic C–H bonds [2.95, 96]. Acetylene reacts to form an acetylide ($C_2H_{(a)}$) species which disproportionates upon heating to gaseous acetylene and $C_{2(a)}$ [2.95]. Other reactions of note are the decomposition of propylene to $CO_{2(g)}$ and $H_2O_{(g)}$, probably via an unstable allyl ($CH_2CHCH_{2(a)}$) species [2.96], and the likely formation of surface trimethylenemethane [$(CH_2)_3C_{(a)}$] from 2-methylpropene (Scheme 2.4) [2.97].

Because hydrocarbon oxidation on Ag(1 1 0) is largely dictated by the acidity of the reacting hydrocarbon, it is possible to achieve extremely selective

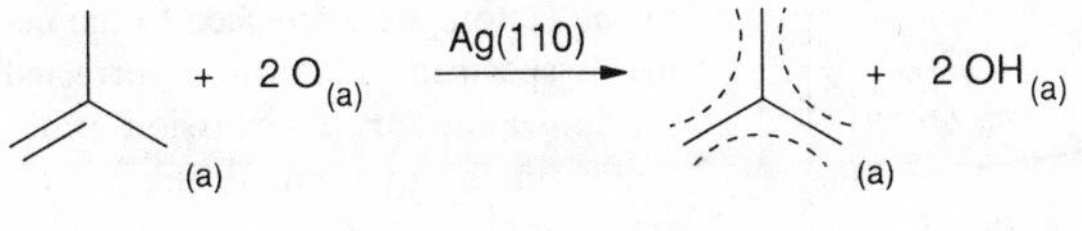
+ 2 O(a)
Ag(110)
+ 2 OH(a)
(a)
(a)

Scheme 2.4. Formation of the surface trimethylenemethane intermediate on Ag(1 1 0) from reaction of isobutylene with atomically chemisorbed oxygen

chemical transformations. As an illustration of this important principle, consider the oxidation of cyclohexene on Ag(1 1 0). Cyclohexene is more acidic than water ($\Delta H_{acid} = 387^{98}$ vs. 395 kcal/mol [2.61]) due to the acidic allylic C–H bond, and so acid-base reactions are anticipated to be important in the oxidation chemistry. Temperature programmed reaction of cyclohexene on Ag(1 1 0) precovered with $O_{(a)}$ shows that acid-base chemistry does indeed occur (Fig. 2.17) [2.99]. The principal oxidation product is benzene, which is the product of four separate C–H bond activation steps. The hydrogen produced from cyclohexene dehydrogenation evolves from the surface as H_2O via hydroxyl disproportionation. Some cyclohexene undergoes total oxidation (combustion) during reaction with $O_{(a)}$, as evidenced by the CO_2 and high temperature features in the H_2O temperature programmed reaction spectrum. Notably, no partially hydrogenated cyclohexene products- 1,3- or 1,5-cyclohexadiene- were detected. HREELS vibrational spectroscopy showed that C–H bond activation begins to occur by 150 K, as evidenced by the appearance of a hydroxyl O–H stretch. Furthermore, benzene formation on Ag(1 1 0) was *complete* by 200 K, with no detectable formation of a surface-stable cyclohexadiene species. These experiments led to the proposal that the slow step in cyclohexene dehydrogenation by $O_{(a)}$ on Ag(1 1 0) is the *initial* C–H bond activation step to form a surface allyl species, as shown in Fig. 2.18. π-Electron delocalization of the cyclohexallyl species is greater than that of cyclohexene

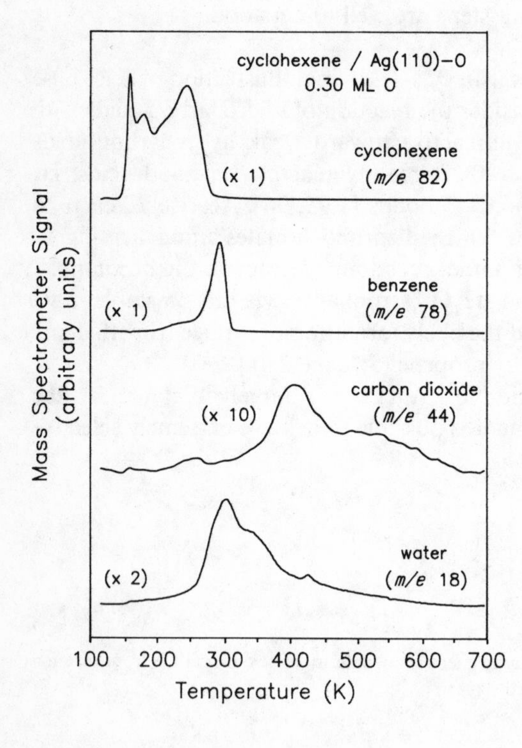

Fig. 2.17. Temperature programmed reaction spectra of cyclohexene multilayers on Ag(1 1 0) precovered with 0.30 monolayers of atomically adsorbed oxygen. Products detected were water (detected as *m/e* 18), carbon dioxide (m/e 44), benzene (*m/e* 78), and cyclohexene (detected as *m/e* 82). Adsorption occurred at 100 K, and the heating rate was approximately 6 K/s. Multiplication factors are referenced to the benzene spectrum and are uncorrected for ionization or transmission probability in the quadrupole mass spectrometer

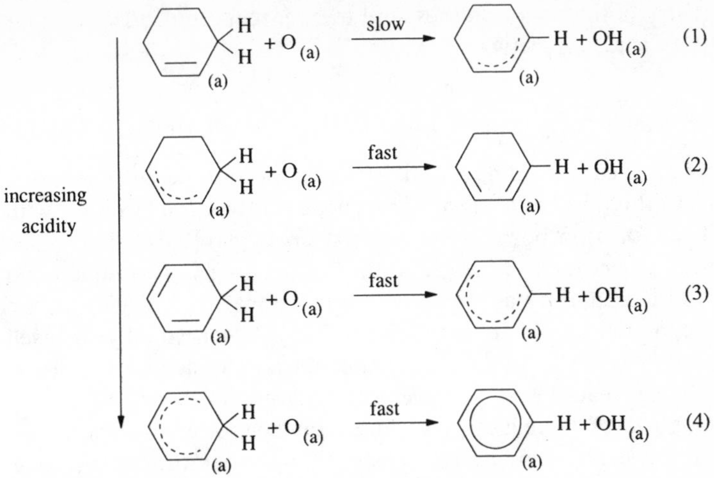

increasing acidity

$C_6H_{10(a)} + O_{(a)} \xrightarrow{\text{slow}} C_6H_{9(a)} + OH_{(a)}$ (1)

$C_6H_{9(a)} + O_{(a)} \xrightarrow{\text{fast}} C_6H_{8(a)} + OH_{(a)}$ (2)

$C_6H_{8(a)} + O_{(a)} \xrightarrow{\text{fast}} C_6H_{7(a)} + OH_{(a)}$ (3)

$C_6H_{7(a)} + O_{(a)} \xrightarrow{\text{fast}} C_6H_{6(a)} + OH_{(a)}$ (4)

Fig. 2.18. Proposed pathway for the dehydrogenation of cyclohexene to benzene by atomically adsorbed oxygen on Ag(1 1 0)

itself, leading to allylic C–H bonds that are more acidic than those in cyclohexene. Thus, the second C–H bond activation step is faster than the first, the third is faster than the second, and the fourth is faster than the third. The pathway indicated in Fig. 2.18 rationalizes the absence of any long-lived cyclohexadiene intermediate in the HREELS vibrational spectra and also the lack of any cyclohexadiene evolution into the gas-phase. Interestingly, vibrational spectroscopy shows that the intermediate which leads to cyclohexene combustion on Ag(1 1 0) is likely an alkoxy ($RO_{(a)}$) species which in the presence of excess $O_{(a)}$ is converted to a phenolate ($C_6H_5O_{(a)}$) or related species [2.100]. The pathway leading to alkoxy formation is not an acid–base reaction but may be related to the oxygen addition reactions described in the next subsection.

Related to the dehydrogenation of cyclohexene is the dehydrogenation of 1-butene to 1,3-butadiene [2.101]. An important difference is observed, however, because, while benzene itself is unreactive toward $O_{(a)}$ [2.42], 1,3-butadiene is reactive and goes on to form furan and furan-derived products by cyclization [2.101], as described in the next subsection.

Concluding remarks on acid-base chemistry. The ability of $O_{(a)}$ to abstract a proton from a chemisorbed molecule on Ag(1 1 0) is defined by the gas-phase acidity of that molecule. These Brønsted acid–base reactions allow for the rational synthesis of an essentially limitless number of surface compounds. The structure of these compounds can then be determined spectroscopically, and their reactions may be examined. We have seen in the present subsection that these surface compounds react via a number of distinct pathways: decomposition via C-H metallation, disproportionation, and addition of excess $O_{(a)}$ via nucleophilic attack. The possibilities for future studies in this area are many. The acid–base chemistry of these molecules is sure to be fascinating and instructive in the principles of surface reactivity.

2.3.3 Addition of Atomic Oxygen to Carbon–Carbon Double Bonds on Silver

The addition of $O_{(a)}$ to carbon–carbon double bonds was only recently reported on silver under ultrahigh vacuum, when norbornene was observed to react with $O_{(a)}$ on Ag(1 1 0) to form norbornene epoxide (Scheme 2.5a) [2.44]. This is the first unambiguous observation of oxygen atom transfer to a C–C double bond on any metallic surface, and it has important implications, in particular for the silver-catalyzed epoxidation of ethylene (Scheme 2.5b). Adsorbed ethylene itself does not react with $O_{(a)}$ or $O_{2(a)}$ on silver under ultrahigh vacuum because it desorbs before it can react [2.102]. Under the high-pressure conditions of a catalytic reactor, however, there is always a finite ethylene surface coverage even at temperatures above the characteristic desorption temperature of ethylene. Indeed, ethylene epoxidation proceeds readily on both Ag(1 1 1) and Ag(11) at appropriately high pressures [2.43, 103]. Oxygen addition to C–C double bonds is to be distinguished from nucleophilic addition. Carbon–carbon double bonds are actually weakly nucleophilic themselves, and would therefore not be expected to undergo nucleophilic attack by $O_{(a)}$. An explanation for the apparent bifunctional nature of $O_{(a)}$ – why it acts both as a nucleophile and an electrophile – will be given below.

The reactions of norbornene with $O_{(a)}$ were examined because norbornene is only weakly acidic (ΔH_{acid} = 402 kcal/mol [2.98]), as distinct from other cyclic olefins like cyclohexene (ΔH_{acid} = 387 kcal/mol [2.98]). We therefore expected that acid-base reactions with $O_{(a)}$ would not contribute in any significant way to the chemistry observed. Norbornene is such a poor acid because the allyl C–H bond is fixed in space by the bridgehead position so that it is nearly orthogonal to the p-orbitals in the C–C double bond. The conjugate base of norbornene is therefore not resonance stabilized, in contrast to other simple olefins where resonance stabilization occurs. The most acidic C–H bonds of norbornene are in fact not the allyl C–H bonds, but are the olefinic C–H bonds [2.98], suggesting that norbornene may be a good model for ethylene (ΔH_{acid} = 406 kcal/mol [2.86]), at least in the context of olefin oxidation on silver. Norbornene was also chosen as an attractive candidate for epoxidation under ultrahigh vacuum

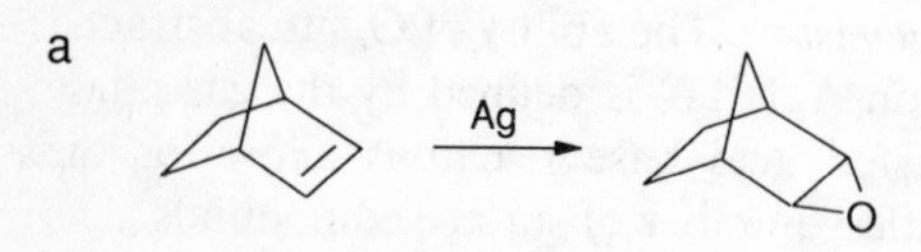

Scheme 2.5. (a) Conversion of norbornene to norbornene epoxide. (b) Conversion of ethylene to ethylene epoxide

because it has a much higher molecular weight than ethylene, and so could remain chemisorbed to temperatures sufficiently high for reaction to occur with $O_{(a)}$.

Temperature-programmed reaction spectra for norbornene reacting with 0.2 monolayers of $O_{(a)}$ are shown in Fig. 2.19. Norbornene oxide evolves at 310 K, well above its characteristic desorption temperature of ≈ 250 K on either clean or oxygen-covered Ag(1 1 0) [2.44]. The activation energy for norbornene oxide evolution therefore reflects the kinetics for epoxide formation, not epoxide desorption. The activation energy (15 kcal/mol) is very close to the activation energy for ethylene epoxidation over an unpromoted silver catalyst [2.1]. The selectivity for norbornene epoxidation, defined as that fraction of norbornene which reacts to form epoxide, is approximately 50%, again very close to the selectivity for ethylene epoxidation over an unpromoted silver catalyst [2.1]. The other 50% forms combustion products – CO_2 and H_2O. Reaction selectivity is relatively independent of both the norborene coverage and the oxygen coverage between 0.05 and 0.50 monolayers.

Norbornene oxide was identified as the norbornene oxidation product by comparison of its mass spectral cracking pattern with that of an authentic norbornene oxide sample (Table 2.3). As recently as five years ago, identification of the norbornene oxidation product would have been exceedingly difficult, but recent advances in instrumentation allow for the use of computer multiplexing

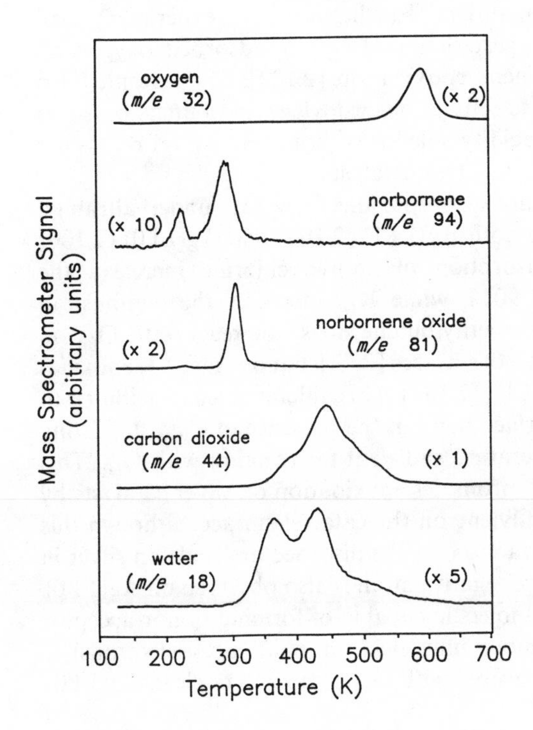

Fig. 2.19. Temperature programmed reaction of norbornene in the presence of excess surface oxygen atoms on Ag(1 1 0). The oxygen coverage was approximately 0.20 monolayers. The heating rate was 12 K/s. Multiplication factors are referenced to the carbon dioxide spectrum and are corrected for degree of fragmentation in the mass spectrometer

Table 2.3. Relative yields of the ions for determination of the norbornene epoxide product

m/e	norbornene oxidation product	pure norbornene epoxide sample
81	100.0	100.0
82	30.6	30.5
95	23.8	20.6
105	1.6	1.2
106	1.5	1.1
109	3.7	4.0
110	6.1	5.8

to simultaneously detect up to 300 different ions during a single TPRS experiment [2.104]. Multiple-ion detection is absolutely essential for the identification of complicated products like norbornene oxide. Furthermore, computer multiplexing allows for the detection of several different products during one experiment, making the quantification of product yields both easier and more accurate.

The epoxidation of norbornene by $O_{(a)}$ on a silver is an important observation for two reasons. First, the nature of the oxygen species which is active for ethylene epoxidation has been a subject of debate for decades [2.1, 43, 105]. This is the first unambiguous demonstration that atomically chemisorbed oxygen is active for epoxidation on a metal surface. Furthermore, the experiments discussed earlier in this chapter suggest that molecularly adsorbed oxygen on Ag(1 1 0) is not active for norbornene epoxidation [2.100]. These results also suggest that the low epoxidation selectivity for propylene and butene on silver catalysts is due to the high olefin acidity. Olefins without acidic C–H bonds are likely to be efficiently oxygenated on silver catalysts.

Other weakly acidic olefins are also epoxidized by $O_{(a)}$ under ultrahigh vacuum. Styrene is epoxidized on both Ag(1 1 1) [2.106] and Ag(1 1 0) [2.100] surfaces. On Ag(1 1 1) the coadsorption of atomic chlorine increases the epoxidation selectivity to above 90%, while $K_{(a)}$ increases the combustion efficiency to 100% [2.106], 3, 3-Dimethylbutene does not react with $O_{(a)}$ on either the Ag(1 1 1) [2.107] or Ag(1 1 0) surfaces [2.42], but on Ag(1 1 1) epoxidation does occur in the presence of $Cl_{(a)}$ [2.107]. This difference was attributed to a stronger 3, 3-dimethylbutene-surface bond in the presence of $Cl_{(a)}$, stabilizing the chemisorbed olefin to a temperature sufficient for reaction with $O_{(a)}$. This result led to the suggestion that Cl promotes epoxidation on silver catalysts by increasing the residence time of ethylene on the catalyst surface, although this conclusion is tentative. The stabilization of a chemisorbed molecule on silver in the presence of a coadsorbed electronegative atom is also observed on Ag(1 1 0). The desorption temperature of a molecule capable of forming donor-acceptor bonds (i.e., a molecule with nonbonding electron pairs or π-electrons) is invariably higher on Ag(1 1 0) precovered with $O_{(a)}$ compared to clean Ag(1 1 0).

For instance, benzene desorbs from Ag(1 1 0) at 255 K, but in the presence of $O_{(a)}$ desorption occurs at 280 K [2.100]. Indeed, it was proposed over ten years ago that electronegative adatoms on silver stabilize chemisorbate bonds by creating Lewis acid sites, leading to enhanced bonding between adsorbates which bond via electron donation to the silver surface [2.53].

Butadiene and $O_{(a)}$ react on Ag(1 1 0), but not to form epoxide adducts [2.101]. The principal products are furan, 2,5-dihydrofuran, CO_2, and H_2O, along with very small amounts of maleic anhydride (Scheme 2.6). Formally, the reaction of butadiene and $O_{(a)}$ to form 2,5-dihydrofuran is a [4 + 2] cycloaddition reaction. Thus, cycloaddition reactions can be considered another reaction class of $O_{(a)}$. Five-membered oxygen containing rings are formed rather than epoxides presumably because their lower ring strain makes them more thermodynamically stable. Butadiene also dimerizes on Ag(1 1 0) via a non-oxidative pathway to form vinylcyclohexene, a Diels-Alder adduct [2.108]. Vinylcyclohexene has acidic allylic C–H bonds, which are dehydrogenated in a fashion analogous to cyclohexene, forming styrene [2.101].

The epoxidation of ethylene over a silver-powder catalyst has recently been investigated using a new transient technique known as Temporal Analysis of Products (TAP) [2.109]. The TAP apparatus is a pulsed microreactor for which contact times between the reactant gases and the catalyst are on the order of several milliseconds, and the time delay between reactant pulses can be varied. All ethylene epoxidation experiments were performed for catalyst temperatures between 475 and 575 K. At 525 K epoxidation occurs even if ethylene is pulsed as late as 500 ms after the initial O_2 pulse (Fig. 2.20). For these temperatures and delay times, dissociation and/or desorption of molecularly adsorbed oxygen should be complete before adsorption of ethylene occurs. Therefore, these

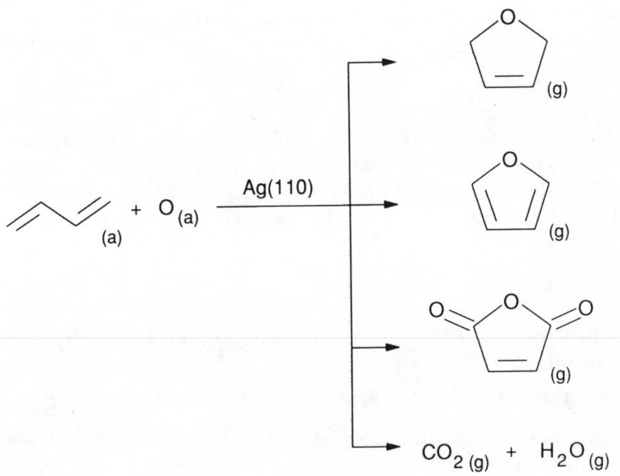

Scheme 2.6. Oxidation of 1,3-butadiene to dihydrofuran, furan, maleic anhydride, and carbon dioxide

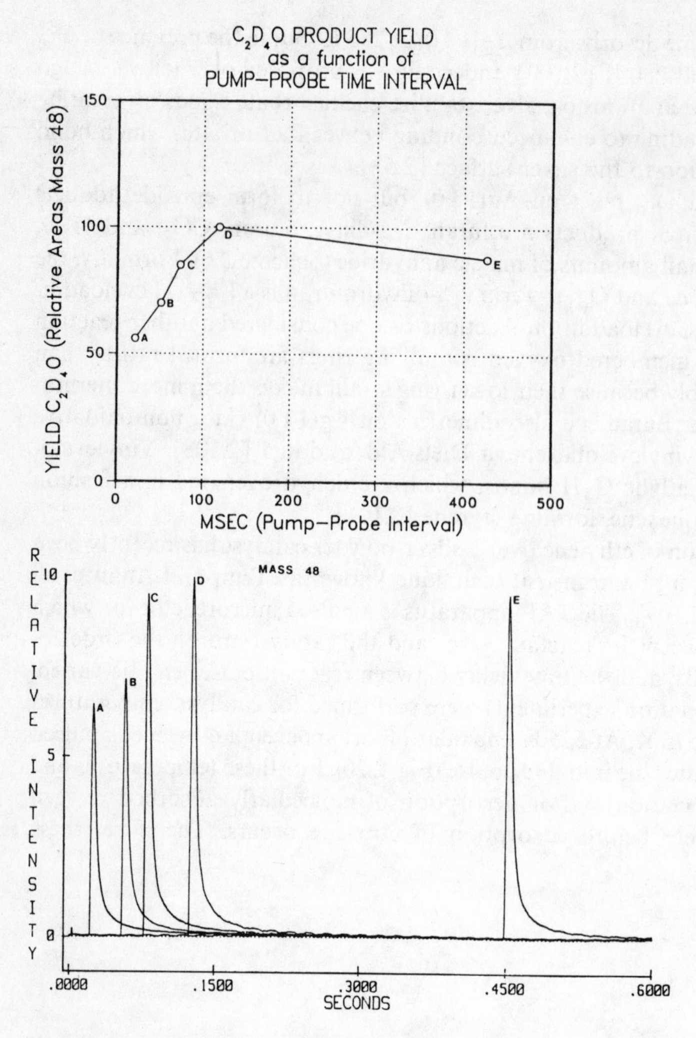

Fig. 2.20. Series of ethylene oxide-d_4 yields in the TAP reactor, with the silver powder temperature held fixed at 525 K. The delay between the initial $O_{2(g)}$ pulse and the reactant ethylene-d_4 pulse is varied: (a) 20 ms, (b) 50 ms, (c) 72 ms, (d) 117 ms (e) 425 (ms). The fact that the yield changes so little with increasing delay time indicates that molecularly adsorbed oxygen is not the active species for ethylene epoxidation

experiments further confirm that $O_{(a)}$, not $O_{2(a)}$ is the active species for ethylene epoxidation over a silver catalyst.

The detailed mechanism by which $O_{(a)}$ adds to olefins to form oxygen-containing rings is still unclear. Particularly mysterious is how $O_{(a)}$ can act as an electron-rich species, participating in both nucleophilic addition and acid–base reactions, and also, at least formally, as an electrophile in olefin epoxidation. The exact solution to this puzzle remains to be discovered but quantum mechanical calculations suggest that two forms of atomic oxygen exist on a silver surface – a closed shell electron-rich species and an oxyradical [2.34, 37]. Two calculations have been performed, both generalized valence bond treatments of $O_{(a)}$ on a small silver cluster. The results of the calculations differ in that one finds the two forms of $O_{(a)}$ are approximately equal in energy [2.34], while the second finds a 10–15 kcal/mol difference, with the closed-shell species the more stable [2.37]. It is possible that the closed-shell species acts as a nucleophile and a base, while the oxyradical adds to C–C double bonds via a radical addition pathway. If the oxyradical is truly less stable than the closed-shell species, then the origin of the activation barrier for olefin epoxidation may be the promotion energy of closed shell $O_{(a)}$ to the oxyradical. The possibility that the oxyradical is the less stable species also rationalizes the fact that cyclohexene undergoes almost exclusive dehydrogenation to benzene during temperature programmed reaction. No competitive epoxidation is observed because all $O_{(a)}$ reacts as a Brønsted base below the temperature at which promotion to the oxyradical might occur. These conclusions are tentative, but the provocative suggestion that $O_{(a)}$ on silver exists both as a closed shell and as a radical species certainly merits further exploration.

2.3.4 Generalization to Other Metals

One might predict $O_{(a)}$ to react with coadsorbed molecules on any metal surface to which $O_{(a)}$ is reversibly bound. In general, this prediction proves to be true. Although no metal surface has been studied as extensively as silver, both partial and/or total oxidative processes have been observed on many transition metal surfaces, among them gold, palladium, platinum, and rhodium. As will be seen, however, the acid-base paradigm, which works so well for silver, cannot always be extended to other metal surfaces.

An unusual class of olefin oxidation reactions has recently been reported on Rh(1 1 1). In contrast to Ag(1 1 0), where C–H bond activation is the dominant oxidative pathway observed for acidic olefins, at oxygen coverages near 0.5 ML oxygen addition processes are observed on Rh(1 1 1). Olefins containing a vinylic C–H bond are oxidized to the corresponding methyl ketone. For example, propene forms acetone and styrene forms acetophenone. Norbornene, 1,3-butadiene, and all isomers of butene are oxidized to their corresponding methylketones. Atomically adsorbed oxygen, not $OH_{(a)}$ or $O_{2(a)}$, is the oxidant in these reaction systems.

The oxidation pathway for olefins on Rh(1 1 1) is exemplified by that of propene (Fig. 2.21). Although no surface intermediates have been spectroscopically identified to date, direct addition of oxygen to C_2 of propylene with concomitant oxametallacycle formation is proposed as the first step in the reaction sequence. Analogous oxametallacycles have been observed in organometallic complexes [2.110]; they were also proposed as intermediates during the oxidation of *t*-butanol on Ag and Cu [2.80, 82]. Net transfer of one hydrogen from C_2 to C_1 of propylene is required for acetone formation. However, reaction

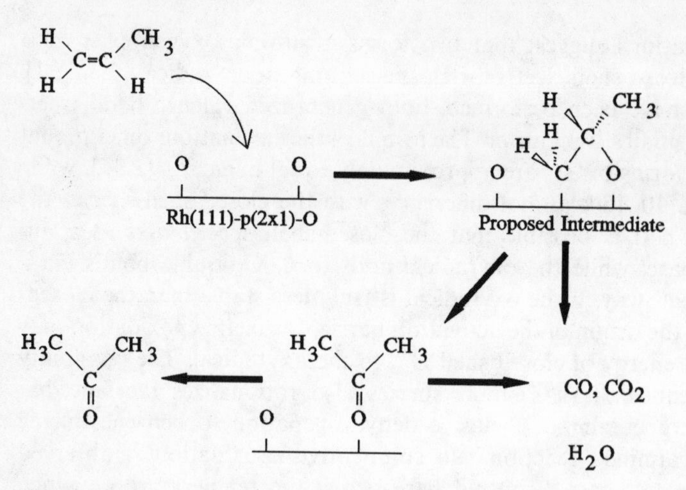

Fig. 2.21. Pathway proposed for the reaction of propene with $O_{(a)}$ on Rh(1 1 1)

of a mixture of propene-d_6 and propene-d_0 produced mixed isotopes of acetone, demonstrating that the hydrogen transfer step is not intramolecular. Transfer can thus occur via the surface or by dispropotionation. Unlike oxidation reactions on Ag(1 1 0), allylic C–H bonds are *not* broken along the pathway for ketone formation. Selective isotopic labeling experiments specifically demonstrated that allylic C–H bond activation does not occur during acetone formation from propene on Rh(1 1 1)-p(2 × 1)–O [2.111]. Specifically, reaction of propene-3, 3, 3-d_3 produced only acetone-d_3. The absence of exchange H–D exchange for the reaction of propene-3, 3, 3-d_3 shows that allylic C–H bonds are not broken along the pathway for acetone formation. The kinetics and selectivity for styrene oxidation on Rh(1 1 1)-p(1 × 1)–O [2.112], which does not contain allylic C–H bonds, are similar to propene, providing corroborating evidence that allylic C–H bond activation does not control olefin oxidation over rhodium.

The vinyl C–H bond of propene is selectively cleaved along the pathway for acetone formation. This observation suggests that oxygen addition precedes hydrogen transfer since addition of oxygen at C_2 is expected to weaken the vinyl C–H bond [2.113]. Elimination of hydrogen bound to the same carbon as oxygen, or β-elimination of an activated C–H bond, is expected to be most facile since the electronegative oxygen lowers the energy for homolytic C–H cleavage. Indeed, the preference for β-C–H bond cleavage on Rh(1 1 1)-p(2 × 1)–O has been specifically demonstrated for the related 2-propoxide intermediate using selective isotopic labeling [2.114], and is expected on the basis of the reactivity of alkoxides established of copper [2.76, 82, 83] and silver [2.70–72, 80] surfaces.

The importance of vinyl C–H bond cleavage in oxygen addition to alkenes on Rh(1 1 1) was further demonstrated on the basis of studies of the oxidation of isobutene. Isobutene does not contain a vinylic hydrogen and, consequently,

β-hydride elimination cannot follow oxygen addition to C_2. As noted above, tertiary alkoxides are very stable, and new reaction channels can be expected to open due to the enhanced kinetic stability of the intermediate. Isobutene is oxidized to *tert*-butoxide on Rh(1 1 1) at oxygen coverages above 0.3 monolayers. The *tert*-butoxide intermediate reacts to form isobutene and *tert*-butanol at approximately 380 K during temperature programmed reaction; the products formed are identical to those from *tert*-butoxide on Cu(1 1 0) [2.82].

It is of obvious importance to establish, if possible, the nature of the transition state which leads to ketone formation. The possibility that a cationic intermediate is formed in the oxidation of olefins to ketones on Rh(1 1 1) was ruled to be unlikely on the basis of studies of norbornene oxidation. Norbornene is oxidized to norbornanone on Rh(1 1 1)-p(2 × 1)–O; there are no products involving skeletal rearrangement. Previously, the oxidation of norbornene on an alumina-supported Rh complex was suggested to occur via a cationic intermediate based on the observation of products which formed via rearrangement of the hydrocarbon skeleton [2.115]. The oxidation of norbornene on the extended Rh(1 1 1) surface is clearly different from that on the supported catalyst.

Surprisingly, oxygen-rich conditions favor partial oxidation on Rh(1 1 1). For all olefins except isobutene, the maximum ketone yield is achieved at oxygen coverages of approximately 0.5 monolayers. At lower oxygen coverages, combustion to CO, CO_2, and H_2O predominate. This unusual behavior is attributed to the inhibition of competitive C–H bond breaking processes by surface oxygen atoms on Rh(1 1 1). On clean Rh(1 1 1), dehydrogenation is facile. For example, propene undergoes extensive C–H bond breaking below 370 K on clean Rh, forming propylidyne, which decomposes to ethylidyne and CH at approximately 300 K [2.116]. Importantly, propylidyne does not produce acetone when reacted with $O_{(a)}$ on Rh(1 1 1); it yields CO, CO_2, and H_2O. As the oxygen coverage is increased on Rh(1 1 1), the fraction of propene which forms propylidyne decreases and the propene desorption yield increases. Inhibition of dehydrogenation by $O_{(a)}$ on Rh(1 1 1) leaves propene intact for oxygen addition and ultimately acetone formation.

The yield of *tert*-butanol from isobutene oxidation is at a maximum for intermediate oxygen coverages. This is attributed to the fact that some of the isobutene must dehydrogenate on clean surface sites to provide a source of hydrogen for *tert*-butanol production. Iostopic labeling experiments demonstrated that one C–H bond and one O–H bond are formed during *tert*-butanol formation from isobutene; there is no reversible C–H bond activation during partial oxidation. Dehydrogenation leads to combustion.

The role of oxygen at high coverages on Rh(1 1 1) is opposite to that on Ag, for example. On the late transition metals, such as Ag and Cu, oxygen promotes C–H bond activation via Brønsted acid-base reactions. Clearly, oxygen on Rh(1 1 1) does not serve as a Brønsted base. The origin of the differences between Rh and the Group IB metals is not yet understood, but is currently the topic of theoretical studies.

On Au(1 1 0), where $O_{(a)}$ recombination to $O_{2(g)}$, occurs at 590 K [2.7] $O_{(a)}$ readily acts as both a nucleophile and as a Brønsted base. Carbon monoxide reacts with $O_{(a)}$ on Au(1 1 0) via nucleophilic oxygen addition to form carbon dioxide which immediately evolves into the gas-phase even for surface temperatures as low as 125 K [2.117]. In contrast to CO_2 on Ag(1 1 0), CO_2 does not suffer attack from $O_{(a)}$ to form surface carbonate ($CO_{3(a)}$) on Au(1 1 0) [2.120]. Nucleophilic addition of $O_{(a)}$ also occurs for formaldehyde (H_2CO) [2.118]. Oxygen addition to formaldehyde results in the formation of the transiently stable η^2-methylenedioxy intermediate which decomposes to surface formate and $H_{(a)}$:

$$O_{(a)} + H_2CO_{(a)} \rightarrow H_2CO_{2(a)},$$
$$H_2CO_{2(a)} \rightarrow HCO_{2(a)} + H_{(a)}. \quad (2.14)$$

$H_{(a)}$ undergoes competitive recombination to $H_{2(g)}$ and reaction with excess $O_{(a)}$ to $H_2O_{(g)}$. Although other molecules subject to nucleophilic attack have not been investigated in any great detail on Au(1 1 0), it is clear that $O_{(a)}$ on Au(1 1 0) has strong nucleophilic character, just as it does on Ag(1 1 0).

The Brønsted acid–base paradigm explained above for $O_{(a)}$ on Ag(1 1 0) also extends to $O_{(a)}$ on Au(1 1 0). Water is dissociated by $O_{(a)}$ via an acid–base reaction below 190 K [2.119]:

$$H_2O_{(a)} + O_{(a)} \rightarrow 2\,OH_{(a)}. \quad (2.15)$$

Hydroxyl disproportionation back to $H_2O_{(g)}$ and $O_{(a)}$ occurs at 215 K. Atomic oxygen also activates the O–H bond in methanol ($\Delta H_{acid} = 382$ kcal/mol [2.65]:

$$CH_3OH_{(a)} + O_{(a)} \rightarrow CH_3O_{(a)} + OH_{(a)}. \quad (2.16)$$

The decomposition of $CH_3O_{(a)}$ on Au(1 1 0) [2.119] is somewhat more complicated than on Ag(1 1 0) [2.71]. At 250 K, methoxy disproportionation occurs to yield gaseous methanol and chemisorbed formaldehyde:

$$2\,CH_3O_{(a)} \rightarrow CH_3OH_{(g)} + H_2CO_{(a)}. \quad (2.17)$$

The formaldehyde product reacts before it can desorb, undergoing nucleophilic attack by another $CH_3O_{(a)}$ and then C–H bond activation, giving gaseous methyl formate ($HCOOCH_{3(g)}$) and $H_{(a)}$, which is largely scavenged by any remaining $CH_3O_{(a)}$ to give $CH_3OH_{(g)}$:

$$CH_3O_{(a)} + H_2CO_{(a)} \rightarrow HCOOCH_{3(g)} + H_{(a)},$$
$$CH_3O_{(a)} + H_{(a)} \rightarrow CH_3OH_{(g)}. \quad (2.18)$$

The O–H bond in formic acid also reacts with $O_{(a)}$ to form surface formate and $OH_{(a)}$ [2.118]. Formate on Au(1 1 0) is stable until 340 K when disproportionation occurs:

$$2\,HCO_{2(a)} \rightarrow CO_{2(g)} + HCOOH_{(g)}. \quad (2.19)$$

In the presence of excess $O_{(a)}$, some H_2O is also produced. The formate-decomposition temperature on Au(1 1 0) is $\approx$ 100 K lower than on either the Ag(1 1 0) [2.53, 55] or Cu(1 1 0) [2.120] surfaces and is attributed to a weaker Au-formate bond. Alkynes also react with $O_{(a)}$ via Brønsted acid-base reactions. Acetylene (ΔH_{acid} = 378 kcal/mol [2.65]), for instance, is unreactive on the clean surface, but in the presence of $O_{(a)}$ reacts to form combustion products, with H_2O evolving into the gas-phase at 205 K and CO_2 at 525 K [2.119]. The reactions of $O_{(a)}$ on Au(1 1 0) with olefins, either acidic or nonacidic, have not been studied. The possibility that olefin dehydrogenation and/or oxygenation might occur on Au(1 1 0) is of some interest and ought to be explored.

The selective reactivity of oxygen adsorbed on silver and gold surfaces is dramatized by the low intrinsic reactivities of the clean surfaces toward organic substrates. Thus, on these surfaces competing reactions between the organic molecule and the metal do not interfere. Palladium and platinum also bind oxygen reversibly, but these metals also have the capacity to activate O–H and C–H bonds in the absence of adsorbed oxygen. Nonetheless, there is some evidence that direct proton transfer to oxygen adsorbed on these surfaces does occur. The reactions of $O_{(a)}$ with formic acid [2.121] and methanol [2.122] have been explored in some detail on Pd(1 0 0). In general, $O_{(a)}$ acts as a Brønsted base on this group VIII metal surface, although some differences from silver can be detected. In contrast to Ag(1 1 0) and Au(1 1 0), formic acid is decomposed on the clean Pd(1 0 0) surface [2.121], as evidenced by temperature programmed reaction mass spectrometry. If formic acid is adsorbed on Pd(1 0 0) precovered with $O_{(a)}$, however, formate is formed readily at 80 K [2.121]. The vibrational spectrum strongly suggests that formate is monodentate with C_s symmetry. Formate formation in the presence of $O_{(a)}$ was further suggested by the temperature programmed reaction spectrum, which showed H_2O evolution from hydroxyl disproportionation at 175 K as well as a sharp CO_2 feature at 265 K. These results suggest that formate formation occurs via a Brønsted acid–base reaction, just as occurred on Ag(1 1 0) and Au(1 1 0). Methanol also reacts with $O_{(a)}$ on Pd(1 0 0) in an acid-base reaction [2.122]. Proton transfer takes place by way of a *direct* transfer mechanism between the acidic methanol proton and $O_{(a)}$:

$$CH_3OH_{(a)} + O_{(a)} \rightarrow CH_3O_{(a)} + OH_{(a)}. \tag{2.20}$$

Temperature programmed reaction of methoxy on Pd(1 0 0) is similar to that on Au(1 1 0) [2.119]. Coincident evolution of formaldehyde, methyl formate, and water occurs at 200 K via a combination of disproportionation and nucleophilic addition reactions. In contrast to Au(1 1 0), methoxy decomposition on Pd(1 0 0) also leads to the formation of some surface formate which decomposes at 265 K.

The oxidation of alcohols by $O_{(a)}$ on Pd(1 1 1) also appears to proceed initially via an acid–base reaction between $O_{(a)}$ and the alcoholic O–H bond [2.123]. Methanol, ethanol, 1-propanol, and 2-propanol all appear to form alkoxy species by $\approx$ 200 K. The decomposition of methoxy on Pd(1 1 1) is essentially identical to that on Pd(1 0 0). Formaldehyde evolves from methoxide C–H bond scission at 240 K. Some formate is also formed, which decomposes to $CO_{2(g)}$ and a small amount of gaseous formic acid at 280 K. Not surprisingly, the products of ethanol, 1-propanol, and 2-propanol oxidation by $O_{(a)}$ on Pd(1 1 1) are more complicated. Reaction of ethanol with $O_{(a)}$ leads to the formation of several products: acetaldehyde, acetic acid, methane, dihydrogen, H_2O, CO_2, and CO. The product distribution from 1-propanol is similar to that from ethanol: propanal, propanoic acid, ethylene, dihydrogen, H_2O, and CO. 2-Propanol is a secondary alcohol, unlike methanol, ethanol, and 1-propanal, so the products from 2-propanol oxidation are somewhat different. Acetone, H_2O, and dihydrogen are the principal product formed, along with small amounts of acetic acid and CO. Overall the reactions appear to be governed by the Brønsted basicity and the nucleophilicity of $O_{(a)}$ on Pd(1 1 1).

The reactions of formic acid on clean and oxygen-precovered platinum surfaces have been investigated in some detail. Surprisingly, formic acid is completely unreactive on the clean Pt(1 0 0) surface, desorbing quantitatively at 200 K [2.124]. In the presence of $O_{(a)}$, formate formation occurs readily below 180 K via the reaction:

$$2\,HCOOH_{(a)} + O_{(a)} \rightarrow 2\,HCO_{2(a)} + H_2O_{(a)} . \tag{2.21}$$

Water desorbs at 180 and 200 K, and $HCO_{2(a)}$ decomposes at 310 K to form $CO_{2(g)}$ and $H_{2(g)}$, which evolve simultaneously into the gas phase [2.124]. In contrast to Pt(1 0 0), some formate formation does occur on clean Pt(1 1 1), as evidenced by TPRS [2.125] and vibrational spectroscopy [2.125a]. In the presence of coadsorbed $O_{(a)}$, however, formate formation is greatly facilitated due to a Brønsted acid–base reaction between formic acid and $O_{(a)}$ [2.125a]. The maximum formate coverage that can be attained on the $O_{(a)}$-precovered surface is approximately six times that which can be attained on clean Pt(1 1 1). The clean Pt(1 1 0) surface is also active for formate formation from formic acid [2.126]. The reaction of $O_{(a)}$ with formic acid on Pt(1 1 0) has not been examined, although formate formation would presumably occur readily. It is not yet clear why the Pt(1 0 0) surface is uniquely inert for O–H bond activation in formic acid, especially since Pt(1 0 0) is considered by some to be the most active platinum crystal face for the electrocatalytic oxidation of formic acid [2.127].

Although our understanding of $O_{(a)}$ on Au, Pd, and Pt does not approach that of $O_{(a)}$ on silver, the available evidence suggests that the acid-base and nucleophilic addition paradigms generally hold true. On Rh, however, the situation is clearly different. The reaction of other acids such as organic nitriles and linear olefins with $O_{(a)}$ on these surfaces would undoubtedly be of great interest. Olefins like norbornene and styrene should also be examined to see if oxygen addition can occur on Au and/or Pd. Finally, oxidation reactions should be explored on metals where less selective oxidation is expected to occur. Some work has been done in this area, concerning, for instance, the oxidation of HCN on Pt(1 1 1) [2.128] and the oxidation of ammonia over Pt, Pd, and Rh foils [2.129]. These and other areas are sure to provide useful insights into the mechanisms of transition metal catalyzed oxidation.

2.4 Conclusion

The past four to five years have seen a rapid increase in our knowledge of the mechanisms of partial oxidation reactions on single-crystal transition-metal surfaces. The wealth of experimental and theoretical tools now at our disposal often allows for a clear understanding of the elementary steps involved in the reactions of simple chemisorbed molecules with both atomically and molecularly chemisorbed oxygen, as well as determination of the kinetics of the rate determining steps. In the case of Ag(1 1 0), those elementary steps are now so well understood that the most likely oxidation pathways of complex molecules can be postulated with confidence, provided that the gas-phase acidity and electrophilicity of the reacting molecule are reasonably well known. Importantly, the Ag(1 1 0) surface has also been shown to epoxidize C–C double bonds in olefins, providing that certain key conditions are met. Ag(1 1 0) under ultrahigh vacuum has therefore proved to effectively model the reactivity of a silver partial oxidation catalyst. These gains in knowledge attest to the importance of recent advances in surface science.

The future holds much promise for continuing progress in study of heterogeneously catalyzed partial oxidation. Although oxidation on Ag(1 1 0) is now reasonably well understood, there is more work to be done, particularly concerning hydrocarbon oxidation by $O_{(a)}$. Furthermore, it is important to compare oxidation on Ag(1 1 0) to oxidation on the closest-packed Ag(1 1 1) surface. Investigations of formic acid, formaldehyde, and methanol on Pd(1 0 0) and Au(1 1 0) suggest that the reactivity of $O_{(a)}$ on these surfaces is similar to that on Ag(1 1 0). These experiments should be extended to include, for instance, reactions of alkenes and nitriles so that oxidation on Ag, Pd, and Au could be more adequately compared and contrasted. Atomically adsorbed oxygen on Rh(1 1 1) appears to react with styrene to form acetophenone, but on Ag(1 1 0)-$O_{(a)}$, styrene is epoxidized. The reason for this marked difference should be investigated. These unsolved problems in partial oxidation are important, and the chemistry revealed during the course of their investigation is bound to prove fascinating. Given the necessary resources, there is every reason to expect that gains over the next five years will match or even eclipse those made over the past.

Acknowledgements. R.J.M. gratefully acknowledges the support of the National Science Foundation for much of the work reported in this chapter.

References

2.1 R.A. Van Santen, H.P.C.E. Kuipers: Adv. Catal. **35,** 265 (1987)
2.2 M.A. Pepera, J.L. Callahan, M.J. Desmond, E.C. Milberger, P.R. Blum, N.J. Bremer: J. Am. Chem. Soc. **107,** 4883 (1985)
2.3 G.E. Keller, M.M. Bhasin: J. Catal. **73,** 9. (1982)
H.D. Gesser, N.R. Hunter, C.B. Prakash: Chem. Rev. **85,** 235 (1985)
K.C.C. Kharas, J.H. Lunsford: J. Am. Chem. Soc. **111,** 2336 (1989)
2.4 A.T. Ashcroft, A.K. Cheetham, J.S. Foord, M.L.H. Green, C.P. Grey, A.J. Murrell, P.D.F. Vernon: Nature **344,** 319 (1990)
2.5 G.N. Lewis, M. Randall: Thermodynamics, 2nd edn (McGraw-Hill, New York, 1961)
2.6 M. Bowker, M.A. Barteau, R.J. Madix: Surf. Sci. **92,** 528 (1980)
M.A. Barteau, R.J. Madix: Surf. Sci. **97,** 101 (1980)
2.7 N.D.S. Canning, D. Outka, R.J. Madix: J. Surf. Sci. **141,** 240 (1984)
A.G. Sault, R.J. Madix, C.T. Campbell: Surf. Sci. **169,** 347 (1986)
2.8 C. Nyberg, C.G. Tengstål: Surf. Sci. **126,** 163 (1983)
J.-W. He, P.R. Norton: Surf. Sci. **204,** 26 (1988)
2.9 N.R. Avery: Chem. Phys. Lett. **96,** 371 (1983)
J.L. Gland, B.A. Sexton, G.B. Fisher: Surf. Sci. **95,** 587 (1980)
2.10 C.T. Campbell: Surf. Sci. **157,** 43 (1985)
C.T. Campbell: Surf. Sci. **173,** L641 (1986)
2.11 A third type of oxygen, subsurface oxygen ($O_{(sub)}$), is also formed on silver. Subsurface oxygen plays an important role in the partial oxidation of ethylene over a silver catalyst, but its effect is mostly to modify silver's electronic structure. $O_{(sub)}$ is not reactive toward chemisorbed molecules, and is therefore beyond the scope of this work.
2.12 C.T. Campbell, M.T. Paffett: Surf. Sci. **143,** 517 (1984)
2.13 H. Albers, W.J.J. Van der Wal, G.A. Bootsma: Surf. Sci. **68,** 47 (1977)
2.14 C. Benndorf, M. Franck, F. Thieme: Surf. Sci. **128,** 417 (1983)
2.15 R.B. Grant, R.M. Lambert: Surf. Sci. **146,** 256 (1984)
2.16 H.A. Engelhardt, D. Menzel: Surf. Sci. **57,** 591 (1976)
2.17 H.A. Engelhardt, A.M. Bradshaw, D. Menzel: Surf. Sci. **40,** 410 (1973)
2.18 W. Heiland, F. Iberl, E. Taglauer, D. Menzel: Surf. Sci. **53,** 383 (1975)
2.19 M. Taniguchi, K. Tanaka, T. Hashizume, T. Sakurai: Surf. Sci. Lett. **262,** L123 (1992)
2.20 P.A. Redhead: Vacuum **12,** 203 (1962)
M. Bowker: Surf. Sci. **100,** L472 (1980)
2.21 See, for instance: E. Tronconi, L. Lietti: Surf. Sci. **199,** 43 (1988)
J.L. Falconer, R.J. Madix: Surf. Sci. **48,** 393 (1975)
2.22 B.A. Sexton, R.J. Madix: Chem. Phys. Lett. **76,** 294 (1980)
C. Backx, C.P.M. de Groot, P. Biloen: Surf. Sci. **104,** 300 (1981)
C. Backx, C.P.M. de Groot, P. Biloen: Appl. Surf. Sci. **6,** 256 (1980)
2.23 T.H. Upton, P. Stevens, R.J. Madix: J. Chem. Phys. **88,** 3988 (1988)
2.24 D.A. Outka, J. Stöhr, W. Jark, P. Stevens, J. Solomon, R.J. Madix: Phys. Rev. B. **35,** 4119 (1987)
J. Stöhr: *NEXAFS Spectroscopy*, Springer Ser. Surf. Sci., Vol. 25 (Springer, Berlin, Heidelberg 1992)
2.25 J. Stöhr, D.A. Outka: Phys. Rev. B. **36,** 7891 (1987)
2.26 The residual intensity recorded for $\boldsymbol{E}$ in the [0 0 1] azimuth is due to incomplete polarization of the light which gives rise to a component of $\boldsymbol{E}$ along the $[1\bar{1}0]$ azimuth even when the major component of $\boldsymbol{E}$ is along [0 0 1]
2.27 F. Sette, J. Stöhr, A.P. Hitchcock: J. Chem. Phys. **81,** 4906 (1984)
J. Stöhr, F. Sette, A.L. Johnson: Phys. Rev. Lett. **53,** 1684 (1984)
2.28 R.L. Redington, W.B. Olson, P.C. Cross: J. Chem. Phys. **36,** 1311 (1962)

2.29 A.F. Carley, P.R. Davies, M.W. Roberts, K.K. Thomas: Surf. Sci. Lett. **238,** L467 (1990)
2.30 T. Matsushima: Surf. Sci. **157,** 297 (1985)
2.31 P.A. Thiel, J.T. Yates Jr., W.H. Weinberg: Surf. Sci. **82,** 22 (1979)
2.32 A. Puschmann, J. Haase: Surf. Sci. **144,** 559 (1984)
2.33 E.M. Stuve, R.J. Madix, B.A. Sexton: Surf. Sci. **111,** 11 (1981)
2.34 E.A. Carter, W.A. Goddard III.: Surf. Sci. **209,** 243 (1989)
E.A. Carter, W.A. Goddard III.: J. Cat. **112,** 80 (1988)
2.35 M.A. Barteau, R.J. Madix: Surf. Sci. **140,** 108 (1984)
2.36 R.J. Madix, J. Stöhr: unpublished results
2.37 T.H. Upton: private communication
2.38 A.J. Capote, J.T. Roberts, R.J. Madix: Surf. Sci. **209,** L151 (1989)
2.39 D.A. Outka, R.J. Madix, G.B. Fisher, C. DiMaggio: J. Phys. Chem. **90,** 4051 (1986)
D.A. Outka, R.J. Madix, G.B. Fisher, C.L. DiMaggio: Langmuir **2,** 406 (1986)
2.40 E.M. Stuve, R.J. Madix, B.A. Sexton: Chem. Phys. Lett. **89,** 48 (1982)
2.41 M.A. Barteau, R.J. Madix: J. Chem. Phys. **74,** 4144 (1981)
2.42 J.T. Roberts, R.J. Madix: unpublished results
2.43 C.T. Campbell, M.T. Paffett: Surf. Sci. **139,** 396 (1984)
2.44 J.T. Roberts, R.J. Madix: J. Am. Chem. Soc. **110,** 8540 (1988)
2.45 X. Guo, A. Winkler, P.L. Hagans, J.T. Yates Jr.: Surf. Sci. **203,** 33 (1988)
2.46 T. Matsushima: Surf. Sci. **123** L663 (1982)
T. Matsushima: Surf. Sci. **127,** 403 (1983)
2.47 W. Ho: private communication
2.48 A.F. Carley, M.W. Roberts: J. Chem. Soc., Chem. Commun. 355 (1987)
2.49 A.F. Carley, M.W. Roberts, S. Yan: J. Chem. Soc., Chem. Commun. 267 (1988)
A.F. Carley, S. Yan, M.W. Roberts: J. Chem. Soc. Farad. Trans. **86,** 2701 (1990)
2.50 A nucleophile reacts with an electrophile via transfer of an electron pair from the nucleophile to an electropositive center on the electrophile. Since oxygen is an electronegative element, chemisorbed oxygen atoms have a high electron density and are therefore good nucleophiles.
2.51 A Brønsted base in a substance capable of proton (H^+) abstraction
2.52 D.A. Outka, R.J. Madix, G.B. Fisher, C. DiMaggio: Surf. Sci. **179,** 1 (1987)
2.53 M.A. Barteau, M. Bowker. R.J. Madix: Surf. Sci. **94,** 303 (1980)
2.54 M.A. Barteau, M. Bowker, R.J. Madix: J. Catal. **67,** 118 (1981)
2.55 E.M. Stuve, R.J. Madix, B.A. Sexton: Surf. Sci. **119,** 279 (1982)
2.56 C. Ayre, R.J. Madix: to be submitted
2.57 M.A. Barteau, R.J. Madix: J. Electron Spect. Relat. Phenom. **31,** 101 (1983)
2.58 R.J. Madix, J.L. Solomon, J. Stöhr: Surf. Sci. **197,** L253 (1988)
2.59 W. Wurth, J. Stöhr: to be published
2.60 M.A. Barteau, R.J. Madix: Surf. Sci. **120,** 262 (1982)
2.61 P.A. Schulz, R.D. Mead, P.L. Jones, W.C. Lineberger: J. Chem. Phys. **77,** 1153 (1982)
2.62 Exceptions to this rule are observed only when the reacting molecule contains a carbonyl carbon, which is highly susceptible to nucleophilic attack by $O_{(a)}$. In acetaldehyde (DH_{acid} = 367 kcal/mol[53]) nucleophilic attack is the preferred reaction pathway, while in acetone both nucleophilic attack and proton abstraction occur.
2.63 J.B. Cumming, P. Kebarle: Can. J. Chem. **56,** 1 (1978)
2.64 S.W. Jorgensen, A.G. Sault, R.J. Madix: Langmuir 1, 526 (1985)
2.65 J.E. Bartmess, J.A. Scott, R.T. McIver Jr.: J. Am. Chem. Soc. **101,** 6046 (1979)
2.66 J.M. Vohs, B.A. Carney, M.A. Barteau: J. Am. Chem. Soc. **107,** 7841 (1985)
2.67 D.F. McMillen, D.M. Golden: Annu. Rev. Phys. Chem. **33,** 493 (1982)
2.68 M. Fujio, R.T. McIver Jr., R.W. Taft, J. Am. Chem. Soc. **103,** 4017 (1981)
2.69 S.W. Benson: Thermochemical Kinetics (Wiley, New York, 1976)
2.70 I.E. Wachs, R.J. Madix: Appl. Surf. Sci. **1,** 303 (1978)
2.71 I.E. Wachs, R.J. Madix: Surf. Sci. **76,** 531 (1978)
2.72 P. Merrill, R.J. Madix: Langmuir **7,** 3034 (1991)
2.73 J.L. Solomon, R.J. Madix: J. Phys. Chem. **91,** 6241 (1987)

2.74 J.L. Solomon, R.J. Madix, J. Stöhr: J. Chem. Phys. **89,** 5316 (1988)
2.75 R.L. Brainard, C.G. Peterson, R.J. Madix: J. Am. Chem. Soc. **111,** 4553 (1989)
2.76 M. Bowker, R.J. Madix: Surf. Sci. **116,** 549 (1982)
2.77 M. Imachi, N.W. Cant, R.L. Kuczkowski: J. Catal. **75,** 404 (1982)
2.78 I.E. Wachs, R.J. Madix: Surf. Sci. **76,** 531 (1978)
2.79 A.H. Janowicz, R.G. Bergman: J. Am. Chem. Soc. **105,** 3929 (1983)
2.80 R.L. Brainard, R.J. Madix: J. Am. Chem. Soc. **109,** 8082 (1987)
R.L. Brainard, R.J. Madix: J. Am. Chem. Soc. **111,** 3826 (1989)
2.81 C.M. Friend, J.T. Roberts: Acc. Chem. Res. **21,** 394 (1988)
2.82 R.L. Brainard, R.J. Madix: Surf. Sci. **214,** 396 (1989)
2.83 M. Bowker, R.J. Madix: Surf. Sci. **95,** 190 (1980)
2.84 T.M. Miller, G.M. Whitesides: Organometallics **5,** 1473 (1986)
H.E. Bryndza, J.C. Calabrese, M. Marsi, D. Roe, W. Tam, J.E. Bercaw: J. Am. Chem. Soc. **108,** 4805 (1986)
D.L. Thorn, R. Hoffmann: J. Am. Chem. Soc. **100,** 2079 (1978)
2.85 A.J. Capote, R. Madix: J. Am. Chem. Soc. **111,** 3570 (1989)
A.J. Capote, R.J. Madix: Surf. Sci. **214,** 276 (1989)
2.86 C.H. DePuy, V.M. Bierbaum, R. Damraurer: J. Am. Chem. Soc. **106,** 4051 (1984)
2.87 A. J. Capote, A.V. Hamza, N.D.S. Canning, R.J. Madix: Surf. Sci. **175,** 445 (1986)
2.88 P.A. Stevens, R.J. Madix, J. Stöhr: J. Chem. Phys. **91,** 4338 (1989)
2.89 A.G. Sault, R.J. Madix: J. Molec. Catal. **52,** 211 (1989)
2.90 E.O. Fischer, P. Stückler, F.R. Kreissl: J. Organomet. Chem. **129,** 197 (1977)
2.91 F.R. Kreissl: Transition Metal Carbene Complexes (Verlag Chemie, Deerfield Beach, FL, 1983) 153
2.92 D.M. Thornburg, R.J. Madix: Surf. Sci. **220,** 268 (1989)
2.93 D.M. Thornburg, R.J. Madix: Surf. Sci. **226,** 61 (1990)
2.94 G.J. MacKay, R.S. Hemsworth, D.K. Bohme: Can. J. Chem. **54,** 1624 (1976)
2.95 M.A. Barteau, R.J. Madix: Surf. Sci. **115,** 355 (1982)
2.96 M.A. Barteau, R.J. Madix: J. Am. Chem. Soc. **105,** 344 (1983)
2.97 C.R. Ayre, R.J. Madix: to be published
2.98 R.E. Lee, R.R. Squires: J. Am. Chem. Soc. **108,** 5078 (1986)
2.99 J. T. Roberts, R.J. Madix: Surf. Sci. **226,** L71 (1990)
2.100 J.T. Roberts, R.J. Madix: to be published
2.101 J.T. Roberts, A.J. Capote, R.J. Madix: to be published
2.102 M.A. Barteau, R.J. Madix: Surf. Sci. **103,** L171 (1981)
2.103 C.T. Campbell: J. Catal. **94,** 436 (1985)
2.104 A.C. Liu, C.M. Friend: Rev. Sci. Instrum. **57,** 1519 (1986)
2.105 S.A. Tan, R.B. Grant, R.M. Lambert: J. Catal. **104,** 156 (1987)
2.106 S. Hawker, C. Mukoid, J.P.S. Badyal, R.M. Lambert: Surf. Sci. **219,** L615 (1989)
2.107 C. Mukoid, S. Hawker, J.P.S. Badyal, R.M. Lambert: Catal. Lett. **4,** 57 (1990)
2.108 A.J. Capote, R.J. Madix: submitted
2.109 J.T. Gleaves, A.G. Sault, R.J. Madix, J.R. Ebner: J. Catal. **121,** 202 (1990)
2.110 J.F. Hartwig, R.G. Bergman, R.A. Andersen: J. Am. Chem. Soc. **112,** 3234 (1990)
J.F. Hartwig, R.G. Bergman, R.A. Andersen: Organometallics **10,** 3326 (1991)
D.P. Kline, J.C. Hayes, R.G. Bergman, R.A. Andersen: J. Am. Chem. Soc. **110,** 3704 (1988)
2.111 X. Xu, C.M. Friend: J. Am. Chem. Soc. **113,** 6779 (1991)
2.112 X. Xu, C.M. Friend: J. Am. Chem. Soc. **112,** 4571 (1990)
2.113 A.H. Janowicz, R.G. Bergman: J. Am Chem. Soc. **105,** 3929 (1983)
R.L. Brainard, R.J. Madix: J. Am Chem. Soc. **111,** 3826 (1989)
2.114 X. Xu, C.M. Friend: Surf. Sci. **260,** 14 (1992)
2.115 J.W. McMillan, H.E. Fischer, J. Schwartz: J. Am. Chem. Soc. **113,** 4014 (1991)
2.116 B.E. Bent, C.M. Mate, J.E. Crowell, B.E. Koel, G.A. Somorjai: J. Phys. Chem. **91,** 1493 (1987)
2.117 D.A. Outka, R.J. Madix: Surf. Sci. **179,** 351 (1987)
2.118 D.A. Outka, R.J. Madix: Surf. Sci. **179,** 361 (1987)

2.119 D.A. Outka, R.J. Madix: J. Am. Chem. Soc. **109,** 1708 (1987)
2.120 D.H.S. Ying, R.J. Madix: J. Catal. **61,** 48 (1980)
2.121 S.W. Jorgensen, R.J. Madix: J. Am. Chem. Soc. **110,** 397 (1988)
2.122 S.W. Jorgensen, R.J. Madix: Surf. Sci. **183,** 27 (1987)
2.123 J.L. Davis, M.A. Barteau: Surf. Sci. **197,** 123 (1988)
2.124 N. Kizhakevariam, E. Stuve: J. Vacuum Sci. Technol. A. **8,** 2557 (1990)
2.125 N.R. Avery: Appl. Surf. Sci. **11/12,** 774 (1982)
N. Abbas, R.J. Madix: Appl. Surf. Sci. **16,** 424 (1983)
2.126 G.E. Gdowski, J.A. Fair, R.J. Madix: Surf. Sci. **127,** 541 (1983)
P. Hofmann, S.R. Bare, N.V. Richardson, D.A. King: Surf. Sci. **133,** L459 (1983)
2.127 R. Parsons, T. Van der Noot: J. Electroanal. Chem. **257,** 9 (1988)
2.128 X. Guo, A. Winkler, I. Chorkendorff, P.L. Hagans, H.R. Siddiqui, J.T. Yates Jr.: Surf. Sci. **203,** 17 (1988)
2.129 T. Pignet, L.D. Schmidt: J. Catal. **40,** 212 (1975)

3. Desulfurization Reactions Induced by Transition Metal Surfaces

C.M. Friend

Catalytic desulfurization reactions are of extreme industrial importance due to the need for upgrading fossil fuel feedstocks with a high sulfur content. Sulfur must be removed from fuel because SO_x combustion products contribute to acid rain and sulfur renders "catalytic converters" in automobile engines ineffective. Fundamental studies of desulfurization reactions induced on single-crystal metal surfaces under ultrahigh vacuum conditions are the focus of this chapter. There have been many excellent studies of desulfurization reactions under well-defined conditions which we use here to establish several general features of this class of surface reactions. While we have attempted to include all of these studies, some fine work has most certainly been overlooked and we apologize in advance for our oversight.

3.1 Background

In practical processes, sulfur is removed from fossil-fuel feedstocks in the presence of a solid metal sulfide catalyst under high pressures of dihydrogen. Volatile hydrocarbons are the desired product. Commercial catalysts are composed of MoS_2 modified by another transition metal, usually cobalt or nickel. The precise nature of the active sites in hydrodesulfurization catalysts and the role of promoters are still a matter of some controversy. It is generally agreed that anion vacancies and edge sites at the surface are the most catalytically active, however [3.1].

Although molybdenum-based catalysts are used in practical processes, other transition-metal sulfides are more effective in inducing hydrodesulfurization. As illustrated in Fig. 3.1, the activity for dibenzothiophene hydrodesulfurization is a maximum for sulfides of Ru and Rh [3.2]. Transition-metal induced sulfur removal reactions have been intensely studied in an effort to better understand the mechanism for the process and the basis of the periodic trends in activity.

The goal of fundamental studies of desulfurization reactions on single-crystal surfaces under ultrahigh vacuum conditions is to ascertain the underlying molecular mechanisms and bonding interactions that are responsible for

Springer Series in Surface Sciences, Vol. 34
Surface Reactions Ed.: R.J. Madix

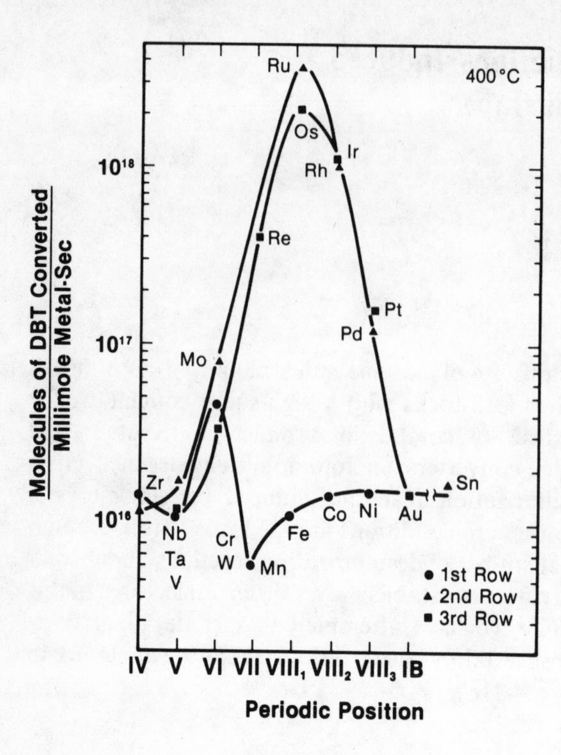

Fig. 3.1. Periodic trends in the rate of hydrodesulfurization of dibenzothiophene taken from the work of *Chianelli* and *Harris* [3.2]

specific chemical transformations. The geometry of the metal atoms on a single-crystal surface is well defined. Furthermore, the surface geometry can be altered in a well-controlled manner by employing different crystallographic orientations of the same metal. Therefore, the dependence of reactivity and selectivity on the types of coordination sites available can be systematically studied. On practical catalysts, there is a distribution of possible reaction sites, and it is often difficult to separate the contributions of different intermediates and sites to the overall reaction process.

In our work, we have focused on the Mo(1 1 0) surface (Fig. 3.2). This is the most thermodynamically stable surface and, hence, is the least prone to reconstruction of the surface. Reconstruction involves significant displacements of the metal atoms in the surface layer. Such displacements are often reversible and depend on the presence or absence of adsorbed species. As a result, the surface structure may change during the course of a reaction due to changes in the surface concentration (coverage) of surface species. The change in the surface structure will contribute to the kinetics of the reaction [3.3, 4] and complicate the analysis of kinetic data. While this is an exceedingly interesting effect, it would render the investigation of complex organosulfur compounds intractable since there is little known about the reaction in the absence of surface reconstruction. Once we have developed an understanding of the mechanisms for

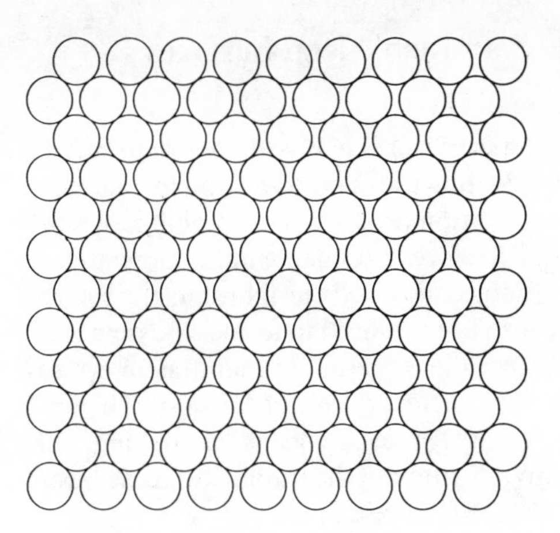

Fig. 3.2. A schematic representation of Mo(1 1 0)

desulfurization on the stable Mo(1 1 0) surface, the effect of surface reconstruction on more open surfaces will be investigated since it may be important in determining reaction selectivity and kinetics.

A major advantage of single-crystal, ultrahigh-vacuum investigations is that a battery of surface spectroscopies can be employed to identify surface intermediates, determine their structure, and measure reaction kinetics. As a result, the mechanistic pathways for desulfurization can be determined in great detail.

Ultrahigh-vacuum conditions also preserve the purity of the metal surface so that the effects of adlayers, such as sulfur, can be specifically studied. Sulfur overlayers play a central role in catalytic desulfurization chemistry; therefore, understanding their effect on chemical reactivity is essential.

There are two major classes of molecules that have been studied to model transition-metal induced desulfurization: S-containing rings and organic thiols. Both classes of reactants are either present in fossil fuel feedstock or are possible intermediates along the pathway to desulfurization. Furthermore, by comparing the reactivity of several related molecules, the steps that control the kinetics and selectivity for desulfurization can be deduced. In a complex reaction system such as this, several bond-breaking and formation steps may be important in determining the kinetics and selectivity. By systematically varying the energetics for cleaving specific types of bonds, for example the C–S bond, the role of that step in determining the rate of reaction and the product distribution is probed.

The experimental approach and accompanying analysis of data is discussed in detail for thiols and cyclic sulfides. We will focus on results from our own group and bring in comparisons with other surfaces and organometallic complexes in order to illustrate general factors governing desulfurization processes induced by metals. This work is not intended as a comprehensive review of the literature so that interested individuals are urged to perform a more comprehensive search.

3.2 The Reactions of Thiols on Transition-Metal Surfaces

The reactions of thiols on transition-metal surfaces is of broad interest in surface chemistry. A major motivation for investigating thiols on transition metals is the potential for modeling the hydrodesulfurization process since thiol linkages are common functionalities in fossil fuels. Therefore, a fundamental understanding of C–S bond activation in this class of molecules yields insight into the related catalytic process. Recently, thiols have also been utilized to form self-assembling monolayers which can be used to improve adhesion and to manipulate optical properties of interfaces [3.5–7]. Thus far, the adsorption of thiols on Group-I metals has been most extensively investigated as a means of making self-assembling monolayers, although the investigation of the stability of these layers on more reactive surfaces is also important for these applications.

Fundamental studies of reaction pathways available on surfaces offer the potential to identify factors that determine reaction selectivity and kinetics. Ideally, studies of reactivity and structure can be coupled with theoretical work to understand a reaction mechanism. Methods for investigating reactivity and identifying surface intermediates are well-developed. However, techniques for determining the structure of adsorbed intermediates and for theoretically describing bonding are still limited. The methodology for investigating complex surface reactions is illustrated for thiol reactions.

3.2.1 Spectroscopic Identification and Characterization

Both chemical and spectroscopic methods are used to probe the nature of surface intermediates and their reactions. A powerful combination of methods is temperature-programmed reaction, X-ray photoelectron and high-resolution electron energy loss spectroscopies used in conjunction with isotopic labeling and exchange experiments. Temperature-programmed reaction spectroscopy measures the kinetics and selectivity for surface reactions. When used in conjunction with isotopic labeling methods, the types of bonds that are broken or formed during the course of a reaction can be identified. In selected cases, kinetic isotope effects give insight into the nature of the rate-limiting step in the reaction. Products must be evolved into the gas phase to be detected by temperature-programmed reaction spectroscopy, however, necessitating the use of other probes for surface-bound species. Spectroscopic probes are also necessary to identify intermediates present on the surface along the reaction pathway.

X-ray photoelectron and vibrational spectroscopies are complementary methods for identifying intermediates that are present on the surface prior to formation of gaseous products. Thus, a reaction pathway can be mapped by determining the nature of intermediates as a function of surface temperature and correlating that information with reactivity data. The reactions of thiols, for example, have been studied in detail using this experimental approach.

The S–H bond in thiols and H_2S are facilely cleaved on almost all transition-metal surfaces investigated. The cleavage of S–H bonds has been demonstrated using X-ray photoelectron and electron energy loss spectroscopies, as described below. On Mo(1 1 0), all of the C_1–C_5 primary thiols [3.8–14] benzenethiol [3.15], and *t*-butanethiol [3.16] form the corresponding thiolates upon adsorption at 120 K. Similarly, methanethiol forms CH_3S- on Cu(1 0 0) [3.17], Cu(1 1 1) [3.18, 19], Ni(1 0 0) [3.18], Ni(1 1 0) [3.20], Pt(1 1 1) [3.21], Fe(1 0 0) [3.22], and W(2 1 1) [3.23]. The exception is methanethiol on Au(1 1 1): the S–H bond remains intact on this surface and molecular desorption is exclusively observed [3.7]. Hydrogen sulfide, H_2S, forms SH and H on Rh(1 0 0) [3.24], Mo(1 0 0) [3.25], Pt(1 1 1) [3.26, 27] Ni(1 1 0), Ag(1 1 0) [3.20, 28,29], and Ru(1 1 0) [3.30]. On Cu(1 1 1), complete dissociation to S and adsorbed hydrogen were reported below 200 K [3.31].

The facility for S–H bond breaking is consistent with its rather weak bond strength and the large thermodynamic driving force to form strong metal–sulfur bonds. Indeed, S–H bond cleavage is analogous to O–H bond breaking in alcohols, commonly induced on surfaces [3.32].

Vibrational data clearly show that the S–H bond of organic thiols adsorbed on Mo(1 1 0) cleaves upon adsorption at 120 K for all coverages studied. The reactions of methanethiol on Mo(1 1 0) are used to illustrate the experimental approach adopted in our work. High-resolution electron energy loss spectra show that the S–H stretch mode is missing for methanethiol adsorbed directly onto the Mo(1 1 0) surface at 120 K (Fig 3.3). Reference spectra for condensed methanethiol contain a peak at 2545 cm^{-1} which is ascribed to the ν(S–H) mode (Fig. 3.3).

The absence of the S–H stretch mode in electron energy loss data is clear evidence that this bond is broken upon adsorption. The S–H stretch should be observable in electron energy loss spectra even if it is oriented parallel to the surface plane since vibrational modes involving displacement of hydrogen are subject to short-range, so-called, impact scattering [3.33]. In contrast, the absence of a mode in reflection infrared spectra for molecules adsorbed on a metal surface could be due either to the absence of that bond or an orientation of that bond parallel to the surface plane. This is because infrared transitions are governed entirely by dipole scattering so that only modes with a component of their dynamic dipole moment perpendicular to the surface can be observed [3.34]. Clearly, electron energy loss spectroscopy is the preferred vibrational technique when posing questions regarding bond breaking. Furthermore, electron energy loss spectroscopy is extremely sensitive, in particular for vibrational energies below 800 cm^{-1}. Infrared spectroscopy has the advantages of higher resolution and well-defined selection rules so that it is preferable in addressing more detailed structural questions as illustrated below for the cases of methyl thiolate and methoxide on Mo(1 1 0).

The absence of the S–H stretch in vibrational data clearly shows that the S–H bond is cleaved, but vibrational data alone do not define the conditions for isolation of an intermediate for detailed study nor do they clearly determine

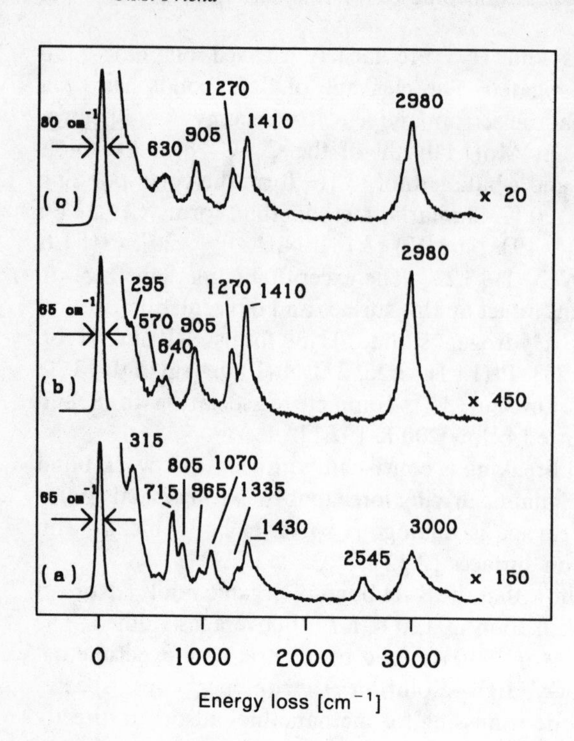

Fig. 3.3. Electron energy loss data for (a) methanethiol multilayers (b) methyl thiolate on Mo(1 1 0), and (c) off-specular data for methyl thiolate [3.8]. The thiolate was formed by heating the condensed thiol to 200 K

whether the C–S bond is intact in the intermediate derived from methanethiol. The ν(C–S) mode in thiols is weak so that it cannot be used for a marker for an intact C–S bond. Hence, X-ray photoelectron spectroscopy is used in complement with the vibrational spectroscopy to determine the conditions required for C–S bond breaking.

X-ray photoelectron data clearly show that the C–S bonds of thiols remain intact when adsorbed on Mo(1 1 0). The C–S bond remains intact up to temperatures where gaseous reaction products are evolved, $\approx$ 250 K for the case of methanethiol. A single C(1s) peak with an energy of 284.7 eV is observed following methanethiol adsorption on Mo(1 1 0) at 120 K (Fig. 3.4). A set of two peaks in the sulfur region correspond to the S($2p_{1/2}$) and S($2p_{3/2}$) states and have binding energies 162.7 and 163.9 eV, respectively (Fig. 3.4). These values are in the range expected for intact C–S bonds and are substantially higher than those measured for atomic sulfur and adsorbed hydrocarbon fragments. Atomic sulfur has S($2p_{1/2}$) and S($2p_{3/2}$) binding energies of 161.3 and 162.5 eV, respectively (Fig. 3.4ii). Hydrocarbon fragments bound to Mo have C(1s) binding energies in the range of 283–284 eV. The higher C(1s) binding energy associated with the thiolate is due to the fact that the carbon is bound to an electronegative sulfur atom. The binding energies measured following adsorption of other thiols to the Mo(1 1 0) surface are similar, indicating in all cases that the C–S bond remains intact.

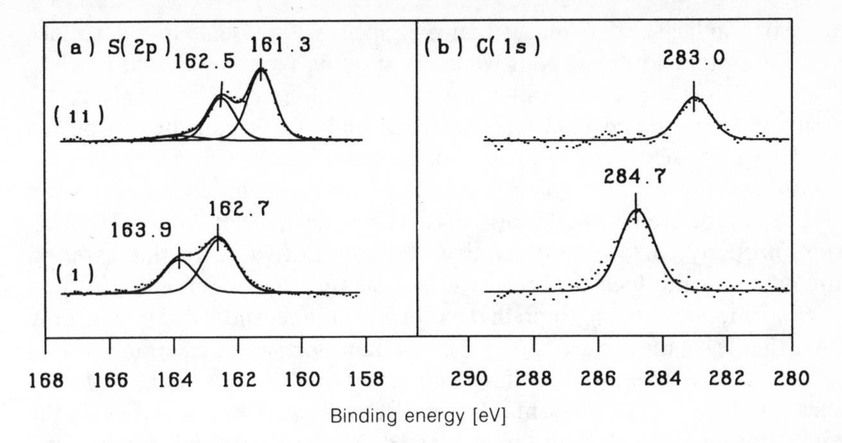

Fig. 3.4. (a) S(2p) and (b) C(1s) X-ray photoelectron spectra of methanethiol condensed on Mo(1 1 0) and heated to (i) 200 K and (ii) 400 K [3.8]

Besides aiding in the identity of a surface intermediate, X-ray photoelectron data can be used to determine the conditions under which a single molecular intermediate is isolated on the surface. Establishing these conditions is important in interpreting other spectroscopic data. For example, a species must be isolated in order to make reliable vibrational assignments. Furthermore, isolation of intermediates is requisite for a structural determination. If a mixture of species is present, a convolution of several different sets of peaks will be observed making assignments and intensity analyses virtually impossible.

Peaks for chemically distinct species may overlap and may have very similar binding energies in X-ray photoelectron spectra. Therefore, the width of a peak must be considered when analyzing such data. In the case of methanethiol on Mo(1 1 0), the peak widths measured following adsorption at 120 K are the same as those measured for a pure atomic sulfur layer and methanethiol multilayers ($\approx$ 1.1 eV). Therefore, the thiolate is the only species present. The widths of the C(1s) and S(2p) peaks increase and intensity is shifted to lower binding energies, as the layer is heated above 250 K, for example, indicating the formation of a second species. Careful analysis indicates that the broadened peaks are best fit by two sets of peaks corresponding to the intact thiolate and atomic sulfur formed when C–S bonds are cleaved. Similarly, the lower C(1s) binding energy peak is attributed to one or more hydrocarbon fragments bound to the surface and formed from C–S bond cleavage. The widths of C(1s) peaks must always be scrutinized since there may be more than one species enveloped in a peak due to the small chemical shifts relative to the resolution of the experiment. For example, the FWHM of the C(1s) peak for methyl thiolate adsorbed on Mo(1 1 0) at 250 K is $\approx$ 1.1 eV, near the instrumental resolution. In contrast, a C(1s) peak width of $\approx$ 2.5 eV is measured following reaction of 2,5-dihydrothiophene on clean Mo(1 1 0) [3.35]. Although X-ray photoelectron spectroscopy cannot definitely identify the different species in this case, the broad peak width clearly indicates that more than one species is present on the surface which must be taken into account when interpreting vibrational data.

X-ray photoelectron intensities can also be quantitatively analyzed to determine the absolute coverages of molecular and atomic constituents on the surface. Hence, the selectivity of a reaction and the coverage of intermediates can be determined. For example, the selectivity for hydrocarbon elimination from methanethiol is derived from the ratio of carbon and sulfur intensities following reaction, properly corrected for relative atomic sensitivities. In this case, no sulfur-containing products leave the surface so that the sulfur intensity is a measure of the amount of thiol that reacts. Methane and H_2 are the only products that leave the surface [3.16]. The methane formation reaction depletes the surface of carbon relative to sulfur. The carbon that remains on the surface is a direct measure of the amount of nonselective reaction. Accordingly, the selectivity for methane formation is $\approx 40\%$ for methanethiol reaction on Mo(1 1 0).

The maximum or saturation coverage of methyl thiolate can also be derived from the S(2p) intensities. By comparing the S(2p) intensity following reaction at saturation coverage to that of the Mo(1 1 0)–p(4 × 1)–S surface, which is known to have a sulfur coverage of 0.5, the saturation coverage is determined to be 0.3 $\pm$ 0.05 methyl thiolates per Mo atom. The saturation coverage for other thiolates is somewhat different and generally correlates with the steric bulk of the alkyl group. For example, phenyl thiolate has a maximum coverage of $\approx$ 0.12 monolayers.

Vibrational and isotopic exchange data are important complements to X-ray photoelectron spectra when identifying surface intermediates and interrogating their structure. For example, the X-ray photoelectron data for methanethiol on Mo(1 1 0) clearly demonstrated that the C–S bond remains intact up to the onset of hydrocarbon production. It does not, however, give detailed information about the integrity of C–H bonds in the molecule or the molecular structure of the intermediate present on the surface.

Electron energy loss data indicate that methyl thiolate is the majority intermediate formed when methanethiol reacts with Mo(1 1 0) at 100 K. This intermediate persists on the surface up to $\approx$ 300 K, the temperature at which methane is formed. Methyl thiolate is identified by comparing the vibrational spectra for condensed methane thiol and the intermediate formed from methane thiol reaction with the surface below 300 K (Fig. 3.3). As noted above, the primary changes in the vibrational spectrum induced upon reaction with the surface are the disappearance of the ν(S–H) and δ(S–H) modes, indicating S–H bond cleavage. Overall, the vibrational spectra for the condensed thiol and the intermediate formed on the surface are qualitatively, similar suggesting that the methyl group remains intact. In addition to the absence of modes involving the S–H bond, there are several other differences induced by bonding to the surface: the ν(C–S) mode is shifted down in energy from 715 cm^{-1} in CH_3SH to 640 cm^{-1} for the thiolate; a new mode is present at 570 cm^{-1} for the adsorbed intermediate, corresponding to the Mo–S stretch; and, there is only one resolved methyl deformation mode at 905 cm^{-1} for the thiolate as opposed to two at 965 and 1070 cm^{-1} for the condensed thiol.

The differences between the thiolate and thiol spectra provide important information about the bonding to the surface. The low ν(C–S) indicates that the C–S bond is weakened as a result of interaction with the surface. Such weakening would facilitate desulfurization and is consistent with the high activity of Mo for sulfur removal. The collapse of the two methyl deformation modes into one suggests an increase in symmetry of the thiolate to C_{3v} which provides valuable structural information, discussed in detail below.

While the vibrational data are compelling evidence for a methyl thiolate intermediate, isotopic labeling data are necessary to unequivocally demonstrate that it is the majority intermediate that produces methane. A partially dehydrogenated species, such as CH_2S, would have a vibrational spectrum quite similar to the methyl thiolate, CH_3S. A CH_2S species could be rehydrogenated to methane. Reaction of methanethiol in the presence of surface deuterium rules out this possibility, however. Only a single deuterium is incorporated into the methane produced in this experiment, ruling out the partially dehydrogenated intermediate in this case. If CH_2S were the intermediate present up to the onset of hydrocarbon formation, up to two deuteriums would be incorporated into the methane product. Since there is a single majority intermediate, the data overwhelmingly supports a thiolate intermediate.

This example illustrates the importance of using complementary chemical and spectroscopic methods in concert when investigating surface reactions. Even for the relatively simple case of methanethiol, several types of experiments were necessary to define the nature of the intermediate leading to methane formation. In more complex, longer-chain thiols the complementarity of these methods assumes greater importance.

In large molecules, such as thiols with longer alkyl chains, intramolecular coupling must be taken into consideration when making vibrational assignments. In the case of methyl thiolate, there are only a few vibrational modes to assign and there is only minimal vibrational coupling in the gas-phase thiol. However, mode coupling is well-documented in the gas phase for long-chain alcohols and thiols because there are often several modes with similar energies and the same symmetry. If several modes are coupled in the free molecule, there will likewise be coupling in the adsorbed state or in closely related intermediates derived from the free molecule.

The vibrational coupling may be severely altered by virtue of the interaction of the molecule with the surface so that mode assignments may not directly follow from the gas phase. Specifically, if the force constant for a vibration involved in the coupling is changed via bonding to the surface, the degree of coupling with other modes will be perturbed. For example, the C–S bond in methyl thiolate on Mo(1 1 0) is shifted by 75 cm^{-1} to lower frequency. Similarly, a decrease of the ν(C–S) stretch frequency in methanethiol has also been reported previously on Fe(1 0 0) [3.22], Cu(1 0 0) [3.22], and Ni(1 1 0) [3.20]. Although coupling is not important in this case, an analogous shift in ν(C–S) for

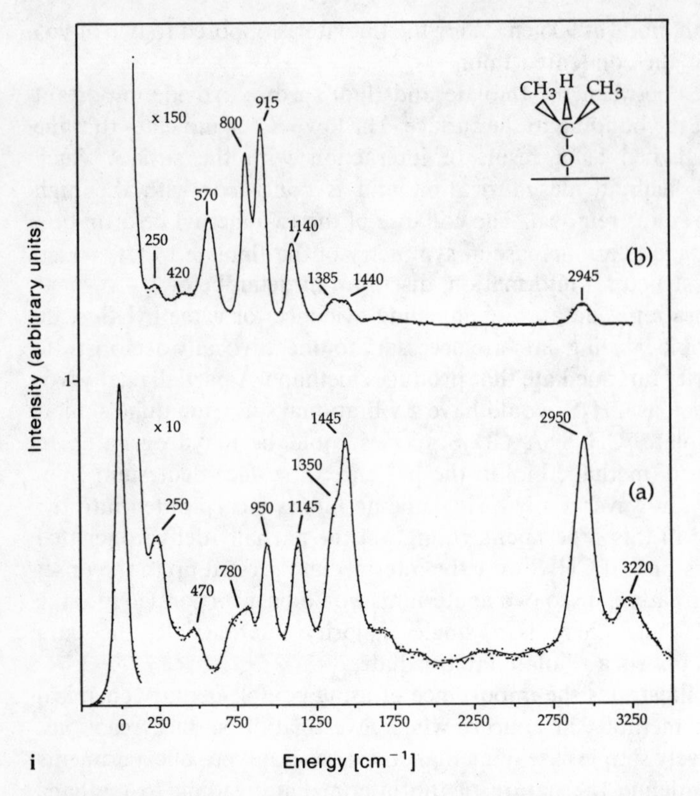

Fig. 3.5. Electron energy loss data for (i) $(CH_3)_2$ CHOH and (ii) $(CD_3)(CH_3)$CHOH and (a–bii) their respective alkoxides

longer-chain thiols will severely alter the vibrational coupling and mode assignments. Since there is a general trend toward C–S bond weakening on metal surfaces, such effects are anticipated for larger thiolates.

Isotopic labeling experiments and modeling of the vibrational spectra are necessary to make definitive assignments for longer-chain organic molecules due to the possibility of alteration in vibrational coupling. Unfortunately, our ability to perform extensive isotopic labeling experiments for thiols and their derivatives is hindered by the lack of commercially-available isotopically labeled thiols. In these cases, analogy with the gas phase and comparison to hydrocarbons with similar skeletons have been used to make assignments. The closely-related case of 2-propoxide derived from 2-propanol on Mo(110) will be used to illustrate the application of isotopic labeling to interrogate changes in vibrational coupling, however.

2-Propanol forms 2-propoxide upon interaction with the surface and deoxygenates to eliminate propene in a process similar to the desulfurization of thiols [3.36]. Intramolecular coupling in gaseous 2-propanol has been well-

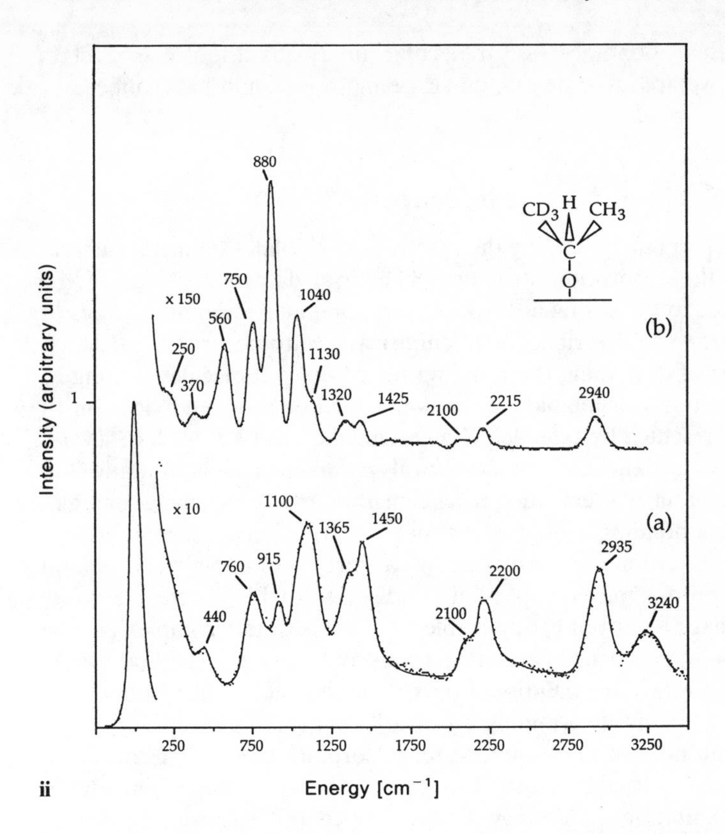

ii

documented, but the extent of coupling is changed when 2-propoxide is formed on the surface because the C–O bond, which is strongly coupled to the C–C stretch and methyl deformations, is weakened due to bonding [3.37]. Vibrational data were obtained for four different 2-propanol isotopomers and were modelled by *ab initio* force field calculations in order to explain the vibrational data. Because the C–O stretch is coupled to the C–C stretch and methyl deformations and is oriented nearly perpendicular to the surface, there are substantial intensity redistributions in the vibrational peaks of the different isotopomers of 2-propoxide (Fig. 3.5). The force field calculations accurately modeled the intensity redistributions and peak shifts due to isotopic labeling [3.38]. Isotopic labeling changed the degree of coupling because it perturbed the symmetry of the molecule and the vibrational energies. The calculations demonstrated that the primary reason for the change in coupling was that the C–O stretch was shifted to lower energy by more than 100 cm^{-1}. Without the isotopic-labeling experiments and force-field calculations, the vibrational modes would have been incorrectly assigned and the C–O bond weakening would not have been evident.

It is clear that a combination of several spectroscopic methods in conjunction with isotopic labeling is necessary to identify surface-bound intermediates and to probe their bonding. For molecular intermediates of the level of complexity that we are interested in, no single method would have sufficed.

3.2.2 Structural Studies of Adsorbed Intermediates

An important step in understanding the reactivity of thiolates on metal surfaces is to determine the adsorption structure. Structural data is necessary if the bonding that leads to the strong adsorption and bond activation in thiolates or other molecules is to be described. Semi-empirical electronic structure calculations are capable of describing the bonding for complex molecules as long as structural information is available. For example, the bonding of ethylene on Pt has been well-represented by extended Hückel calculations [3.39, 40]. In several other cases, however, the lack of structural parameters is a limitation in describing the bonding. The extended Hückel method may yield similar energies for several different bonding configurations of an intermediate. If there is little or no information regarding the structure of adsorbed intermediates, several bonding configurations may be essentially indistinguishable. There are no *ab initio* methods that are currently applicable to systems of this complexity. The only alternative is to perform cluster calculations. In this case, structural data is still necessary as a test of the validity of the calculation since edge or finite-size effects may play a role in determining the theoretical result.

Currently, the molecular orientation of adsorbates can be inferred from vibrational data or near edge X-ray absorption fine structure experiments. In selected cases exhibiting long-range order, low-energy electron diffraction [3.41] or He diffraction [3.42, 43] may also be used. In our work, we have employed both vibrational spectroscopy and near edge X-ray absorption fine structure measurements to infer molecular orientation, as illustrated for methyl and phenyl thiolate in the following section.

Vibrational intensities can be used to determine the disposition of the C–S bond with respect to the surface in selected cases. Symmetry arguments may be used to infer the orientation of a specific functional group in some molecular adsorbates. The orientation is only as reliable as the vibrational assignments, however, underscoring the importance of accurate assignments. Each case will be different so that no general arguments about how to derive orientation from vibrational data can be made. The cases of methyl thiolate and 2-propoxide will be used to illustrate two different ways of using vibrational intensities to derive orientation.

Symmetry arguments were made to deduce the nearly perpendicular orientation of the C–S bond in methyl thiolate on Mo(1 1 0) [3. 8]. In this case, there is an increase in the symmetry of the methyl thiolate bound to the Mo(1 1 0) surface compared to free methanethiol. The increase of the symmetry of the methyl group in methyl thiolate can be achieved if the C–S bond is nearly perpendicular to the surface. The increase in symmetry manifests itself in the methyl deformation modes which collapse into a single resolvable peak at

905 cm^{-1} in the thiolate but which are non-degenerate (1070, 965 cm^{-1}) in the thiol (Fig. 3.3). As the C–S bond is tilted toward the normal, the symmetry approaches C_{3v} and the splitting in the symmetric and asymmetric modes decreases. On most other surfaces, both methyl rocking modes are detected in vibrational spectra, indicating a relatively low symmetry. For example, on Cu(1 0 0) [3.17], Pt(1 1 1) [3.44], Fe(1 0 0) [3.22], and Au(1 1 1) [3.7], both symmetric and asymmetric $\rho(CH_3)$ modes are observed. Only one peak attributed to a methyl rock was resolved on Ni(1 1 0) [3.20], although this may be due to the lack of resolution in the data.

This particular symmetry argument relies on the absence of a mode in the vibrational spectrum. The absence of this mode could, in principle, also be due to a low intensity, rendering it difficult to detect. The symmetry argument is plausible because the intensities of the non-degenerate methyl deformation modes in the thiol itself are very similar and the methyl deformation mode in methyl thiolate has been determined to be impact scattered. The absolute intensity of an impact scattered mode is insensitive to molecular orientation. Impact scattering is a short-range effect and is identified based on the weak dependence of the vibrational intensity on the electron scattering angle [3.33]. Dipole scattering, in contrast, is characterized by a sharp decrease in the vibrational intensity as the detector is moved off of the specular scattering angle. Only modes with a component of the dynamic dipole perpendicular to the surface are allowed for dipole scattering. The methyl rock mode at 905 cm^{-1} for methyl thiolate is not very sensitive to the scattering angle, indicating a strong contribution of impact scattering (Fig. 3.3) [3.8].

Usually modes that are predominantly impact-scattered cannot be used to determine molecular orientation. In this case, impact scattering was essential in making the argument regarding the thiolate symmetry. First of all, if a dipole-scattered mode is absent from a vibrational spectrum it may be due to an orientational effect [3.33]. Hence, the observation of only one methyl rocking mode would not necessarily indicate an increase in symmetry if they were dipole scattered. Since the methyl rock modes were experimentally determined to be mainly impact scattered, they should be observable independent of the orientation of the methyl group and the symmetry argument can therefore be made. Furthermore, the rocking modes would not be observable for a perpendicular orientation of the methyl group if these vibrations were dipole scattered because the dynamic dipole moment for this vibration would lie parallel to the surface plane. This example underscores the importance of angular-dependent measurements when performing electron energy loss experiments so that the scattering mechanism can be defined.

More commonly, the intensities of dipole-scattered vibrational modes are related to the orientation of a specific functional group in a molecule. For example, we have used the relative intensities of the symmetric and asymmetric methyl deformation modes to determine the orientation of the methyl group in several alkoxides on Mo(1 1 0) [3.45]. In these cases, the methyl deformations were experimentally determined to be dipole scattered [3. 37].

The relative intensities of the two different methyl deformations is related to the orientation of the methyl group since they have different polarizations and are both dipole-scattered. The doubly degenerate asymmetric mode has a dynamic dipole moment perpendicular to the C_3 axis of the methyl group whereas the dipole moment of the symmetric deformation lies along the symmetry axis (Fig. 3.6). Consequently, only the symmetric methyl deformation would be observed if the C_3 axis of the methyl group were oriented perpendicular to the surface plane. Conversely, only the asymmetric modes would be detected if the C_3 axis were oriented exactly parallel to the surface. Both modes would be detected for intermediate orientations.

The methyl groups of 2-propoxide were determined to be oriented at an angle of 67° with respect to the surface normal based on an analysis of the relative intensities of the methyl deformations. The disposition of the methyl groups is consistent with a C–O bond orientation near the surface normal, assuming that there is no significant change in the bond angles of 2-propoxide compared to free 2-propanol. This assumption is justified on the basis of the *ab initio* force field calculations [3.38].

Infrared spectroscopy is an alternate method for relating molecular orientation to vibrational intensities. Only dipole active modes are observed in infrared spectroscopy and the spectral resolution is higher. Consequently, the C–H stretch region can be scrutinized in detail in order to probe for tilting of the terminal methyl groups in thiolates and alkoxides, for example. If the C_3 axis of an alkoxide or thiolate is oriented along the surface normal, only the symmetric C–H stretch would be observed in their infrared spectra. The asymmetric C–H stretches are not allowed in dipole scattering since their dipole moment would be parallel to the surface plane. For a tilted geometry, both the symmetric and asymmetric ν(C–H) modes are dipole-allowed. Furthermore, degeneracy of the asymmetric stretches would be broken due to a lowering of the molecular symmetry in a tilted state. Consequently, three distinct C–H stretch peaks are expected for a tilted geometry. Only infrared spectroscopy can be applied this way since the splitting is smaller than the instrumental resolution and impact scattering predominates for C–H stretch modes in electron energy loss spectroscopy (Fig. 3.7).

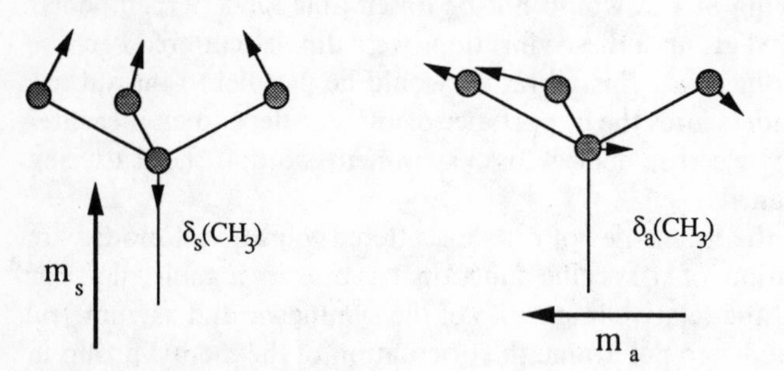

Fig. 3.6. Schematic representation of the symmetric and asymmetric deformation modes of a terminal methyl group in an alkoxide

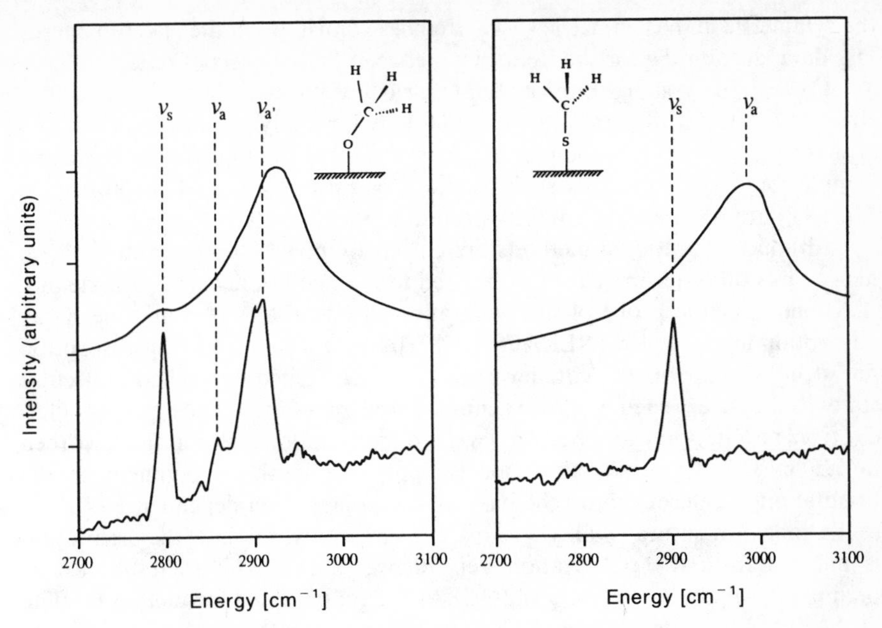

Fig. 3.7. Comparison of the C–H stretch region of (i) infrared reflection and (ii) electron energy loss spectra of (a) CH_3S [3.91] and (b) CH_3O [3.46] adsorbed on Mo(1 1 0). The symmetric and asymmetric C–H stretch modes are shown in the inset

The ν(C–H) stretch modes are split for methoxide bound to Mo(1 1 0), demonstrating that it is tilted away from the surface normal [3.46]. This tilting is corroborated by NEXAFS data [3.46]. The C–H stretch region for methoxide is compared in electron energy loss and infrared spectra to illustrate the superior resolution of the infrared method and the splitting of the degeneracy of the asymmetric stretches (Fig. 3.7). Furthermore, electron energy loss and infrared spectra are compared to illustrate the superior resolution of the infrared method and the splitting of the degeneracy of the asymmetric stretches (Fig. 3.7). Furthermore, electron energy loss spectroscopy is sensitive to both the symmetric and asymmetric modes independent of orientation since C–H stretch modes have a large component of impact scattering.

The symmetric C–H stretch mode dominates the infrared spectrum of methyl thiolate on Mo(1 1 0), confirming the nearly perpendicular orientation of the CH_3 group. These data are consistent with the symmetry arguments made on the basis of the electron energy loss data and serve as an important consistency check [3.47]. There is a low-intensity feature attributed to the ν_a(C–H) mode suggesting a small tilt angle (Fig. 3.7). The small intensity of this mode is consistent with a nearly, but not exactly, perpendicular orientation. The electron energy loss and infrared data can be reconciled by allowing for a small tilt ($\approx 10°$) which would lead to a splitting of the symmetric and asymmetric rock modes, of a magnitude lower than the resolution of the experiment. Both the symmetric and asymmetric C–H stretches contribute to the electron energy loss data, accounting for the frequency between the two sets of data.

Careful analysis of vibrational data clearly yields a substantial amount of structural data. At this stage, the methodology for applying selection rules and symmetry arguments to complex systems is in the developmental stages and complementary experiments such as those performed for methyl thiolate on Mo(1 1 0) are necessary to confirm structural assignments.

Although vibrational methods are extremely powerful structural tools, in many cases other techniques must be used to map additional structural details. A second versatile probe of the orientation of adsorbates is near edge X-ray absorption fine structure (NEXAFS) [3.48a]. In the near edge X-ray absorption fine structure experiment, core level electrons are excited to unfilled molecular states as they are ejected into the vacuum. These excitations can be thought of as electronic excitation from the core levels to antibonding states in the adsorbed molecule. The dipole selection rule is applicable in this experiment so the orientation is determined on the basis of the polarization dependence of these excitations. For saturated alkyl groups, such as methyl thiolate, the orientation is inferred from the polarization dependence of the C–S σ^* resonance. A preliminary analysis of C–K edge NEXAFS data obtained for methyl thiolate on Mo(1 1 0) indicates that the C–S bond is perpendicular to the surface plane, corroborating the conclusions based on electron energy loss and infrared data. For alkyl groups with C–C bond unsaturation, such as phenyl thiolate, the disposition of the alkyl group is derived from the polarization dependence of π^* and σ^* resonances in the alkyl moiety. The tilt angle can usually be more accurately determined for intermediates with C–C bond unsaturation since the C(1s) $\rightarrow \pi^*$ resonances are usually well separated from other peaks and lower in energy than the X-ray absorption threshold, and hence, less subject to variations in the curve-fitting parameters.

Taking phenyl thiolate as an example, the sharp resonances near the C(1s) X-ray absorption threshold are the π^* transitions whereas the broad peaks at higher energy are due to σ^* resonances (Fig. 3.8). The electric-field vector of the light acts as a search light so that resonances have maximum intensity when $\boldsymbol{E}$ is oriented along the transition moment. For a phenyl ring, the π^* resonances would have maximum intensity when the electric field vector is oriented perpendicular to the ring whereas the σ^* intensity will be greatest when $\boldsymbol{E}$ lies in the plane of the ring. Thus, the orientation of the ring in phenyl thiolate adsorbed on Mo(1 1 0) was estimated to be $\approx 20°$ with respect to the surface normal [3.48b]. Visual inspection of the data in Fig. 3.8 reveals that the ring is more closely aligned to the surface normal than with the Mo(1 1 0) plane itself since the π^* resonances are more intense at normal photon incidence [3.49]. The ring is clearly not exactly perpendicular to the surface, however, since the σ^* resonance does not vanish at normal photon incidence.

While there are slight variations in molecular geometry for different surfaces and alkyl groups, the C–S bonds of most thiolates studied to date are nearly perpendicular to the surface plane at high coverages. For example, methyl

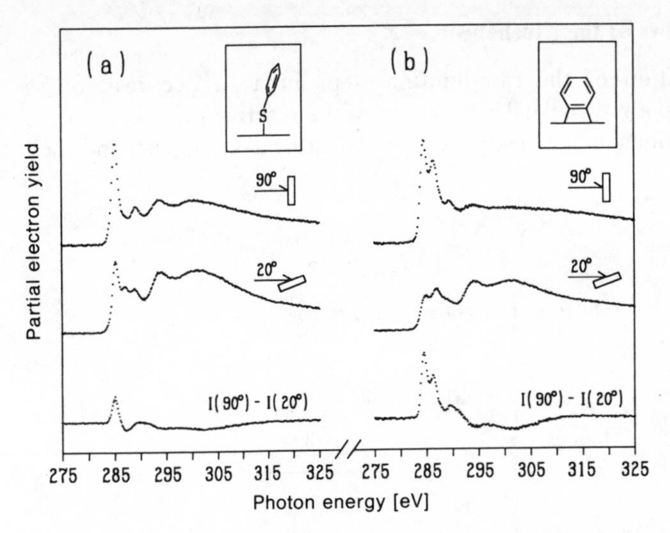

Fig. 3.8. Carbon K-edge NEXAFS data for phenyl thiolate on Mo(1 1 0)

thiolate is essentially perpendicular to the surfaces of Cu(1 0 0) [3.17], Cu(1 1 1) [3.18, 19], Ni(1 0 0) [3.18], and Mo(1 1 0) [3.37]. The presence of a second, somewhat tilted geometry has been proposed for CH_3S- adsorbed on Cu(1 0 0) [3.17] and Pt(1 1 1) [3.21], based on vibrational and NEXAFS data, respectively. Similarly, phenyl thiolate is nearly perpendicular to Mo(1 1 0) at high coverage [3.48]. These geometries are very similar to the orientations of long-chain alkyl thiols which form self-assembled monolayers on Au surfaces [3.5].

The current state-of-the-art in structural measurements is the determination of molecular orientation. In principle, vibrational spectra can be used to infer the symmetry of the adsorption site for molecular intermediates although it is not clear how sensitive the vibrational spectra are to the symmetry of the surface. In the case of methyl thiolate, for example, the symmetry derived from the vibrational data indicates that the spectrum is insensitive to the surface symmetry [3.8]. An independent determination of the coordination site of methyl thiolate, for example, is hence necessary.

In the future, synchrotron radiation methods, namely surface EXAFS, will be used to interrogate the coordination sites and bond distances for selected thiolates so that more detailed modeling of their structure and bonding will be possible. In selected cases, Low Electron Energy Diffraction (LEED) is also a promising method for the study of short-chain thiolates. It has successfully been applied to structural studies of small hydrocarbons, such as ethylene and its derivatives [3.41], and is capable of determining bond distances and coordination site for adsorbates. LEED requires long-range order, however, and detailed calculations to analyze the intensity-energy relationships in the diffraction pattern. As a result, it cannot be applied to all systems of interest. The development of structural techniques is one of the challenges facing the field of surface chemistry.

3.2.3 Chemical Probes of the Mechanism

Chemical interrogation of the rate-limiting steps in a surface reaction is necessary to develop a molecular-level picture of the reactive processes. In the case of thiolate reactions on metal surfaces, several pathways compete, and each

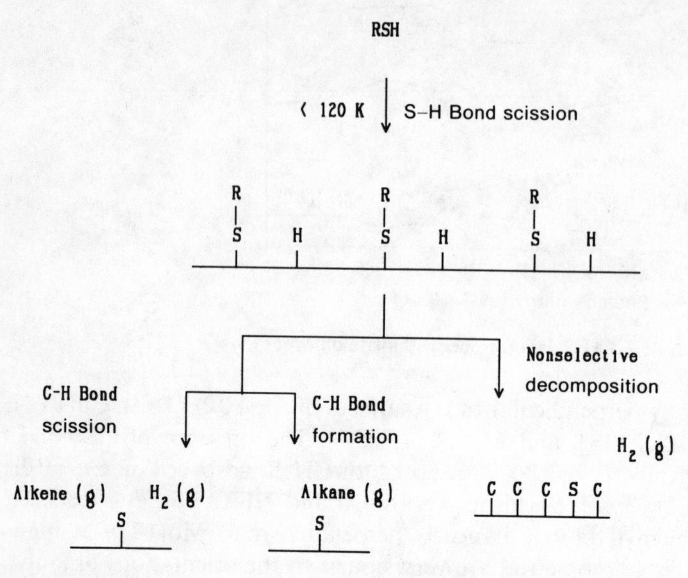

Fig. 3.9. General reaction scheme for thiols on Mo(1 1 0)

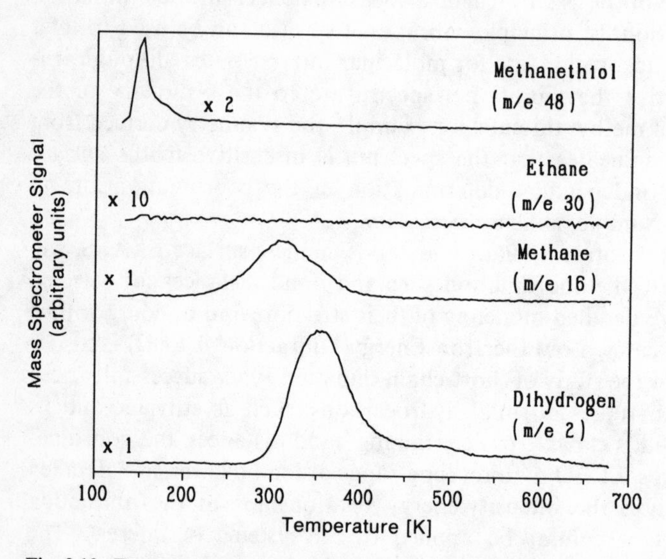

Fig. 3.10. Temperature programmed reaction of methanethiol on Mo(1 1 0). The parent ions are shown in each case and are uncorrected for fragmentation

pathway involves several bond breaking and/or formation steps. By investigating a series of related reactants, the character of the rate-limiting step for a reaction pathway can be probed.

Thiolates react by a general mechanism on Mo(1 1 0) (Fig. 3.9), and this mechanistic investigation serves as an excellent illustration of our experimental approach. Based on a series of experiments, we are able to predict the reaction products and relative selectivity for hydrocarbon formation from thiols with different alkyl substituents when they react on Mo(1 1 0).

There are two competing pathways for hydrocarbon production from thiolates: C–S bond hydrogenolysis to alkanes, and C–S bond breaking accompanied by dehydrogenation to produce alkenes. The third pathway is undesirable; nonselective decomposition affords atomic constituents and gaseous H_2.

Methanethiol forms methane from C–S bond breaking and C–H bond formation on all of the surfaces investigated, including Mo(1 1 0) (Fig. 3.10) [3.8]. Nonselective decomposition, which involves C–S and C–H bond breaking, is also observed. There is no stable gas-phase product of concomitant C–S and C–H bond breaking in methyl thiolate and methylene radicals are expected to be trapped on the surface, if formed. This pathway would ultimately lead to decomposition products on most surfaces. On Cu(1 0 0), reaction of two methyl groups also produces ethane, indicating that dehydrogenation is not facile.

We have proposed that alkene formation occurs via selective scission of the C–H bond γ to the metal (β to the sulfur). γ elimination is the most direct pathway for alkene elimination from thiolates since no hydrogen migration would be required. Furthermore, phenyl thiolate decomposes to adsorbed benzyne, which is the expected product of γ elimination (Fig. 3.11).

Isotopic exchange experiments demonstrate that there is no *reversible* C–H bond activation during desulfurization, that is, once a C–H bond is cleaved, it is *not* reformed. No deuterium is incorporated into the alkene product during reaction of ethanethiol in the presence of surface deuterium atoms, for example. In addition, only ethane-d_1 and ethane-d_0 were formed in the same experiment, indicating that a single C–H(D) bond is formed during hydrogenolysis. Similar results have been obtained for all other thiols investigated on Mo(1 1 0), indicating that the lack of reversible C–H bond scission holds in general for this class of reactions.

Ideally, isotopic labeling would be used to probe for selective γ–C–H bond cleavage. This approach has been successfully used to demonstrate that alkenes are eliminated from alkoxides via C–O and γ–C–H bond cleavage on Mo(1 1 0) [3.36]. For example, 2-propoxide that has been selectively deuterated at the methyl positions reacts to exclusively form propene-d_5 (Fig. 3.12). Similarly, 2-propoxide that is deuterated only at the 1-position produces propene-d_1 (Fig. 3.12). There is no reversible C–H bond activation during deoxygenation of alkoxides on Mo(1 1 0) so that the selective labeling experiments unequivocally demonstrate selective γ elimination. If reversible C–H bond breaking occurred, the isotopic distribution of products would be a convolution of processes and not subject to straightforward interpretation. Unfortunately, analogous ex-

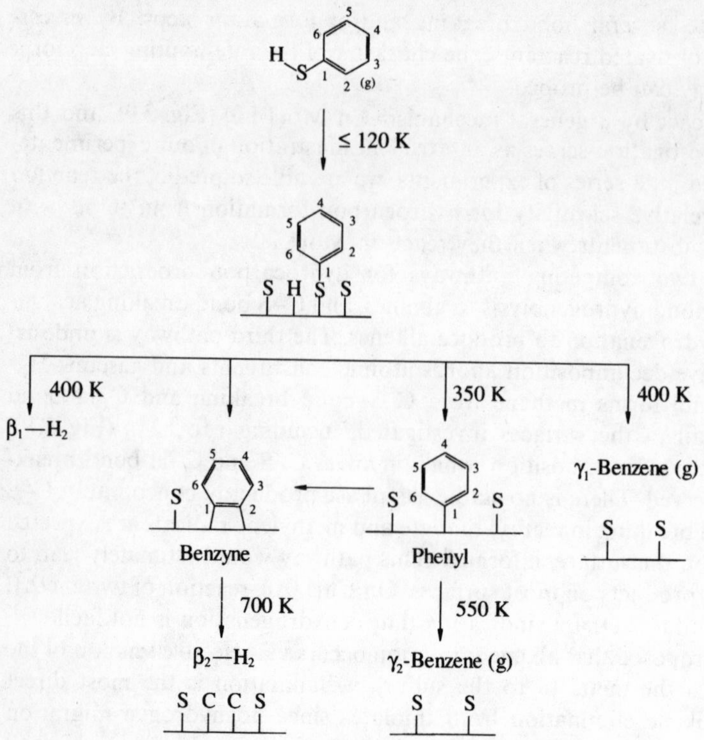
H
S
≤ 120 K
S H S S
400 K
350 K
400 K
β1—H2
S
Benzyne
S
S
Phenyl
γ1-Benzene (g)
S S
700 K
550 K
β2—H2
γ2-Benzene (g)
S C C S
S S

Fig. 3.11. Reaction scheme for benzenethiol on Mo(1 1 0) at high coverage

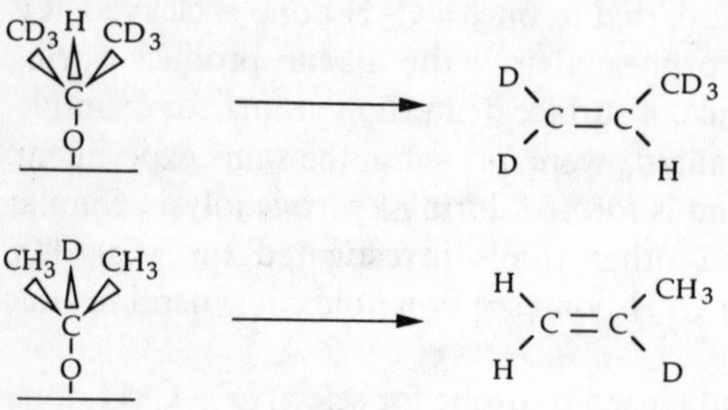
CD3 H CD3
C
O
D
CD3
C = C
D
H
CH3 D CH3
C
O
H
CH3
C = C
H
D

Fig. 3.12. Reactions of isotopically labeled 2-propoxide on Mo(1 1 0). Propene is formed via selective γ-dehydration

periments could not be performed on thiolates because the selectively deuterated isotopomers are not commercially available. However, our model was tested by comparing measured and predicted product distributions for several different thiols on Mo(1 1 0). In particular, the specific alkane and alkene isomers that are formed from longer-chain thiols are correctly predicted by our model.

Support for our proposal that selective γ–C–H bond breaking produces alkenes during the reaction of thiolates is derived from our studies of *tert*-

butanethiol on Mo(1 1 0). All of the hydrogens are equivalent in *tert*-butyl thiolate and at the γ position. Hence, isobutene and isobutane should be produced during desulfurization. As with other thiols, vibrational spectroscopy demonstrates that the S–H bond is broken upon adsorption at 120 K and that the thiolate persists on the surface up to the onset of hydrocarbon formation.

t-Butanethiol, indeed, reacts to form isobutane and isobutene products. The formation of the isobutene isomer supports the proposed mechanism whereby the γ–C–H bond is selectively cleaved without rearrangement.

In order to determine the specific isomer formed in a reaction, the fragmentation patterns of products formed during temperature programmed reaction must be quantitatively measured using a 20 eV ionization voltage in our mass spectrometer. For ionization voltages of 70 eV, there are no measurable differences in the fragmentation patterns of the various isomers of butane and butene. At 20 eV, significant differences in the fragmentation patterns arise and the specific isomers can be readily distinguished (Table 3.1) [3.16].

Thiolate hydrogenolysis involves both C–S bond breaking and C–H bond formation; however, the rate of hydrogenolysis correlates only with the gas-phase C–S bond strengths. The C–S bond strength can be experimentally varied by changing the alkyl substituent on the thiol. All primary thiols, with the exception of methanethiol, have similar C–S bond strengths. Accordingly, their respective thiolates all form the corresponding alkanes with essentially the same kinetics: the line shapes and temperatures of their peaks in temperature programmed reaction data are all indistinguishable [3.10]. In contrast, *t*-butanethiol [3.16] reacts more rapidly than the primary thiols while benzenethiol reacts more slowly [3.15]. The thiols that react most rapidly, also react most selectively. This trend in reaction rate correlates with the stability of the corresponding alkyl radicals or cations that would be formed from C–S bond breaking: Phenyl is the least stable whereas *t*-butyl is one of the most stable [3.50].

Our proposal that C–S bond breaking limits the rate of desulfurization and that dehydrogenation or hydrogenation rapidly follows is also consistent with the mechanism proposed for alkene elimination from thiolates. If dehydrogenation preceded sulfur elimination, the rate of alkene formation would be expected to correlate with C–H bond strength. The weakest C–H bond in a thiolate would be the one on the carbon adjacent to sulfur. Hence, primary thiols, which contain a hydrogen at this position, should dehydrogenate more rapidly than *tert*-butanethiol. In fact, the opposite is observed, indicating that C–S bond scission also precedes dehydrogenation in the alkene elimination

Reactions of 2-Methyl-2-propanethiol on Mo(1 1 0)

Table 3.1. Fragmentation patterns for butane isomers at 27 eV ionization energy

Molecule	*m/e* 41 : *m/e* 56	Molecule	*m/e* 41 : *m/e* 56
1-butane	1.99 ± 0.06	Isobutene	1.47 ± 0.04
cis-2-butene	1.29 ± 0.03	Butene product	1.55 ± 0.06

pathway. Similar trends have also been observed for alkoxides on Mo(1 1 0) [3.37].

The proposal that C–S bond breaking precedes hydrogenation or dehydrogenation is similar to the reaction scheme specifically proposed for methanethiol reaction on Fe(1 0 0) and qualitatively similar to the other surfaces studied. In the case of Fe(1 0 0), methyl, produced from C–S bond breaking in the thiolate, is proposed to yield methane via subsequent hydrogenation.

The dependence of the hydrogenolysis kinetics on C–S bond strength is different, however, for aryl thiolates on Ni(1 1 0) where a similar experimental approach has been recently used [3.51]. These data have been interpreted in terms of a concerted, hydrogen-assisted C–S bond scission step during hydrogenolysis of the thiolate [3.51].

3.2.4 Coverage Dependence of Reactivity

The reactivity of thiols is strongly dependent on their coverage on Mo(1 1 0) and other transition-metal surfaces. This effect is interesting since it indicates that intermolecular interactions can have a strong effect on reactivity and, perhaps, bonding structure. The differences in thermal stability is important from the viewpoint of catalysis and applications involving self-assembled monolayers.

On Mo(1 1 0), thiolates react more rapidly at low coverage. For example, phenyl thiolate decomposition commences at ≈ 225 K at saturation coverage compared to ≈ 150 K at a coverage of 10% of saturation. X-ray photoelectron data indicate that all C–S bonds are cleaved in phenyl thiolate by 250 K at the lower coverage compared to ≈ 500 K at saturation coverage.

The reaction selectivity is also strongly dependent on thiolate coverage. Typically, only nonselective decomposition to surface carbon and sulfur and gaseous H_2 occurs at low coverage. As the coverage is increased, the kinetic stability of the thiolate is increased and hydrocarbon production commences. A similar trend has been reported for benzenethiol on Ni(1 1 0) [3.51].

The differences in selectivity and kinetics may be related to the structure of the thiolates on the surface. As noted above, the alkyl groups of thiolates are generally oriented away from the surface at high coverage. Unfortunately, there is little structural data available for lower coverages. However, it is possible that metal-alkyl interactions occur at lower coverages due to the availability of more coordination sites on the surface. Such interactions may lead to a tilted geometry and perhaps more facile dehydrogenation and C–S bond activation.

3.3 Desulfurization of Cyclic Sulfur-Containing Molecules

The desulfurization of cyclic sulfur-containing molecules has been widely studied on transition-metal surfaces as a model for fossil-fuel reforming. Thiophenic functional groups are prevalent in sulfur-containing fossil fuels so that thiophene has been most extensively explored. Since ring hydrogenation of thiophene derivatives may compete with sulfur extraction, the reactivity of sulfur heterocycles with less C–C bond unsaturation, such as 2, 5-dihydrothiophene and tetrahydrothiophene, is also of interest. The study of these molecules is a means of gaining insight into the steps important in the competing hydrogenation paths since these pathways are not favored under ultrahigh vacuum conditions due to the low H_2 partial pressure.

The reactions of thiophene have been widely studied on transition metal surfaces. Based on investigations of high surface area catalysts, several models for thiophene hydrodesulfurization have been proposed. In one scheme, partial or full hydrogenation of the ring precedes desulfurization [3.52]. Nucleophilic attack at the 2-position by hydride has been specifically proposed as the first step, based on model studies of homogeneous transition metal complexes (Fig. 3.13). Alternatively, hydrogenolysis of carbon–sulfur bonds may occur directly, forming 1, 3-butadiene [3.53]. Finally, thiophene desulfurization may proceed without hydrogen via β-hydrogen elimination to produce diacetylene which would be subsequently hydrogenated [3.54].

On most transition metal surfaces thiophene nonselectively decomposes to surface carbon, sulfur and gaseous H_2 under ultrahigh-vacuum conditions. In cases where no gaseous hydrocarbon products are formed from thiophene decomposition, detailed mechanistic information is not derived. The predominance of nonselective reaction under ultrahigh-vacuum conditions is, at least in part, due to the relatively low coverage of hydrogen on these metal surfaces at temperatures sufficiently high to favor hydrogenation. Second, C–S bond breaking must be more rapid than molecular desorption in order for reaction to occur under ultrahigh vacuum conditions. Therefore, nonselective decomposition is favored on surfaces where C–S bond scission is facile but C–H bond formation is slow relative to hydrogen atom recombination: Re(0001) [3.55], W(211) [3.23], Mo(100) [3.56], Mo(110) [3.57], Ru(0001) [3.58], Ni(111) [3.59], and Pt(100) [3.60]. The slow rate of C–H bond formation under ultrahigh-vacuum conditions on these surfaces precludes the formation of

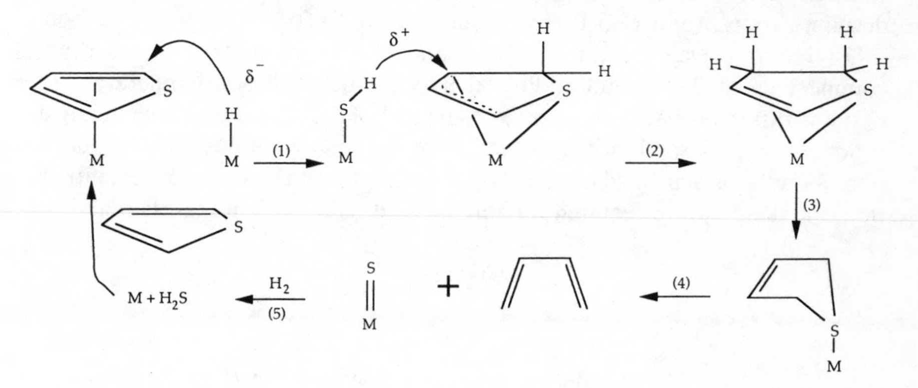

Fig. 3.13. Reaction scheme for thiophene hydrodesulfurization proposed in [3.72]

volatile hydrocarbon products as well as direct hydrogenation of the ring to form partially unsaturated products.

The fact that nonselective decomposition is the exclusive reaction path on many surfaces under ultrahigh-vaccum conditions is an example of the "pressure gap" in comparing ultrahigh-vacuum studies to catalytic reactions that typically occur at atmospheric pressure or above. While hydrogenation does not effectively compete with C–S and C–H bond breaking under ultrahigh-vacuum conditions, such processes may be important under high H_2 pressures such as those used in "real" catalytic processes. For high pressures of H_2, the rate of hydrogenation is increased because of the higher steady-state coverage of hydrogen on the catalyst surface. For example, butadiene is produced from thiophene reaction on Mo(1 0 0) under high pressures of hydrogen whereas no volatile hydrocarbons are produced during thiophene reaction under ultrahigh vacuum conditions [3.61].

Hydrogenation products are produced from thiophene reaction under ultrahigh-vacuum conditions on a few surfaces which are known to have a high activity for hydrogenation. Butadiene is formed from thiophene decomposition on Pd(1 1 1) [3.62,63], Pt(1 1 1) [3.60,64], Ru(0 0 0 1) [3.65], and Rh(1 1 1) [3.66] under ultrahigh-vacuum conditions. This is probably due to the ability of these surfaces to promote rapid C–H bond formation. Therefore, partial hydrogenation of either thiophene itself or of hydrocarbon fragments, formed from sulfur excision from thiophene, competes favorably with dehydrogenation even under ultrahigh-vacuum conditions. Since no ring hydrogenation products are reported in these studies, C–S bond scission is probably still more rapid than desorption or ring hydrogenation under ultrahigh-vacuum conditions.

Surface spectroscopy and selective isotopic labeling experiments have provided the basis for several mechanistic proposals for thiophene decomposition. Sulfur replacement by a surface metal atom to yield a five-membered metallacycle (Fig. 3.14) was first proposed for Pt(1 1 1) based on near edge X-ray absorption fine structure and vibrational data [3.64]. Subsequently, multiple-metal metallacycles, such as $\equiv C{-}CH_2CH_2C \equiv$ bound to the Pt surface through both carbons, were proposed as alternative models for the thiophene desulfurization product on Pt(1 1 1) and Pt(1 0 0) [3.60]. On Pt(1 1 1), carbon–sulfur bond cleavage in thiophene occurs over a broad temperature range, commencing at 290 K and ending at 470 K. Such a broad temperature for desulfurization probably indicates that the rate of desulfurization is changing as products are deposited on the surface. Sulfur probably inhibits the desulfurization. As sulfur is deposited on the surface during the initial stages of desulfurization, the temperature required for thiophene desulfurization increases and the

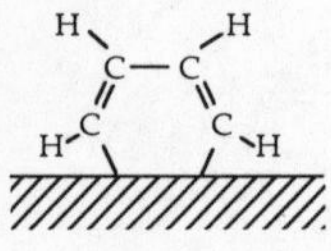

Fig. 3.14. Model of metallacycle proposed for thiophene desulfurization on Pt(1 1 1)

remaining thiophene is kinetically more stable. This effect is well-documented on Mo(1 1 0), for example [3.67].

Butadiene is produced during thiophene reaction on Pt(1 1 1) and a two-step hydrogenation of the C_4H_4 intermediate could afford butadiene. No isotopic exchange experiments were reported that would have provided insight into the details of the hydrogenation steps, however. Isotopic exchange experiments are necessary to probe for possible reversible C–H bond activation which would indicate a more complex mechanism for the butadiene formation reaction. For example, dehydrogenation of thiophene may precede C–S bond breaking, resulting in a C_4H_3 intermediate on the surface which may be subsequently hydrogenated to butadiene. A similar metallacycle intermediate was proposed as the product of thiophene reaction on Ru(0 0 0 1) [3.65]. As on Pt, butadiene is formed, although the exact mechanism for the hydrogenation was not defined.

There is evidence for C–H bond activation during the reaction of thiophene on surfaces which do not promote hydrogenation under ultrahigh vacuum conditions. Formation of a metallacycle-like intermediate from thiophene decomposition has also been proposed for Ni(1 0 0) [3.68] and Ru(0 0 0 1) [3.65] on the basis of vibrational studies. On Ni(1 0 0), a C_4H_3 metallacycle is proposed which ultimately decomposes to surface carbon and H_2 [3.68].

Selective isotopic labeling experiments have shown that there is a preference for cleavage of the C–H bonds adjacent to the sulfur in thiophene. The *ortho* C–H bonds are cleaved more rapidly than at the other positions in thiophene adsorbed on Ni(1 0 0) [3.68], Mo(1 1 0) [3.57], and Mo(1 0 0) [3.56]. The more facile dehydrogenation of the *ortho* carbons is consistent with a lowering of the homolytic C–H bond energy by the adjacent electronegative sulfur atom similar to that proposed for alkoxides [3.69–71].

A surprising difference in the reactivity of Ni(1 1 1) and Ni(1 0 0) underscores the importance of using complementary methods to probe surface intermediates. On Ni(1 0 0) molecular thiophene could not be isolated since desulfurization proceeds upon adsorption at 90 K [3.68]. In contrast, thiophene is purported to be molecularly bound and perpendicular to the Ni(1 1 1) surface at room temperature based on infrared reflection data [3.59]. Since the infrared studies were used without corroborating spectroscopic evidence, it is possible that they actually probed the orientation of a cyclic decomposition product, such as the metallacycle proposed for Ni(1 0 0), since the ring modes of thiophene and such a metallacycle would be similar. The possibility of C–S bond breaking below room temperature was not considered. Ideally, X-ray photoelectron spectroscopy would have been used to complement the infrared data in order to search for C–S bond breaking. Unfortunately, no such experiments were performed.

The geometry that thiophene assumes on the metal surface may also play a role in determining the relative rates of C–H bond scission at the different ring positions as well as the facility for C–S bond breaking. Thiophene may bond to metal surfaces either via the sulfur or the π system. Pure S-bonding would result in a ring orientation nearly perpendicular to the surface, whereas the ring would assume a parallel disposition if there were solely π bonding. Both of these bonding configurations have been observed for transition metal complexes. On surfaces, a combination of S- and π-bonding for thiophene is commonly observed.

The *ortho* hydrogens would be more sterically accessible to the surface when the thiophene ring is oriented perpendicular to the surface, for example. Indeed, thiophene orients perpendicular to the Mo(1 0 0) surface for coverages greater than 0.3 and undergoes selective *ortho* dehydrogenation [3.56]. π-bonding of the ring may also lead to selective activation of the *ortho* C–H bonds, however, as has been demonstrated in organometallic complexes of thiophene [3.72].

The alignment of the thiophene ring typically depends on both its coverage and the presence of decomposition products, such as sulfur. At low coverage, thiophene is π-bound and parallel to the surface plane on all surfaces where the orientation was specifically probed: Cu(1 0 0) [3.73], Mo(1 0 0) [3.56], and Rh(1 1 1) [3.66].

A compressed layer of thiophene, in which the ring assumes a more upright geometry as the coverage is increased, has been reported at low temperature for several surfaces. The compressed state sacrifices π bonding in order to increase the coverage of thiophene on the surface. Since S-metal bonds are typically stronger than metal-π bonds, the increase in coverage even with the loss of π-bonding is still apparently energetically favorable. On Cu(1 0 0) the ring tilts away from the surface as the thiophene coverage increases, ultimately aligning near to the vertical [3.73]. Similarly, a nearly perpendicular orientation has also been identified for high thiophene coverages on Mo(1 0 0) [3.56]. The transition to a perpendicular geometry is not universally observed, however, indicating that the thermodynamic balance between S- and π-bonding is delicate. Thiophene is oriented parallel to the Rh(1 1 1) surface independent of coverage, for example [3.66]. Thiophene is estimated to be tilted $\approx 40°$ with respect to the surface normal on Pt(1 1 1) at low temperature and high coverages. This tilted state is the precursor to formation of the C_4H_4 metallacycle. In addition, the thiophene ring is oriented roughly perpendicular to Mo(1 0 0) for all thiophene coverages studied when there is a high coverage of sulfur [3.56].

There is no clear correlation between the reactivity and molecular orientation of thiophene on surfaces. Thiophene on Cu molecularly desorbs without reaction at all coverages, indicating that the orientation does not affect reactivity in this case. Thiophene molecularly desorbs from Mo(1 0 0) at high coverages of either thiophene or sulfur, the conditions for perpendicular geometry [3.56,74]. The reactivity of thiophene on Rh(1 1 1) and Pt(1 1 1) is very similar despite the reported difference in orientation.

Because of the possibility of hydrogenation of the thiophene ring, understanding the reactivity of hydrogenated sulfur-containing rings on metal surfaces is important. The reactions of 2, 5-dihydrothiophene have been investigated on Mo(1 1 0) because 2, 5-dihydrothiophene was proposed as an intermediate along the pathway for thiophene desulfurization in organometallic complexes [3.75]. 2, 5-Dihydrothiophene has not been investigated on other surfaces. 2, 5-Dihydrothiophene intramolecularly eliminates butadiene with $\approx 70\%$ selectivity on Mo(1 1 0). The kinetics for butadiene elimination are very rapid but depend on the presence of sulfur. 2, 5-Dihydrothiophene reacts similarly on Mo(1 1 0)–p(4 × 1)–S ($\theta_s = 0.50$) but with slower kinetics and higher selectivity [3.67].

High-resolution electron energy loss spectroscopy [3.67] was used to show that the orientation of 2, 5-dihydrothiophene with respect to the Mo(1 1 0)–p(4 × 1)–S surface depends on its coverage. The intensity of the HC = out-of-plane bending mode was used to infer the approximate orientation. NEXAFS data was necessary to quantitatively determine the disposition of the C–C = C–C plane [3.8, 76]. At high coverage, the C–C = C–C plane is oriented $\approx 8°$ away from surface normal. At lower coverage the C–C = C–C plane tilts towards the surface plane.

The reactivity of 2, 5-dihydrothiophene on Mo(1 1 0)–p(4 × 1)–S at high coverage is remarkably similar to that on a mononuclear Ru complex [3.77]. 2, 5-Dihydrothiophene bound to the Ru complex is nonplanar with a dihedral angle of $\approx 26.2°$ (Fig. 3.15). Such a nonplanarity may account for the small ring tilt of 2, 5-dihydrothiophene on Mo(1 1 0)–p(4 × 1)–S. NEXAFS is, in principle, capable of determining the disposition of the C–S–C plane and, therefore, the dihedral angle. Unfortunately, there are two possible transitions with different initial-state symmetries to the orbital associated with the C–S bonds. Depending on the initial state, the ring may be bent or flat. A calculation of transition intensities is necessary to determine the dihedral angle based on NEXAFS. The mode of adsorption for intact 2, 5-dihydrothiophene on clean Mo(1 1 0) was not identified because reaction commenced upon adsorption at 120 K so that the intact molecule could not be isolated. However, only sulfur-bound, 2, 5-dihydrothiophene has been observed in transition metal complexes [3.78–80] so that it is the most likely configuration on the clean surface as well.

The overall reactivity of 2, 5-dihydrothiophene on Mo(1 1 0) is substantially different from that of thiophene. Thiophene only undergoes complete decomposition to H_2, surface S, and surface C on Mo(1 1 0) below 120 K [3.57]. In contrast, although a significant amount of 2, 5-dihydrothiophene also undergoes desulfurization upon adsorption, approximately two-thirds of the irreversibly-absorbed 2, 5-dihydrothiophene ultimately reacts to selectively form gaseous

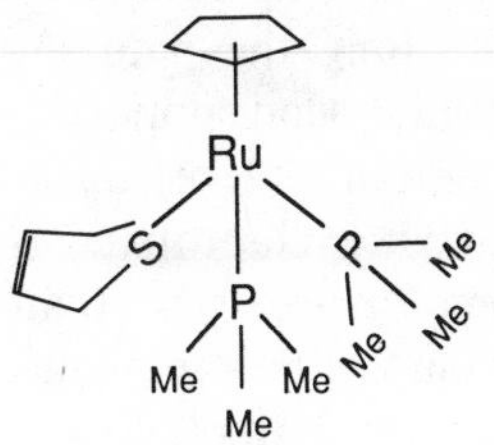

Fig. 3.15. The proposed structure of 2,5-dihydrothiophene in a ruthenium complex [3.77]

butadiene. In addition, approximately twice as much sulfur is deposited during the reaction of 2,5-dihydrothiophene as for thiophene (0.23 vs. ≈ 0.1 monolayers).

The high selectivity for gaseous butadiene formation from 2,5-dihydrothiophene arises, at least in part, from the fact that it can desulfurize to form a hydrocarbon product in a direct intramolecular elimination reaction whereas thiophene cannot. The elimination of butadiene from 2,5-dihydrothiophene is analogous to the elimination of a C_4H_4 hydrocarbon which has been proposed as the metallacycle on Pt(1 1 1). A minimum of two C–H bonds must be formed to make butadiene from thiophene, for example, which is not favored on Mo.

The rapid elimination of butadiene from 2,5-dihydrothiophene suggests two possible pathways for the formation of butadiene from thiophene on Pd(1 1 1) and Ru(0 0 0 1). Thiophene itself could be initially hydrogenated to dihydrothiophene, which would facilely eliminate butadiene, or a C_4H_4 intermediate may be formed first followed by hydrogenation. Studies of 2,5-dihydrothiophene on Pd(1 1 1) or Ru(0 0 0 1), for example, would test for this mechanism. Unfortunately, the reactivity of 2,5-dihydrothiophene has not been investigated on either of these surfaces.

Dihydrothiophene itself may also be hydrogenated in the catalytic process, so the fully hydrogenated ring, tetrahydrothiophene, is also of interest. Indeed, a series of studies of fully hydrogenated cyclic sulfides with varying ring size on Mo(1 1 0) yields insight into the rate-limiting step in desulfurization.

The fully hydrogenated ring cyclic sulfides react via three competing pathways on Mo(1 1 0): intramolecular elimination, C–S bond hydrogenolysis to a thiolate, and nonselective decomposition (Fig. 3.16). Molecular desorption of the cyclic sulfides is also a competing process. The thiolate, formed from C–S bond hydrogenolysis reacts to produce hydrocarbons via C–S bond hydrogenolysis and combined C–S and C–H bond cleavage.

The relative contribution of the possible reaction channels for cyclic sulfides on Mo(1 1 0) depends on the ring strain in the cyclic sulfides indicating that C–S bond breaking controls the rate and selectivity for desulfurization. Direct, intramolecular elimination is favored for the highly strained cyclic sulfides, ethylene sulfide, C_2H_4S, and trimethylene sulfide, C_3H_6S. In fact, intramolecular elimination of ethylene from ethylene sulfide is the predominant path [3.81]. There is also a minor amount of nonselective decomposition to carbon, sulfur and H_2. There is no thiolate formation for ethylene sulfide.

Intramolecular elimination is favored when there is a high ring strain and when a minimal amount of reorganization is required along the path for elimination. Theoretical [3.82] and related experimental studies lend support to the proposal that the ease of elimination is related to the amount of reorganization required in going from reactant to product. Although cyclopropane is eliminated from trimethylene sulfide, the kinetics for elimination are slower than for ethylene sulfide [3.83]. As a result, other reaction channels become competitive; in particular, C–S bond hydrogenolysis to *n*-propyl thiolate and

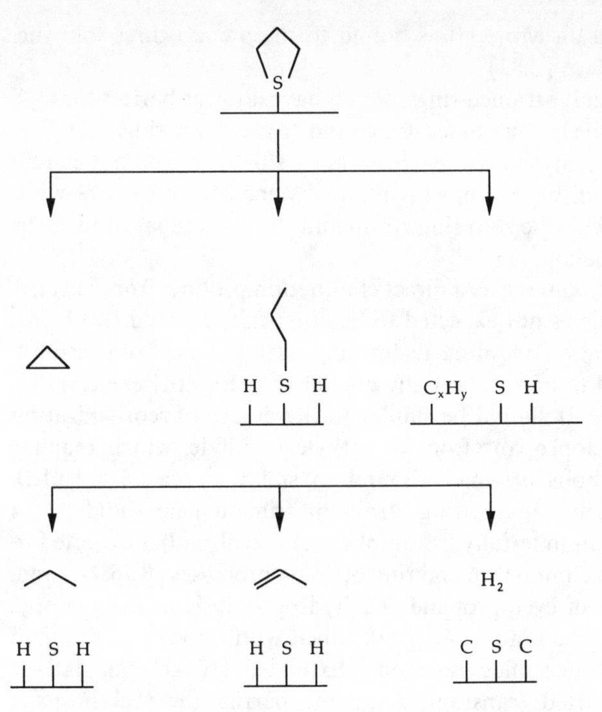

Fig. 3.16. The competing reaction pathways for cyclic sulfide desulfurization on Mo(1 1 0). The specific case shown is for trimethylene sulfide [3.12]

nonselective decomposition. The carbon centers must move ≈ 1.1 Å[12] to form the C–C σ bond in the cyclopropane product.

The ethylene that is formed from ethylene sulfide is evolved directly into the gas phase at 120 K. This is a surprising result since ethylene itself does not desorb from clean and S-covered Mo(1 1 0) up to ≈ 220 K. We have rationalized this observation by proposing that the ethylene is formed with a large component of its momentum perpendicular to the surface. The elimination reaction is exothermic by ≈ 40 kcal/mol due to the formation of a strong Mo–S bond [3.81]. If a significant fraction of the energy is imparted to the leaving ethylene, it would directly evolve into the gas phase as long as there is very little energy transfer to the surface. If the ethylene is formed with a large component of its momentum perpendicular to the surface, there would be little opportunity for energy transfer to the surface. A concerted transition state, where C–C π bond formation C–S bond breaking occur smoothly, would produce ethylene perpendicular to the surface for S-bound ethylene sulfide.

Extended Hückel calculations show that a concerted transition state is plausible for ethylene elimination from ethylene sulfide. Electron donation from a single orbital, the b_1, to the metal d-band leads to C–S bond weakening and C–C π bond formation. Furthermore, the lowest energy configuration for ethylene sulfide adsorbed on Mo(1 1 0) is bound to the pseudo-three fold site, perpendicular to the surface [3.82].

In contrast to the highly strained rings, which have a lower barrier for C–S bond breaking, the relatively unstrained four- and five-carbon rings, tetrahydrothiophene and pentamethylene sulfide, react more slowly and do not exhibit intramolecular elimination. Indeed, most pentamethylene sulfide desorbs without reaction, consistent with the zero ring strain and the large separation of the carbon centers in the reactant.

The formation of cyclobutane as a direct elimination product from tetrahydrothiophene, for example, is not expected to be kinetically favored [3.12, 75]. The degree of reorganization required in forming cyclobutane from tetrahydrothiophene is expected to be substantially greater than for ethylene elimination from ethylene sulfide. It should be similar to the degree of reorganization required for forming cyclopropane from trimethylene sulfide, which requires displacement of the carbons originally bound to sulfur by ≈ 1.1 Å [3.12]. However, the substantially greater ring strain in trimethylene sulfide (19.4 kcal/mol) compared to that in tetrahydrothiophene (1.7 kcal/mol) is expected to lower the barrier for both elimination and ring opening processes [3.75, 84] and, indeed, both elimination of cyclopropane and hydrogenolysis to form propyl thiolate occur during its reaction on Mo(1 1 0), albeit with slower kinetics than for the analogous ethylene sulfide reactions. Extended Hückel calculations suggest that for a concerted transition state, the barrier for cyclopropane elimination from trimethylene sulfide is ≈ 0.4 eV, substantially higher than the barrier for ethylene elimination from ethylene sulfide and in good agreement with experiment [3.82]. Taken together, the absence of ring strain in tetrahydrothiophene and the substantial amount of reorganization required in the transition state for a concerted elimination appear to render the intramolecular elimination pathway kinetically unfavorable for larger rings.

In the cases where elimination is not kinetically favored, pathways involving C–H bond activation become important. Hydrocarbon production occurs in the larger C_3–C_5 rings, via C–S bond hydrogenolysis, to form the corresponding thiolate intermediates. As discussed in detail in earlier subsections, the thiolate undergoes competing hydrogenolysis, to form alkanes, and C–S bond breaking accompanied by C–H bond breaking, to form either alkenes or nonselective decomposition products. The relative importance of these three channels depends on the kinetics for thiolate formation and the coverage of hydrogen. The relationship between the ring strain and rate of desulfurization is strong evidence that C–S bond breaking controls the rate of desulfurization in cyclic sulfides. As the ring strain increases, C–S bond breaking becomes more facile and the rate of reaction on Mo(1 1 0) increases.

The intramolecular elimination pathway is important in putting the 2, 5-dihydrothiophene reactivity in context. While the reaction of 2, 5-dihydrothiophene is significantly different from that of either thiophene or tetrahydrothiophene, it is analogous to ethylene sulfide [3.11]. The qualitative difference in the reactivities of tetrahydrothiophene and 2, 5-dihydrothiophene can be rationalized in terms of the degree of reorganization required for a direct elimination process for the two reactants. The absence of thiolate formation in the reaction of 2,5-dihydrothiophene suggests that direct elimination to form butadiene and atomic sulfur is kinetically more favorable than C–S bond hydrogenolysis. This is also supported experimentally by the fact that butadiene is eliminated from 2,5-dihydrothiophene at a considerably faster rate than tetrahydrothiophene is desulfurized on Mo(1 1 0) [3.11, 12, 81, 82]. However, both tetrahydrothiophene and 2,5-dihydrothiophene have minimal ring strain, 1.7 kcal/mol [3.84], so that their differing reactivities cannot arise from this factor.

Cyclic sulfides have not been widely investigated on other surfaces. The only other study of saturated cyclic sulfides is tetrahydrothiophene on Cu(1 0 0) [3.85]. On Cu(1 0 0), no reactivity or desorption studies were reported. Tetrahydrothiophene is oriented nearly perpendicular to the Cu(1 0 0) plane based on ultraviolet photoemission data.

The case of cyclic sulfide desulfurization exemplifies how the study of a homologous series of molecules yields insight into the molecular-level control of surface reactions. If only tetrahydrothiophene, for example, had been studied, the relationship between reactivity and ease of C–S bond breaking would not have been established. Furthermore, the intramolecular elimination pathway would not have been discovered.

3.4 Conclusions

The study of desulfurization reactions on single-crystal metal surfaces exemplifies the application of surface chemistry as a means of inferring the mechanistic details of important surface processes. By investigating the reactivity of molecules that model different steps of the desulfurization pathway, key aspects of the mechanism have been inferred. The desulfurization of cyclic sulfides and thio-hydrocarbons is largely controlled by the facility of C–S bond breaking. The driving force for these reactions is the strong metal–sulfur bond on transition metal surfaces.

The fundamental spectroscopic and mechanistic studies that have been described in this chapter have laid the ground work for a more detailed understanding of desulfurization reactions. Next, the nature of the transition state for desulfurization needs to be probed. The trends in reactivity that have been observed indicate that C–S bond breaking limits the rate of reaction but cationic and radical-like transition states are both plausible. A methodology must be developed to address these issues.

Structural studies are extremely important to developing a broad understanding of surface reactions and must be further developed. Currently, there are several complementary methods for determining the orientation of specific bonds in an adsorbed intermediate. In the future, methods for determining the adsorption site geometry and bond distances must be developed so that accurate structural models can be made. Once structures are known for surface intermediates, their bonding can be probed with semi-empirical electronic structure calculations. Furthermore, the structural data can serve as a consistency test for higher-level calculations on clusters or extended solids and will thus promote the development of theoretical methods that can be used to predict bonding and reactivity.

One of the primary challenges that remain in this area of surface chemistry is to understand the role of surface impurities, such as sulfur, and promoters, such as Co, in determining the bonding and reactivity of sulfur-containing molecules on transition metal surfaces. There is clear evidence that impurities modify reactivity of both thiols and cyclic sulfides but little is known of the bonding in these systems. Furthermore, thin metal films form a variety of interesting structures [3.86–88] but little is known about their reactivity.

The study of complex surface reactions is still in its infancy. However, the tools that are now available to us will make it possible to meet the challenge of predicting structure, reactivity and bonding for surface reactions.

Acknowledgements. I gratefully acknowledge the support of the U.S. Dept. of Energy. Office of Basic Energy Sciences under grant No. DE-FG02-84ER13289, which provided funding for most of the work reviewed in this chapter contribution. I would also like to thank Dr. J. Stöhr. Dr. D. Outka, Dr. Wiegand and Dr. Per Uvdal for use of their figures. I am also grateful to Dr. Per Uvdal, M. Weldon, and Dr. Han Xu for their input as well as R. Pachtman, for editorial assistance.

References and Notes

3.1 S.J. Tauster, T.A. Pecoraro, R.R. Chianelli: J. Catal. **63,** 515 (1980)
3.2 S. Harris, R.R. Chianelli: J. Catal. **86,** 400 (1984)
3.3 A. Horlacher Smith, R.A. Barker, P.J. Estrup: Surf. Sci. **136,** 327 (1984)
3.4 S. Ladas, R. Imbihl, G. Ertl: Surf. Sci. **198,** 42 (1988)
3.5 E.B. Troughton, C.D. Bain, G.W. Whitesides, R.G. Nuzzo, D.L. Allara, M.D. Porter: Langmuir **4,** 365 (1988)
3.6 L.H. Dubois, B.R. Zegarski, R.G. Nuzzo: Proc. Natl Acad. Sci. USA **84,** 4739 (1988)
3.7 R.G. Nuzzo, B.R. Zegarski, L.H. Dubois: J. Am. Chem. Soc. **109,** 733 (1987)
3.8 P.C. Uvdal, B.C. Wiegand, C.M. Friend: Surf. Sci., (1992)
3.9 P.A. Stevens, R.J. Madix, C.M. Friend: Surf. Sci. **205,** 187 (1988)
3.10 J.T. Roberts, C.M. Friend: J. Am. Chem. Soc. **109,** 4423 (1987)
3.11 J.T. Roberts, C.M. Friend: J. Am. Chem. Soc. **108,** 7204 (1986)
3.12 J.T. Roberts, C.M. Friend: J. Am. Chem. Soc. **109,** 3872 (1987)
3.13 J.T. Roberts, C.M. Friend: Surf. Sci. **198,** L321 (1988)
3.14 J.T. Roberts, C.M. Friend: J. Phys. Chem. **92,** 5205 (1988)
3.15 J.T. Roberts, C.M. Friend: J. Chem. Phys. **88,** 7172 (1988)
3.16 B.C. Wiegand, P. Uvdal, C.M. Friend: J. Phys. Chem. **96,** 4527 (1992)
3.17 B.A. Sexton, G.L. Nyberg: Surf. Sci. **165,** 251 (1986)
3.18 S. Bao, C.F. McConville, D.P. Woodruff: Surf. Sci. **187,** 133 (1987)
3.19 D.L. Seymour, S. Bao, C.F. McConville, M.D. Crapper, D.P. Woodruff, R.G. Jones: Surf. Sci. **189/190,** 529 (1987)

3.20 D.R. Huntley: Surf. Sci. **240,** 13 (1990)
3.21 X.-Y. Zhu, M.E. Castro, S. Akhter, J.M. White, J.E. Houston: Surf. Sci. **207,** 1 (1988)
3.22 M.R. Albert, J.P. Lu, S.L. Bernasek, S.D. Cameron, J.L. Gland: Surf. Sci. **206,** 348 (1988)
3.23 J.B. Benziger, R.E. Preston: J. Phys. Chem. **89,** 5002 (1985)
3.24 R.I. Hegde, J.M. White: J. Phys. Chem. **90,** 296 (1986)
3.25 J.L. Gland, E.B. Kollin, F. Zaera: Langmuir **4,** 118 (1988)
3.26 R.J. Koestner, M. Salmeron, E.B. Kollin, J.L. Gland: Chem. Phys. Lett. **125,** 134 (1986)
3.27 R.J. Koestner, M. Salmeron, E.B. Kollin, J.L. Gland: Surf. Sci. **172,** 668 (1986)
3.28 C.T. Campbell, M.T. Paffett: Surf. Sci. **139,** 396 (1984)
3.29 C.T. Campbell, M.T. Paffett: Surf. Sci. **177,** 417 (1986)
3.30 G.B. Fisher: Surf. Sci. **87,** 215 (1979)
3.31 C.T. Campbell, B.E. Koel: Surf. Sci. **183,** 100 (1987)
3.32 C.M. Friend, X. Xu: Annu. Rev. Phys. Chem. **42,** 251 (1991)
3.33 H. Ibach, D.L. Mills: *Electron Energy Loss Spectroscopy and Surface Vibrations* (Academic, New York, 1982)
3.34 Y.J. Chabal: Surf. Sci. Rpts. **8,** 211 (1988)
3.35 A.C. Liu, C.M. Friend: J. Am. Chem. Soc. **113,** 820 (1991)
3.36 B.C.Wiegand, P.E. Uvdal, J.G. Serafin, C.M. Friend: J. Am. Chem. Soc. **113(17),** 6686 (1991)
3.37 P. Uvdal, B.C. Wiegand, J.G. Serafin, C.M. Friend: J. Chem. Phys. **97,** 8727–8735 (1992)
3.38 P. Uvdal, A. McKerral, C.M. Friend, M.P. Karplus: In preparation
3.39 J. Silvestre, R. Hoffmann: Langmuir **1,** 621 (1985)
3.40 Y.T. Wong, R. Hoffmann: J. Chem. Soc., Faraday Trans. **86,** 4083 (1990)
3.41 R.J. Koestner, M.A. Van Hove, G.A. Somorjai: J. Phys. Chem. **87,** 203 (1983)
3.42 Marcel den Nijs, Eberhard K. Riedel, E.H. Conrad, T. Engel: Phys. Rev. Lett. **55,** 1689 (1985)
3.43 T. Engel, K.H. Rieder: Surf. Sci. **109,** 140 (1981)
3.44 R.J. Koestner, J. Stöhr, J.L. Gland, E.B. Kollin, F. Sette: Chem. Phys. Lett. **120,** 285 (1985)
3.45 P. Uvdal, B.C. Wiegand, C.M. Friend: In preparation
3.46 P. Uvdal, M.K. Weldon, H. Xu, J. Stöhr, C.M. Friend: Unpublished results
3.47 P. Uvdal, M.K. Weldon, C.M. Friend: Unpublished results
3.48 J. Stöhr, D.A. Outka: Phys. Rev. B **36,** 7891 (1987).
J. Stöhr: *NEXAFS Spectroscopy*, Springer Ser. Surf. Sci., Vol. 25 (Springer, Berlin, Heidelberg 1992)
3.49 The electric field vector for linearly polarized light is perpendicular to the direct of propagation of the light
3.50 M.-C. Tsai, C.M. Friend, E.L. Muetterties: J. Am. Chem. Soc. **104,** 2539 (1982)
3.51 D.R. Huntley: J. Phys. Chem., **96,** 4550 (1992)
3.52 J. Kraus, M. Zdrazil: React. Kinet. Catal. Lett. **6,** 45 (1977)
3.53 J.M.J.G. Lipsch, G.C.A. Schuit: J. Catal. **15,** 179 (1969)
3.54 S. Kolboe: Can. J. Chem. **47,** 352 (1969)
3.55 D.G. Kelly, J.A. Odriozola, G.A. Somorjai: J. Phys. Chem. **91,** 5695 (1987)
3.56 F. Zaera, E.B. Kollin, J.L. Gland: Surf. Sci. **184,** 75 (1987)
3.57 J.T. Roberts, C.M. Friend: Surf. Sci. **186,** 201 (1987)
3.58 R.A. Cocco, B.J. Tatarchuk: Surf. Sci. **218,** 127 (1989)
3.59 G.R. Schoofs, R.E. Preston, J.B. Benziger: Langmuir **1,** 313 (1985)
3.60 J.F. Lang, R.I. Masel: Surf. Sci. **183,** 44 (1987)
3.61 A.J. Gellman, M.H. Farias, G.A. Somorjai: J. Catal. **88,** 546 (1984)
3.62 A.L. Johnson: Ph.D. Thesis, LBL Report No. 20964 (1986)
3.63 T.M. Gentle: Energy Res. Abstracts **9,** 28229 (1984). (Abstract)
3.64 J. Stöhr, J.L. Gland, E.B. Kollin, R.J. Koestner, A.L. Johnson, E.L. Muetterties, F. Sette: Phys. Rev. Lett. **53,** 2161 (1984)
3.65 W.H. Heise, B.J. Tatarchuk: Surf. Sci. **207,** 297 (1989)
3.66 F.P. Netzer, E. Bertel, A. Goldmann: Surf. Sci. **201,** 257 (1988)
3.67 H. Xu, C.M. Friend: J. Phys. Chem. **97,** 3584–3590 (1993)
3.68 F. Zaera, E.B. Kollin, J.L. Gland: Langmuir **3,** 555 (1987)

3.69 X. Xu, C.M. Friend: Surf. Sci. **260,** 14–22 (1992)
3.70 M. Bowker, R.J. Madix: Surf. Sci. **95,** 190 (1980)
3.71 M. Bowker, R.J. Madix: Surf. Sci. **116,** 549 (1982)
3.72 N.N. Sauer, R.J. Angelici: Organometallics **6,** 1146 (1987)
3.73 B.A. Sexton: Surf. Sci. **163,** 99 (1985)
3.74 A.J. Gellman, M.H. Farias, M. Salmeron, G.A. Somorjai: Surf. Sci. **136,** 217 (1984)
3.75 C.M. Friend, J.T. Roberts: Accts. Chem. Res. **21,** 394 (1988)
3.76 H. Xu, P. Uvdal, J. Stöhr, C.M. Friend: In preparation
3.77 S.C. Huckett, L.L. Miller, R.A. Jacobson, R.J. Angelici: Organometallics **7,** 686 (1988)
3.78 N.N. Sauer, R.J. Angelici: Inorg. Chem. **26,** 2160 (1987)
3.79 J.H. Eekhof, H. Hogeveen, R.M. Kellog, E. Klei: J. Organometal. Chem. **161,** 183 (1978)
3.80 J.H. Eekhof, H. Hogeveen, R.M. Kellog: J. Organometal. Chem. **161,** 361 (1978)
3.81 J.T. Roberts, C.M. Friend: Surf. Sci. **202,** 405 (1988)
3.82 M. Calhorda, R. Hoffmann, C.M. Friend: J. Am. Chem. Soc. **112,** 50 (1990)
3.83 J.T. Roberts, C.M. Friend: J. Am. Chem. Soc. **109,** 7899 (1987)
3.84 S.W. Benson: *Thermochemical Kinetics*, 2nd edn. (Wiley-Interscience, New York 1976)
3.85 T.M. Thomas, F.A. Grimm, T.A. Carlson, P.A. Agron: J. Electron Spectrosc. Relat. Phenom. **25,** 159 (1982)
3.86 R.A. Demmin, S.M. Shivaprasad, T.E. Madey: Langmuir **4,** 1104 (1988)
3.87 P.J. Berlowitz, D.W. Goodman: Langmuir **4,** 1091 (1988)
3.88 J-W. He, D.W. Goodman: Surf. Sci. **245,** 29 (1991)

4. Tricyclisation and Heterocyclisation Reactions of Ethyne over Well-Defined Palladium Surfaces

R.M. Lambert and *R.M. Ormerod*

The unusual cyclisation reactions of ethyne on Pd surfaces are described and discussed with reference to analogies with the chemistry of transition metal clusters. Detailed spectroscopic and kinetic characterisation of the adsorbed reactant, intermediate and product species permits the molecular pathway to be constructed in considerable detail. A C_4H_4 tilted metallopentacycle is found to be the crucial surface intermediate for benzene formation and for the formation of heterocycles. Single-crystal data enable useful predictions to be made about the behaviour of practically supported metal catalysts and in regard to new areas of catalytic chemistry. These ideas are borne out in practice, and the effects on cyclisation chemistry of modifying the metal surface by promoters, poisons and alloying are described and discussed.

4.1 Background

The Diels-Alder cyclisation of alkynes in solution is a standard and very well known procedure in synthetic organic chemistry [4.1, 2]. Similarly, the high-temperature gas-phase homogeneous tricyclisation of ethyne to benzene was discovered long ago by *Berthelot* [4.3] who merely heated the reactant to ≈ 650 K in a glass vial. The corresponding *homogeneously metal-catalysed* process was first reported by *Reppe* et al. [4.4] who used $Ni(CO)_4$ in solution as the catalyst. Since then, many other examples of such homogeneously catalysed cyclisations of alkynes have been found for which both mononuclear and polynuclear transition metal cluster compounds are active and selective catalysts [4.5, 6]. Furthermore, it has often been proposed [4.7, 8] that such organometallic cluster compounds can be regarded as useful model systems by means of which the structure, bonding and reactivity of adsorbates on extended metal surfaces may be rationalised. In this sense, studies on model single-crystal systems may be regarded as providing a bridge between cluster chemistry and the catalytic chemistry of high-area supported metal catalysts; the bridge has, of course, nothing to do with the relative sizes of the metal ensembles in the three systems, rather it is related to our ability to carry out detailed measurements, especially structural observations, with the three types of system. In this connection, a particularly interesting case is that of ethyne tricyclisation. Here,

Springer Series in Surface Sciences, Vol. 34
Surface Reactions Ed.: R.J. Madix

the cluster/single-crystal/real catalyst linkage appears to hold in a way which is both revealing and practically useful.

Low-temperature tricyclisation of ethyne to benzene on polycrystalline palladium black was first reported by *Kojima* and co-workers [4.9]. Subsequently, the occurrence of this same reaction over single-crystal Pd(1 1 1) was demonstrated almost simultaneously by *Tysoe* et al. [4.10] and by *Sesselman* et al. [4.11]; these observations being subsequently confirmed by other workers [4.12, 13]. The reaction can be studied over an extremely wide range of pressure – from ultrahigh vacuum conditions to one atmosphere pressure [4.13], a remarkable feature of the process being the way in which benzene evolution into the gas phase can occur at very low temperatures (≈ 200 K). It also exhibits a number of interesting analogies with well-known organometallic cluster chemistry of ethyne and its derivatives. Thus, it provides a unique opportunity to study a useful and unusual heterogeneous catalytic process over a range of conditions which permit reliable and detailed characterisation of the various surface species involved in the overall reaction. A wide range of physical methods have been used to identify and study these species, including LEED, AES, XPS, ARUPS, HREELS and NEXAFS [4.14–20].

The tricyclisation reaction and some closely related processes are the subject of this chapter which summarises recent work aimed at elucidating the mechanistic chemistry of ethyne cyclisation reactions in as much detail as possible, exploring the analogies with known cluster chemistry and extrapolating the single-crystal data to predict the behaviour of practical supported catalysts; at the same time, one might hope to open up new areas of catalytic chemistry. We shall consider such matters as determination of the bonding of reactants, products and intermediates at the metal surface, identification and isolation of the crucial reaction intermediate and elucidation of the molecular pathway by a combination of kinetic and spectroscopic studies, including isotope tracing. In addition, reference will be made to other cyclisation reactions which occur on metal surfaces. For the most part, these are related to ethyne tricyclisation and in some cases have provided additional useful mechanistic information. We also review salient aspects of ethyne adsorption on transition metal surfaces. These are of relevance to an understanding of the apparently almost unique properties of Pd(1 1 1) in ethyne cyclisation.

4.2 Mechanistic Studies of Ethyne Tricyclisation

4.2.1 Molecular Beam Results, Temperature-Programmed Reaction and Isotope Labelling: Molecular Formula of the Reaction Intermediate

Tysoe et al. carried out a molecular beam study [4.14] of the reaction kinetics of benzene formation from ethyne; a C_4 species was detected in the gas phase and the results indicated that this species was the precursor to benzene formation.

The chemical identity of this precursor could not, however, be established by the molecular beam data alone.

Patterson and *Lambert* [4.15] used deuterium isotope tracing to demonstrate that the molecular pathway from ethyne to benzene involves no cleavage of C–C or C–H bonds. Figure 4.1 shows the results of multiplexed temperature-programmed reaction spectra obtained following the coadsorption of a mixture of C_2H_2, C_2HD and C_2D_2 on Pd(1 1 1). These data exhibit two striking features. Firstly, at sufficiently high surface coverages, benzene is evolved in two very different temperature regimes: clearly resolved desorption maxima occur at $\approx$200 K and $\approx$500 K. This immediately raises the question – are there two different surface processes which lead to benzene formation or just one? Secondly, it is apparent that benzene evolution can occur at very low temperatures – what makes the process so efficient?.

It can be seen from Fig. 4.1 that all possible deuterobenzenes are formed, and all appear in the two peaks at $\approx$200 and $\approx$500 K. Table 4.1 lists the observed relative yields of the various benzenes (corrected for mass-spectrometer fragmentation) and the calculated yields for both associative and dissociative reaction mechanisms. It is at once apparent that the data for both peaks agrees with the calculated values for an associative mechanism (i.e., no C–C or C–H cleavage). Closer inspection of Table 4.1 reveals that the results for the low-temperature peak are in very good agreement with the associative mechanism, whereas those for the high-temperature peak show small systematic deviations from the calculated distribution: the observed yields of those benzene molecules

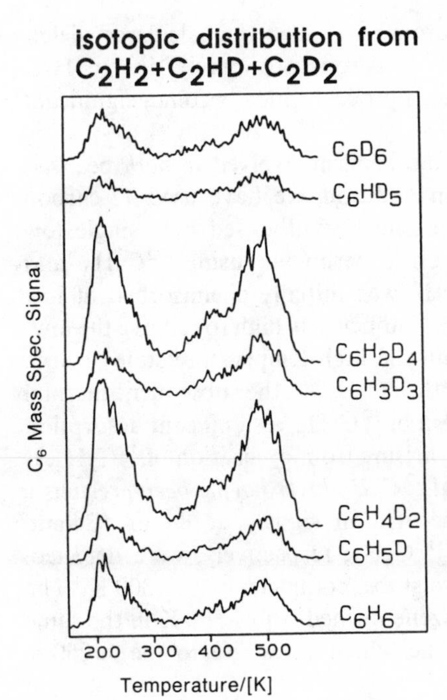

Fig. 4.1. Benzene from ethyne on Pd(111). Multiplexed TPR spectra (raw data) showing formation of all possible benzene species resulting from reaction within a $C_2H_2 + C_2HD + C_2D_2$ adsorption layer

Table 1 Isotopic distribution of benzenes in the low-temperature desorption peak

$(A + B + C)^3 = A^3 + 3A^2B + \cdots$ A, B, C $= C_2H_2, C_2HD, C_2D_2$

Benzene	Observed	Calculated
C_6H_6	10.7	10.7
C_6H_5D	2.9	4.1
$C_6H_4D_2$	32.0	32.0
$C_6H_3D_3$	8.7	8.0
$C_6H_2D_4$	31.7	31.3
C_6HD_5	4.0	4.0
C_6D_6	10.1	10.0

Isotopic distribution of benzenes in the high temperature desorption peak

Benzene	Observed	Calculated
C_6H_6	8.2	10.7
C_6H_5D	4.6	4.1
$C_6H_4D_2$	28.7	32.0
$C_6H_3D_3$	11.5	8.0
$C_6H_2D_4$	31.3	31.3
C_6HD_5	6.3	4.0
C_6D_6	10.0	10.0

containing an odd number of D atoms are greater than the calculated yields. Control experiments indicate that these deviations are merely due to H/D scrambling between the product molecules, a process which becomes significant at higher temperatures.

The conclusion therefore is that *all* the benzene evolved in *both* peaks is formed by an associative mechanism. In addition, we have used 13-carbon labelling to demonstrate that all the benzene is synthesised in a single low temperature process [4.21]: coadsorption experiments using $^{12}C_2H_2$ and $^{13}C_2H_2$ were performed, in which $^{12}C_2H_2$ was initially chemisorbed at low temperature, the sample annealed to 300 K – sufficiently high to remove the low-temperature desorption state but maintain the high-temperature state – before cooling to low temperature and dosing with $^{13}C_2H_2$. In the subsequent thermal desorption sweep, in addition to desorption of $^{13}C_6H_6$, a significant desorption signal at 78 amu, associated with $^{12}C_6H_6$ arising from cyclisation of $^{12}C_2H_2$, is observed below 300 K, *despite the original* $^{12}C_2H_2$ *layer having been previously annealed to* 300 K. Furthermore, no desorption signals at 80 or 82 amu, corresponding to $^{12}C_4{}^{13}C_2H_6$ and $^{12}C_2{}^{13}C_4H_6$, respectively, were detected. Thus we conclude that tricyclisation must be complete below 300 K. The $\approx$500 K peak is due to desorption of benzene formed below 300 K in the same low temperature process as is the benzene which gives rise to the $\approx$250 K desorption peak. All unreacted species remaining at 300 K (ethyne/vinylidene) are *inactive* towards benzene formation.

Deuterium isotope tracing was also used by *Patterson* and *Lambert* [4.16] to demonstrate that the surface intermediate involved in benzene formation has the molecular formula C_4H_4. This was achieved by using dissociative chemisorption of cis-3, 4-dichlorocyclobutene (DCB) as a reagent with which to seed the surface with a C_4H_4 species: $C_4H_4Cl_2 \rightarrow C_4H_4(a) + 2Cl(a)$. Control experiments established that the Cl atoms co-deposited by this procedure exerted no substantial effect on the kinetics and selectivity of benzene formation in the temperature range 200–600 K. Cl desorption from Pd(1 1 1) becomes significant at $\gtrsim 600$ K by which point all the interesting hydrocarbon chemistry has occurred; the Pd(1 1 1)/Cl system was previously studied in detail by *Tysoe* and *Lambert* [4.22].

It was shown [4.16] that this C_4H_4 species reacted with coadsorbed C_2D_2 forming benzene with essentially the same kinetic behaviour as that observed with ethyne alone as a reactant. In these experiments the *only* products formed were $C_6H_4D_2$ (from $C_4H_4 + C_2D_2$) and C_6D_6 (from $3C_2D_2$). This result clearly indicates that the molecular formula of the C_4 reaction intermediate is indeed C_4H_4; furthermore it suggests that it is benzene desorption which is the rate determining step for the evolution of benzene into the gas phase, in agreement with the conclusion from our ^{13}C labelling experiments. This point is very strongly confirmed by the observation that coadsorption of C_6H_6 and C_2D_2 leads to the evolution of C_6H_6 (by simple desorption) and C_6D_6 (formed by reaction, followed by desorption) *with identical kinetics* (Fig. 4.2) [4.15].

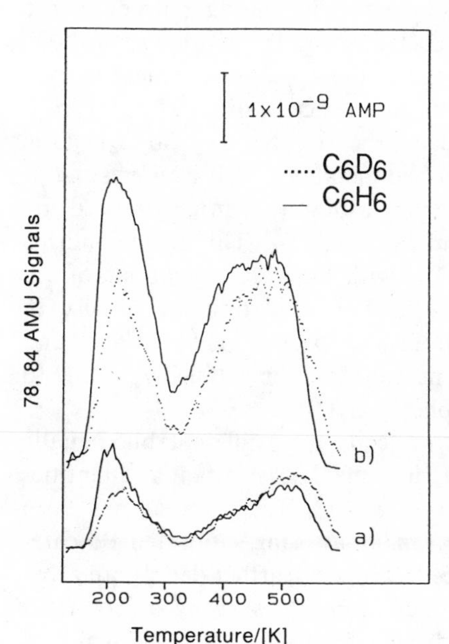

Fig. 4.2. $C_6H_6 + 3C_2D_2 \rightarrow C_6H_6 + C_6D_6$. Desorption of reactively formed benzene (C_6D_6) and adsorbed benzene (C_6H_6) showing identical desorption kinetics. (a) 0.25 C_6H_6 + 1.25 C_2D_2, 150 K. (b) 0.75 C_6H_6 + 1.25 C_2D_2, 150 K

We may therefore conclude that benzene desorption is desorption rate controlled and *not* reaction rate controlled, for both the low-temperature and high-temperature desorption peaks. A final important point remains to be made about these temperature programmed reaction spectra, namely that there is a distinct threshold value for ethyne coverage below which benzene formation does not occur (see figures and Sect. 4.2.3). This threshold coverage corresponds to that required for the formation of a $(\sqrt{3} \times \sqrt{3})$ R30° LEED structure [4.14]; at lower coverages the reactant merely undergoes fragmentation, trimerisation being completely suppressed. The question of adsorbate packing density and its effect on catalytic reactivity is an important one and we shall return to it in Sect. 4.2.3. For the present, it is sufficient to note that with initially clean Pd(1 1 1) the maximum conversion of ethyne to benzene in a TPR experiment is $\approx 25\%$ [4.23], the remainder of the adsorbate undergoing fragmentation to yield hydrogen and (ultimately) carbon. Later, it will be seen that the presence of a suitable co-adsorbate can greatly enhance the benzene yield.

The presence of a reactive C_4H_4 intermediate *under tricyclisation reaction conditions* may be satisfyingly demonstrated by inducing this species to undergo alternative competing reactions. This has, in fact, been achieved, both with single-crystal model catalysts and with high-area supported catalysts, using coadsorbed oxygen or hydrogen to provide competing reaction channels. Thus with preoxygenated Pd(1 1 1), ethyne chemisorption and reaction leads to the production of gaseous furan (C_4H_4O) along with benzene and the products of total oxidation ($CO_2 + H_2O$) (4.24) (Sect. 4.7). It is most unlikely that this partial oxidation of ethyne to furan would have been predicted on the basis of previous knowledge: transition metal surfaces exposed to oxygen and hydrocarbon generally yield $H_2O + CO_2$, the ease of reaction increasing in order alkane < alkene < alkyne. Ethyne would be expected to be the most combustion-prone of all! However, given our knowledge of the apparently almost unique ability of Pd(1 1 1) to generate substantial surface concentrations of C_4H_4 from C_2H_2, the scavenging of the former by coadsorbed oxygen to yield C_4H_4O is exactly what one would expect for an adsorbed metallocycle. It is interesting to note that furan may be generated in a similar fashion (but in larger yield) by coadsorbing DCB ($\rightarrow C_4H_4(a) + Cl(a)$) and oxygen [4.24]. This again confirms that DCB is a reagent for seeding Pd(1 1 1) with high concentrations of the relevant reaction intermediate – providing further validation for its use in spectroscopic studies. Analogous observations have been made by *Gellman* and co-workers [4.25, 26] who used both ethyne and DCB on sulfided Pd(1 1 1) to generate small quantities of gaseous thiophene (C_4H_4S).

Coadsorption of hydrogen with ethyne leads to a different but equally revealing scavenging reaction for the C_4H_4 intermediate: substantial quantities of butadiene ($CH_2 = CH - CH = CH_2$) are formed. This process has been demonstrated under *steady state turnover conditions* using supported Pd catalysts [4.27] contacted with C_2H_2 + surface hydrogen. Further details are given in Sect. 4.3; for the present we note that once again, the scavenged product is exactly what one would expect for the hydrogenation of C_4H_4(a): the structure and bonding of butadiene nicely confirm the structure and bonding proposed for C_4H_4(a), see below.

4.2.2 Characterisation of the C_4H_4 Intermediate

HREELS

The use of DCB as a reagent for producing high surface concentrations of the relevant C_4H_4 reaction intermediate may be exploited in order to determine the chemical identity and chemisorption geometry of this species. High-Resolution Electron Energy Loss Spectroscopy (HREELS) carried out as a function of coverage, temperature and scattering geometry showed [4.17] that the C_4H_4 species is a metallopentacycle. This conclusion was achieved by comparison of the HREEL spectra with infrared reference spectra of known relevant compounds. Furthermore, by comparing the HREEL spectra collected at specular geometry with those collected at off-specular geometry, it was shown that this metallopentacycle is strongly tilted with respect to the metal surface. This conclusion is based on the selection rules which operate in HREELS spectroscopy: vibrations characterized by a large dynamic dipole perpendicular to the metal surface appear only in specular scattering geometry; parallel vibrations appear in both specular and off-specular geometry. Thus a vertical metallocycle with its C_2 axis parallel to the surface normal would exhibit a strong ν(C–C) mode around 1400 cm^{-1}. No such vibration is observed (Fig. 4.3a). However, the loss at $\approx 1220\ cm^{-1}$ may be assigned to an A_1 ν(C–C) mode of a strongly tilted metallocycle bonded to the metal surface by π, σ interaction. This is in

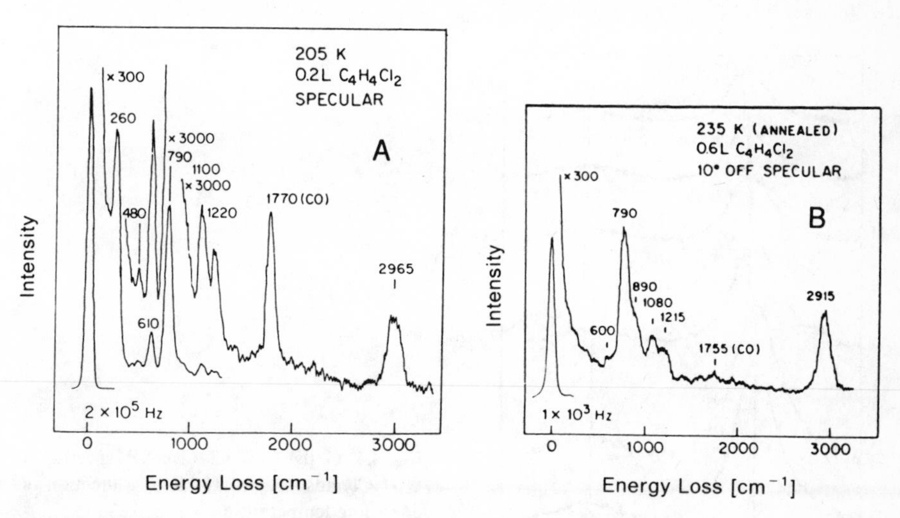

Fig. 4.3. HREEL spectra taken after chemisorption of cis-1,2 dichlorocyclobutene on Pd(1 1 1) at low temperatures. Spectrum obtained in specular geometry; spectrum obtained in off-specular geometry

good agreement with observed infrared spectra of, for example, $Fe_2(CO)_6(C_4H_4)$. The $\approx 1220\ cm^{-1}$ loss may also contain a contribution from $B_1\ \delta$(C–H) mode while the losses at 1100 and 890 cm^{-1} may be assigned to $A_1\ \delta$(C–H) modes of the molecule. It is significant for the analysis that the 890 cm^{-1} loss is observed only in non-specular geometry (Fig. 4.3b). Notice also that a comparison of Figs. 3a and b shows that the strongly dipole active loss at 260 cm^{-1} is quenched in the non-specular spectrum. This feature is reliably assigned to a Pd–Cl stretching mode due to the presence of chlorine adatoms resulting from dissociative chemisorption of the original $C_4H_4Cl_2$. These conclusions are confirmed by our NEXAFS and photoemission studies described below. The HREELS data also indicate [4.17] that the C_4 metallocycle transforms to another species between 250 and 350 K, possibly vinylidene.

ARUPS, XPS, NEXAFS

Here we summarise the results obtained by ARUPS [4.23], XPS [4.28] and NEXAFS [4.28] in the further characterisation of the C_4H_4 intermediate deposited on Pd(1 1 1) by use of DCB, as described earlier. Figure 4.4 displays C (1s) and Cl (2s) XP spectra as a function of annealing temperature following adsorption of DCB at 132 K. Initially, both the C and Cl binding energies (285.4 and 271.0 eV, respectively) are characteristic of multilayers of DCB, as would be expected. Between 132 and 173 K substantial intensity loss occurs associated with the desorption of physisorbed DCB, and a further reduction in intensity occurs on annealing to 213 K; this is accompanied by a shift in the binding energy of both the carbon and chlorine peaks to 284.7 and 270.2 eV, respectively. A close examination of the C (1s) emission suggests that a state character-

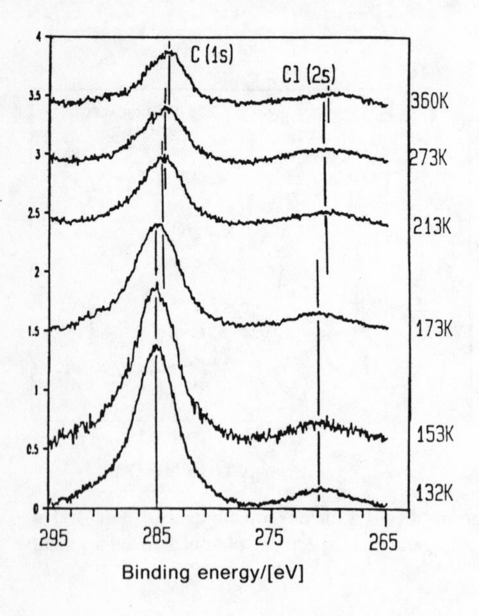

Fig. 4.4. C (ls) and Cl (2s) XP spectra of $C_4H_4Cl_2$ adsorbed at 132 K as a function of annealing temperature ($h\nu$ = 350 eV)

ised by a binding energy of $\approx$284.9 eV begins to be formed at 173 K and has essentially disappeared by 213 K. Subsequently, the spectrum remains unchanged until 360 K when further a shift in C (1s) emission to 284.3 eV occurs. The intensity of the C (1s) and Cl (2s) emissions remain constant in the temperature interval 213–360 K.

These observations may be readily interpreted in the following terms. Multilayers of DCB are present up to 173 K; above this temperature desorption of the physisorbed layers leaves a single non-dissociatively chemisorbed layer of molecular DCB in contact with the metal surface. This chemisorbed layer undergoes C–Cl bond scission below 213 K to form C_4H_4 and chemisorbed atomic chlorine; the dissociation process is characterised by a shift to lower binding energy of the Cl (2s) emission. The C_4H_4 species is then stable on the surface until $\approx$270 K. By 360 K transformation to a different surface species has occurred. All this behaviour is in excellent agreement with the HREELS observations described above, and as mentioned there, we assign this final species as vinylidene [4.17], a conclusion which is supported by independent measurements which show that the C (1s) binding energy found here (284.3 eV) is in agreement with that reported for vinylidene prepared by other means (284.0 eV [4.20, 20]). The small discrepancy between these two values may be ascribed to the presence of atomic chlorine on the surface in the case of the DCB experiments.

Figure 4.5 exhibits the variation in He I normal emission spectra ($\psi = 60°$) with annealing temperature after dosing Pd(1 1 1) with 10 L DCB at 155 K. The

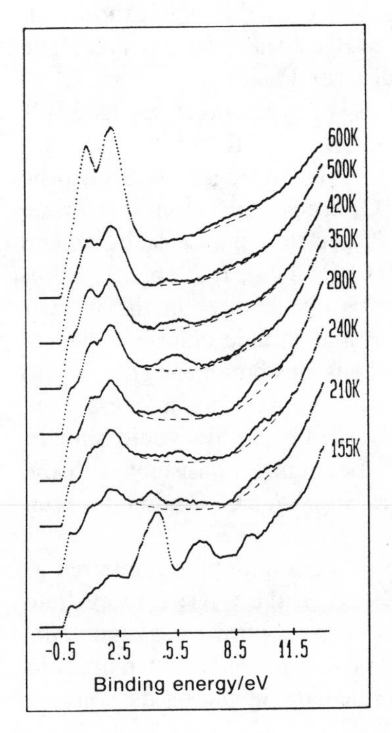

Fig. 4.5. He I UP spectra of $C_4H_4Cl_2$ adsorbed at 155 K as a function of annealing temperature; $\psi = 60°$, $\theta = 0°$

initial spectrum clearly corresponds to DCB multilayers which are subsequently removed below 200 K; the 240 and 280 K spectra are essentially identical but differ significantly from the 210 K spectrum, suggesting that the latter corresponds to molecular DCB whilst the former are due to the presence of both C_4H_4 and chemisorbed chlorine atoms. It is known that emission from the Cl 3p bands occurs at 5.2 eV (p_z) and 5.7 eV ($p_{x,y}$) when chlorine is adsorbed on clean Pd(1 1 1) [4.29]: these emission features are present for the first time in the 240 K spectrum, in agreement with the above proposal; they are characteristic of Cl adatoms. In $C_4H_4Cl_2$ the Cl valence orbitals contribute to the MO structure of the organohalide and are not essentially atomic in character, as they are on a metal surface. Furthermore, spectra taken as a function of photon incidence angle indicate that the Cl 3p emission exhibits the polarisation dependency that one would expect for adsorbed atomic chlorine [4.23] confirming that C–Cl bond scission has indeed occurred. By 350 K there is a further change in the spectrum in accord with the HREELS [4.17] and XPS observations [4.28]. This spectrum does show some similarities to that of vinylidene [4.14] taking into account the emission due to atomic chlorine at ≈5.4 eV. At higher temperatures still, vinylidene undergoes decomposition and the UP spectrum is characterised by weak broad emission centered around 7.5 eV associated with adsorbed CH.

The species present at 210 and 240 K (assigned to adsorbed DCB and C_4H_4 + Cl, respectively) were further studied by examining their UP spectra as a function of photon incidence angle and electron exit angle [4.23]. The 210 K spectrum exhibited almost no angular dependence whilst the 240 K spectrum showed variations characteristic of atomic chlorine as described above; an overall z-polarisation of the band centered at ≈10 eV was also observed. The significance of this latter observation will be discussed below.

Figure 4.6 depicts the angle-integrated UP spectrum taken using 30 eV photons following exposure of Pd(1 1 1) to 2 L DCB at 205 K. It is known from both LEED [4.16] and ARUPS [4.23] that at low coverages DCB adsorbs dissociatively at this temperature to form the C_4 species and atomic chlorine. Also indicated are the peak positions in the UP spectrum of *gaseous* butadiene [4.30] and of *adsorbed* atomic chlorine on Pd(1 1 1) [4.29]. It is apparent that there is quite good agreement between the emission features in the present spectrum and those associated with butadiene and atomic chlorine, the exception being the lowest binding energy peak of gaseous butadiene [the $1b_g$ (π) level]. However, if butadiene were bonded to the surface via a π interaction, which seems likely, then the π level of the adsorbed molecule would shift to higher binding energy. Thus the observed UP spectrum is consistent with the formation of a π-bonded butadiene species suggesting that a C–C bond has been broken in addition to the C–Cl bond.

We have also used NEXAFS to investigate all the surface phases referred to above, namely physisorbed DCB, chemisorbed DCB, the C_4H_4 intermediate and the 350 K vinylidene species, in an attempt both to confirm the assignments made on the basis of HREELS, XPS and ARUPS measurements, and in order to determine the precise nature of the C_4H_4 intermediate, including its bonding

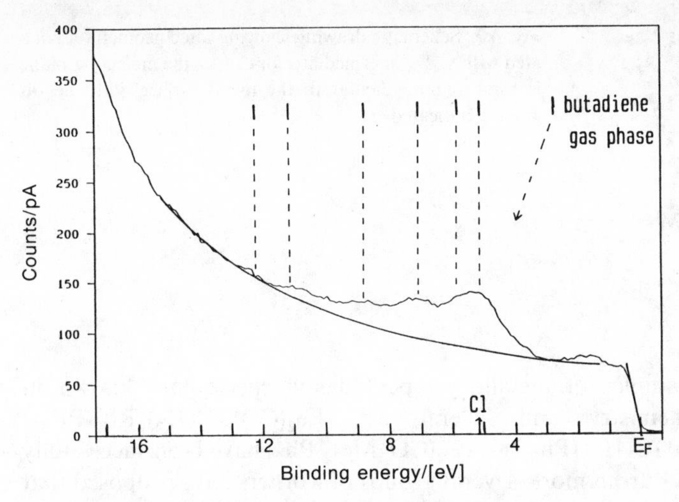

Fig. 4.6. Angle-integrated UP spectra of $C_4H_4Cl_2$ adsorbed to low coverage at 205 K; $h\nu = 30$ eV. Also shown are the peak positions in the gas phase spectrum of butadiene [4.30] and adsorbed chlorine [4.29]

geometry with respect to the surface. As expected, the NEXAFS spectrum of multilayer DCB shows no angular dependence. Similarly, the monolayer NEXAFS observed at 200 K exhibited very little variation in the intensity of these same resonances with photon incidence angle, suggesting an orientation for the chemisorbed molecules close to the magic angle (55°). In contrast, the NEXAFS spectra of the C_4H_4, species (210 K) *do* show a pronounced angular variation. A sharp peak at ≈ 284 eV is observed which corresponds to the $1s \rightarrow \pi^*$ transition. This has maximum intensity at grazing incidence but is still present at normal incidence, suggesting a species which is tilted with respect to the metal surface. Preliminary analysis of the angular dependence of this π^* resonance suggests that the molecular plane is tilted with respect to the surface by $\approx 35°$.

These observations are consistent with the ARUPS data which showed essentially no angle-dependence for chemisorbed DCB and only small angular variations for the C_4H_4 species. In the light of the combined photoemission and NEXAFS results it seems most likely that the C_4H_4 species is a tilted palladia-pentacycle which undergoes some additional π interaction with the metal surface – in good agreement with the conclusions that were reached on the basis of the HREELS observations [4.17]. Figure 4.7 is a schematic drawing showing the adsorption geometry of the C_4H_4 intermediate. From the point of view of reaction mechanism (ethyne $\rightarrow C_4H_4 \rightarrow$ benzene), this geometry for the intermediate provides a very satisfactory basis for rationlising the high efficiency of benzene formation from C_4H_4: the third C_2H_2 molecule can approach the tilted C_4H_4 in a configuration which is optimal for 2 + 4 cycloaddition. Likewise, it is consistent with our conclusion (Sect. 4.2.3) that the benzene product is itself formed in a tilted configuration.

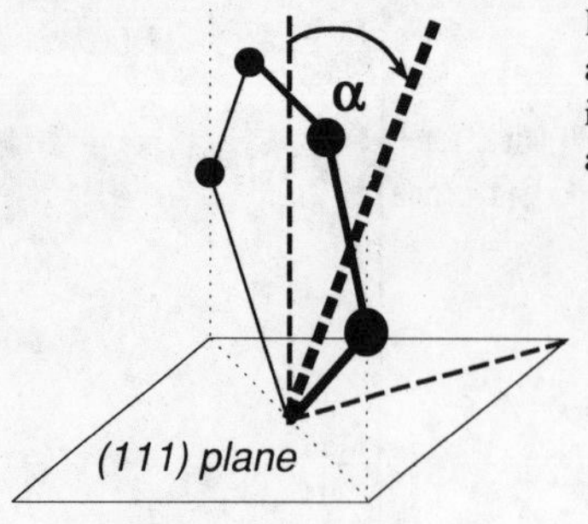

Fig. 4.7. Schematic drawing showing tilted geometry associated with C_4H_4 intermediate: for clarity, the molecular plane is shown perpendicular to the metal surface with the tilt angle α indicated

Analogous examples of metallocyclopentadienyl species are known in organometallic chemistry, and several, e.g., $Fe_2(CO)_6(C_2H_2)_2Rh_2(PF_3)_5(C_2(CO_2Me)_2)_2$ and $(C_5H_5)(Ph_3P)Co(C_4(CO_2Me)_2Ph_2)$ have been successfully isolated [4.31–33]. Furthermore, several groups of workers have proposed that a $\sigma + \pi$ bonded cyclopentadienyl species is the crucial intermediate in the homogeneously catalysed tricyclisation of ethyne [4.34–38]: the close similarity with the present heterogeneous reaction is striking. Furthermore, *Whitesides* and *Ehmann* have used isotopically labelled but-2-yne ($CH_3C_2CD_3$) to show that the $Co_2(CO)_8$ catalysed tricyclisation reaction does *not* proceed via a metal-complexed *cyclobutadiene* intermediate [4.39], whilst *Nakamura* and *Hagihara* demonstrated that the *cyclobutadiene* complex, $Co(C_5H_5)(C_4Ph_4)$ does *not* react with the alkyne, dimethylethyne-dicarboxylate [4.40]. *Thus it seems that, as in the heterogeneously catalysed reaction on Pd surfaces, a cyclopentadienyl species rather than cyclobutadiene is the intermediate formed during cyclisation.* Again, the correspondence with our conclusions on the identity of the key intermediate on Pd(1 1 1) illustrates the close correspondence between the homogeneous and heterogenous reactions: it appears that organometallic cluster compounds *are* good models for extended metal surfaces in the case of alkyne trimerisation.

4.2.3 The Reactively Formed Benzene is Tilted: Effect of Surface Packing Density on the Conformation, Yield and Desorption Kinetics of Benzene Formation

As we shall see later, the adsorption geometry of benzene on Pd appears to be a sensitive function of surface coverage. At low coverage the molecule lies flat; at high coverage it tilts. This fact is crucial to an understanding of the kinetics of benzene evolution. This hypothesis that a high molecular packing density in the reacting adsorbed layer leads to formation of tilted, weakly bonded benzene molecules is very strongly supported by the results of experiments involving two-dimensional compression of ethyne overlayers on Pd(1 1 1) by addition of a chemically inert "spectator" molecule.

The original work of *Tysoe* et al. [4.10] showed that there was a critical ethyne coverage threshold of ≈ 0.3 monolayers below which benzene formation does not occur – presumably because the reactant molecules are too far apart.

This critical coverage is characterised by the appearance of a $(\sqrt{3} \times \sqrt{3})$ R30° LEED structure [4.14]. *Patterson* and *Lambert* performed coadsorption experiments with C_2H_2 and C_2D_2 to demonstrate that although benzene formation does not occur until the $\sqrt{3}$ structure is complete, the reactively formed benzene molecules *do* contain ethyne molecules that were initially chemisorbed into the $\sqrt{3}$ sites, [4.15] i.e., it is not a second, more weakly bound, ethyne species which forms once the $\sqrt{3}$ structure is complete that is responsible for benzene formation. Instead the results can be explained by the requirement for sufficiently close approach of three chemisorbed ethyne molecules for reaction to occur.

When two species are coadsorbed on a surface, broadly speaking there are two possibilities. They may either form a mixed adsorption layer with the two adsorbates distributed essentially randomly within a single adsorbed phase. Alternatively, if the lateral interactions between the adsorbed species are significantly different, phase separation can occur with the formation of separate islands of each of the two adsorbates. Our LEED and TDS studies clearly indicate that coadsorption of NO and C_2H_2 on Pd(1 1 1) leads to such islanding behaviour [4.18], and this may be exploited in order to test the compression/tilting hypothesis referred to above. The choice of NO as spectator species was dictated by the following criteria.

1. Pure NO and pure ethyne form quite distinct and different surface structures which are readily distinguishable by their LEED patterns (this is not true of CO and ethyne, for example). One may therefore readily distinguish between the cases of homogeneous mixed adsorption layers and island formation.
2. NO desorbs from Pd(1 1 1) at temperatures above those at which benzene formation occurs.
3. The N(KLL) Auger and N(1s) core level spectra provide convenient spectroscopic monitors of NO coverage in the presence of the organic species (again this would not be true of CO).

It is found that when NO and ethyne are coadsorbed on Pd(1 1 1) the effects of compression on the ethyne phase lead to dramatic changes in the threshold coverage required for formation and desorption of benzene at low temperatures. These effects are independent of the order of dosing of the adsorbates, as indeed one would expect if separate islands are being formed. This behaviour is illustrated in Fig. 4.8 and 4.9 which show the effect operating in both modes. As can be seen from Fig. 4.8, a 0.6 L dose of ethyne on clean Pd(1 1 1) leads to virtually no C_6H_6 desorption: no benzene is observed in the low-temperature peak and only a very small amount in the high-temperature peak. This ethyne dose corresponds to a surface coverage close to the critical surface coverage required for benzene formation. If however the surface is precovered with a fixed amount of NO it is immediately apparent that *low temperature benzene desorption commences for very low doses of adsorbed ethyne.* By 0.65 L there is effectively 100% conversion of ethyne to benzene *and all the benzene is desorbed at low*

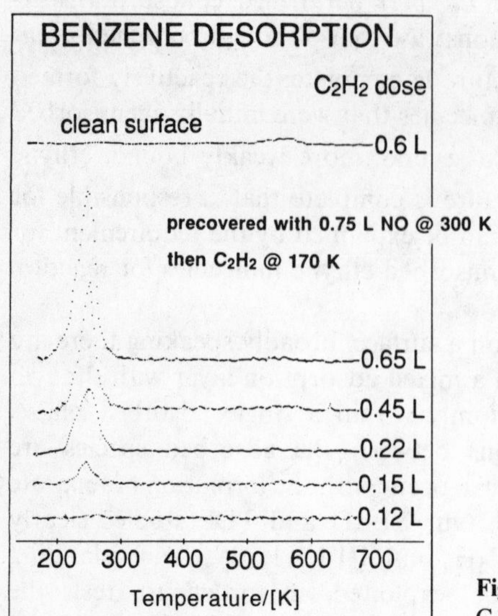

Fig. 4.8. Showing effect of preadsorbed NO on $C_2H_2 \rightarrow C_6H_6$ reaction

temperature- this is to be compared with the upper trace in Fig. 4.8 where on the clean Pd(1 1 1) surface essentially no benzene formation occurs at all for the same initial coverage of ethyne. The inference is that the NO islands maintain sufficient compression on the reacting islands of ethyne such that all benzene formation and desorption occurs from a high density organic phase. Figure 4.9 exhibits the complementary experiment in which the surface was precovered with varying amounts of C_2H_2 followed by a fixed 1.0 L dose of NO. Notice in this case that low temperature evolution of benzene into the gas phase is detectable at extremely small doses of ethyne (0.06 L) dramatically illustrating the effects of compression by NO: although the total number of ethyne molecules on the surface is very small, they are compressed into islands in which the local density exceeds the critical value for benzene formation and highly selective ethyne → benzene conversion occurs for ethyne loadings which would not yield any benzene at all in the absence of NO. Figure 4.10 provides a graphical summary of the results and illustrates how the ethyne coverage threshold for benzene formation is dramatically reduced by coadsorbed NO. These observations confirm nicely the idea that steric crowding leads to reactively formed benzene being generated in a tilted geometry followed by subsequent low temperature evolution into the gas phase.

We have noted in Sect. 4.1.1 (Fig. 4.2) that coadsorption of C_6H_6 and C_2D_2 leads to the evolution of *chemisorbed* C_6H_6 and *reactively formed* C_6D_6 with identical kinetics, i.e. surface crowding of chemisorbed benzene on Pd(1 1 1) by ethyne molecules, leads to desorption of benzene at low temperature, consistent with the formation of a tilted benzene species at high surface coverages.

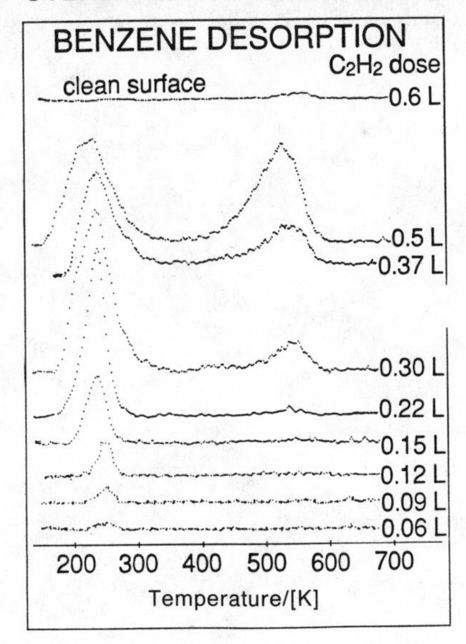

Fig. 4.9. Showing effect of NO on $C_2H_2 \rightarrow C_6H_6$ reaction for a series of different C_2H_2 precoverages

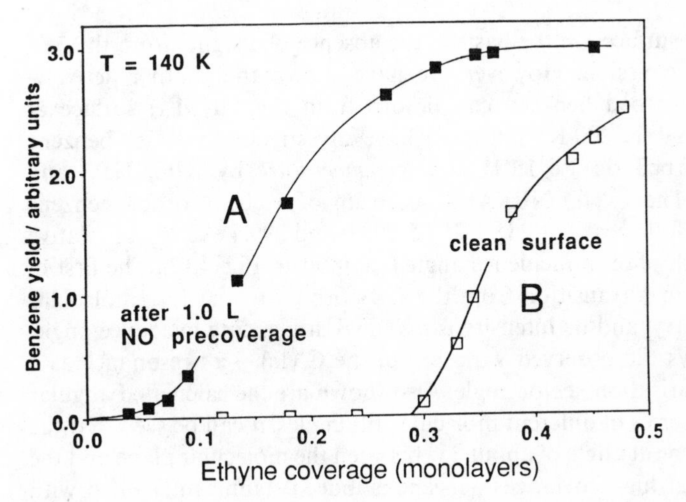

Fig. 4.10. Benzene yield as a function of ethyne coverage for clean and NO precovered Pd(1 1 1)

When benzene is chemisorbed on Pd(1 1 1) at low temperatures, desorption initially occurs at ≈550 K. Increasing the surface coverage leads to a drastic reduction in adsorption strength, with a continuous lowering in the onset of benzene desorption, falling from ≈550 to ≈220 K at saturation coverages (Fig. 4.11) [4.15]. *Netzer* and *Mack* studied benzene chemisorption on Pd(1 1 1) at room temperature by ARUPS [4.41] and concluded that the molecular plane

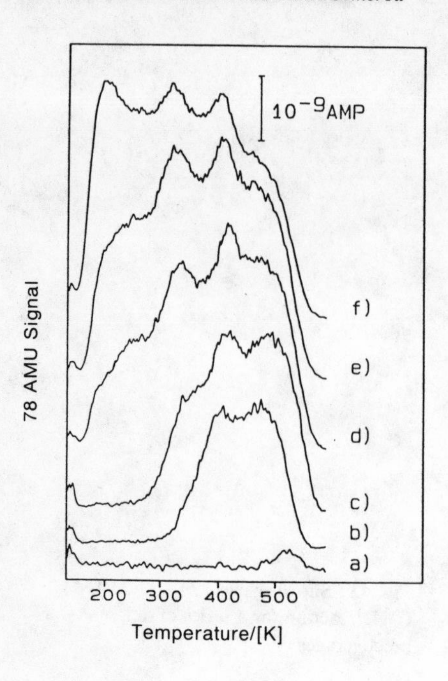

Fig. 4.11. Temperature programmed desorption of C_6H_6 chemisorbed on Pd(1 1 1) at 150 K as a function of C_6H_6 initial coverage

was parallel to the surface, on the basis of the absence of a signal from the $3e_{2g}$ orbital at normal emission. However, we have shown that both reactively-formed and chemisorbed benzene can desorb from the Pd(1 1 1) surface at temperatures as low as 200 K. Thus, we have investigated a dense benzene overlayer chemisorbed on Pd(1 1 1) at *low temperature* by XPS, UPS and NEXAFS [4.19]. The C (1s) NEXAFS spectrum of a chemisorbed benzene monolayer reveals four peaks at 285.4, 288.5, 292.0 and 299.4 eV, whose relative intensities vary with photon incidence angle (see inset to Fig. 4.12). The first of these corresponds to a transition from the C 1s orbital to the $1e_{2u}$ molecular orbital of π symmetry, and its intensity is maximal at glancing incidence angle. Figure 4.12 displays the observed variation of the C (1s) $\rightarrow \pi^*$ resonance as a function of the polarisation vector angle. Also shown are the calculated angular dependences for a series of different molecular tilt angles; it can be seen that the data correspond to a tilt angle of about 33° between the molecular plane and the metal surface, i.e., at high coverages benzene is indeed significantly tilted with respect to the Pd(1 1 1) surface. Quantification of the C (1s) peak intensity in XPS leads to a saturation coverage of benzene on Pd(1 1 1) at $\approx$200 K of 0.24, that is C/Pd = 1.44. This is consistent with significant molecular crowding on the palladium surface at low temperatures.

The tilted geometry obtained by NEXAFS is in *apparent* disagreement with the conclusion reached by *Netzer* and *Mack* that chemisorbed benzene on Pd(1 1 1) lies flat [4.41]. However, the data in [4.41] were obtained for benzene adsorbed at *room temperatures*, and XPS measurements have demonstrated that there is *a* $\approx$35% reduction in adsorbate coverage when a 200 K benzene layer is

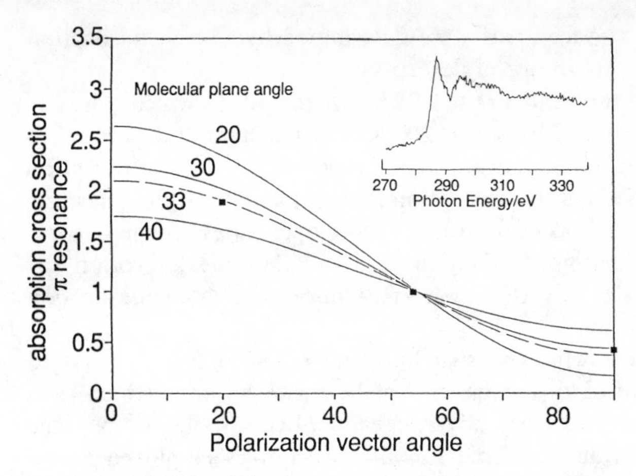

Fig. 4.12. NEXAFS of titled benzene on Pd(1 1 1). Dependence of π resonance intensity on polarisation vector angle for a benzene monolayer chemisorbed on Pd(1 1 1) at 200 K. Inset shows NEXAFS spectrum of benzene for photon incidence angle of 20°

warmed to 300 K [4.19, 23], and thermal desorption data show that significant benzene desorption occurs below room temperature [4.15].

Thus the results are consistent: at low temperatures and high surface coverages, steric effects result in the adsorption of a weakly-bound species with tilted geometry; the necessary surface pressure can be provided either by the benzene molecules themselves (as in the above case) or by coadsorbed species, whilst reduced surface coverages, such as those obtained by room temperature chemisorption, lead to a flat-lying geometry [4.41], with a corresponding increase in adsorption enthalpy [4.14, 15].

4.3 Studies at High Pressures

In the preceding section it was shown how a combination of electron spectroscopic and kinetics studies may be used to isolate and identify the crucial reaction intermediate, determine the bonding of the reactant, products and intermediate, and elucidate the molecular pathway. In order to establish whether such studies performed on well characterised single-crystal surfaces under UHV conditions yield information which is relevant to an understanding of the properties of practical high area supported metal catalysts, we carried out a parallel investigation of the behaviour of practical supported Pd catalysts under high-pressure conditions [4.23, 27].

Catalysts containing between 2 and 10% w/w Pd dispersed on silica, alumina, titania and charcoal supports were used in a single-pass quartz tube microreactor with mass spectroscopic detection. Following pretreatment in flowing oxygen at 473 K, samples were reduced in hydrogen at 470 or 770 K, before purging in helium at 770 K and cooling in helium to the reaction temperature. Palladium surface areas were determined by CO chemisorption followed by temperature programmed desorption.

Figure 4.13 displays the results of a TPRS experiment, in which a freshly reduced Pd/Al_2O_3 sample was cooled to 200 K under helium and exposed to ethyne for twenty minutes. Significant low-temperature ethyne desorption occurs along with a single benzene desorption maximum at $\approx$410 K indicating that cyclisation does indeed occur following low-temperatures ethyne adsorption on dispersed oxide-supported palladium catalysts. Benzene desorption was essentially complete by 470 K, with only a small amount of ethene desorption being detected.

For temperatures above 480 K, switching the gas feed from helium to ethyne immediately resulted in a high rate of benzene synthesis which then decreased to a substantially lower steady-state value. Below 480 K no benzene was detected. This behaviour occurred for all the samples employed and a typical reaction profile for a Pd/Al_2O_3 sample is shown in Fig. 4.14. In addition to benzene, *butadiene and butene* are detected as products. Both exhibit the same time profile as benzene (that of butadiene is shown in Fig. 4.14) but can be identified with certainty on the basis of their mass spectral fragmentation patterns. No C_3 or C_5 products were detected, and the fall from the initial high rate of benzene production to its steady-state level correlates with a rise in the

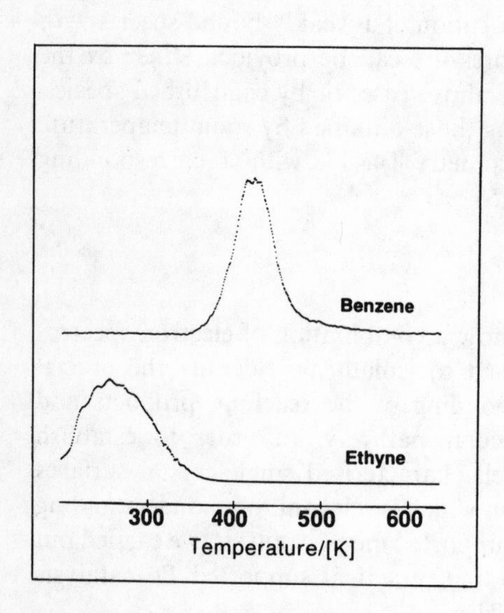

Fig. 4.13.

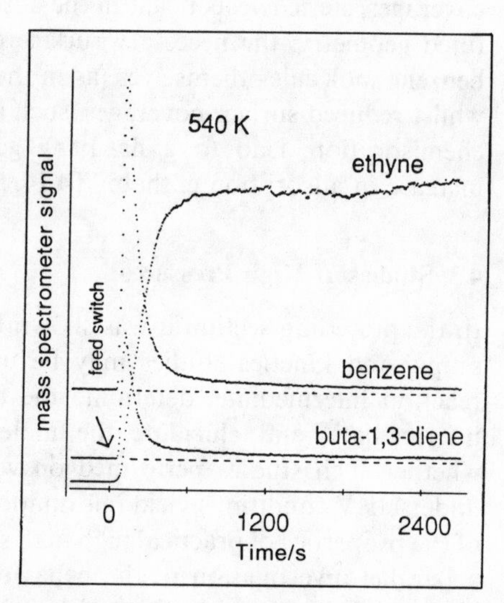

Fig. 4.14.

Fig. 4.13. Desorption spectra following exposure of a Pd/Al_2O_3 catalyst to ethyne at 200 K

Fig. 4.14. Time variation in benzene, buta-1,3-diene and ethyne production on switching gas feed from He to ethyne at 540 K: Pd/Al_2O_3 catalyst

amount of unreacted ethyne to its maximum level (Fig. 4.14). In addition to benzene and the C_4 products, ethene was also formed, its production being observed at significantly lower reaction temperatures to those required for benzene formation. Correspondingly, the transient activity for ethene production was also higher to the extent that initially all the ethyne was consumed [4.27]. This effect was accompanied by a marked decrease in reactant flow rate which recovered in a manner that exactly mirrored the amount of unreacted ethyne.

CO chemisorption measurements carried out by both frontal chromatography and temperature programmed desorption methods indicate that CO chemisorption is completely inhibited once the catalyst has reached steady state performance, whereas considerable chemisorption occurs on the freshly reduced catalyst, as can be seen in Fig. 4.15a [4.27]. Both the CO chemisorption properties and the very high initial level of benzene formation could be restored by an O_2/H_2 oxidation/reduction cycle, extensive CO_2 evolution occurring during the first stage of this process (Fig. 4.15b). This demonstrates that on the freshly reduced catalyst, ethyne tricyclisation competes with irreversible adsorption of ethyne, and that the reduction in benzene yield is related to the substantial amount of unreacted ethyne.

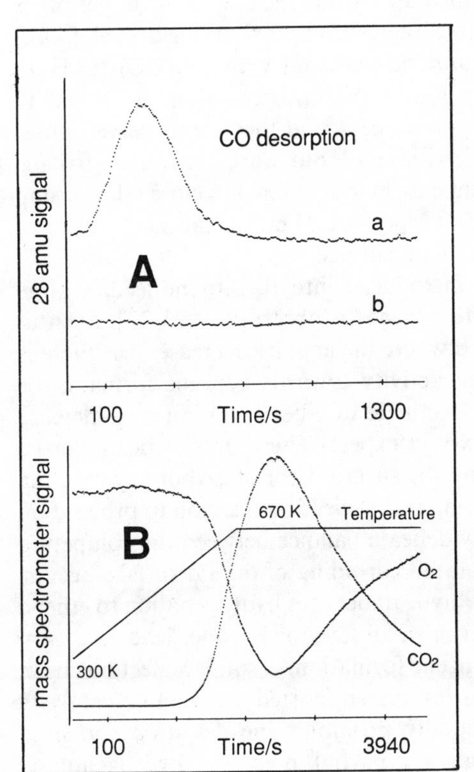

Fig. 4.15. **A** CO thermal desorption spectra in the interval 200–500 K (temperature ramp continued to 720 K): Pd/SiO_2 catalyst. a: Freshly reduced catalyst; b: after exposure to ethyne at 540 K. **B** Temperature programmed oxidation of Pd/Al_2O_3 catalyst after exposure to ethyne at 540 K. Temperature ramp 300–670 K at 10 K/min then maintained at 670 K for 1 h under oxygen flow

Taken together these observations indicate that the very high transient level of benzene synthesis can be attributed to efficient ethyne cyclisation occurring on the clean Pd surface, and the reduction in this activity corresponds to a gradual loss of exposed Pd metal area due to carbon deposition resulting from ethyne adsorption and dehydrogenation. It is this latter process which leads to the initially high level of ethene formation. Thus the much smaller level of steady state conversion of ethyne to benzene actually occurs on an extensively carbided Pd surface.

Most importantly, the synthesis of butadiene and butene, and the complete absence of any C_3 or C_5 products, provides direct evidence for *the presence of a* C_4 *species on the surface under steady state reaction conditions.* We propose that this C_4 species is precisely the C_4H_4 metallocycle identified by the single crystal work summarised above [4.16, 17, 23, 24, 28]: butadiene and butene are exactly what one would expect as primary and secondary hydrogenation products, respectively, of this key intermediate. Thus the tilted metallocycle surface intermediate identified and characterised by isotope tracing, HREELS, ARUPS and NEXAFS is indeed the key player in determining the catalytic chemistry of ethyne on Pd surfaces. Under practical conditions, chemisorbed hydrogen is necessarily present (adventitiously or deliberately introduced) so that butadiene and butene formation compete with benzene formation. Under different conditions, as we shall see, the C_4H_4 intermediate can be induced to form yet other interesting products. It is of considerable interest to note that *Osella* et al. [4.42] have observed *similar homogeneous catalytic chemistry* with $Co_2(CO)_6(C_2H_2)$]. At high ethyne concentrations they found that tricyclisation of ethyne to benzene becomes the dominant catalytic process, and they too observed butadiene and butene formation. The similarities with our work are rather striking, and strongly suggest that, in the present case at least, there is a close relationship between the bonding modes, intermediate species and molecular mechanisms of cluster compounds and the extended metal surface.

When a small amount of H_2 is introduced into the ethyne feed, a large increase in the steady-state turnover to benzene is observed [4.23, 27]. Eventually a C_2H_2/H_2 composition is reached where the activity decreases, i.e., there is an optimum C_2H_2/H_2 ratio at which activity towards benzene formation is maximal. The partial hydrogenation of ethyne to ethene continues to increase with increasing H_2 content, as one would expect. These observations can be rationalised in terms of the H_2 keeping the surface freer of carbon, by reacting with ethyne to form ethene, thus allowing the cyclisation reaction to proceed on the Pd surface. Clearly there is a fairly delicate balance between the competing processes; in the absence of H_2 fairly rapid carbiding of the Pd surface occurs, whilst at high H_2 levels almost all the ethyne undergoes hydrogenation to ethene before cyclisation can occur. However, a small level of H_2 does lead to a very substantial increase in steady-state benzene formation. A similar effect has been observed with cyclohexane/H_2 mixtures on supported ruthenium catalysts [4.43]. where the *presence* of H_2 actually promotes the *dehydrogenation* of cyclohexane to benzene, but at high H_2 partial pressures hydrogenolysis predominates leading to straight chain alkane formation: there is an optimum cyclohexane/H_2 ratio for benzene formation [4.44].

In Section 4.6 we discuss the very pronounced surface specificity of ethyne tricyclisation, and in particular its strong structural sensitivity to the (1 1 1) plane of Pd. Because of this, the formation of benzene on high-area supported Pd particles provides strong evidence that these particles consist of crystallites whose surfaces are dominated by (1 1 1) planes – as might be anticipated, given the relatively high metal loadings employed. Furthermore, catalyst samples reduced at 770 K gave a significantly higher benzene TPD yield than those reduced at 470 K (4.45), consistent with the much higher activity of the (1 1 1) plane towards ethyne cyclisation. This again illustrates the strong reaction sensitivity towards the presence of (1 1 1) planes. Treatment at 770 K sinters the catalyst more than does treatment at 470 K. The total metal area *decreases* but the cyclisation activity *increases*, in contrast with the more usual behaviour. Measurements made with different metal loadings also suggest that the reaction depends markedly on metal surface geometry. Thus high metal loadings which result in larger particle sizes and proportionally *lower* metal areas were found to lead to a *higher* steady-state level of benzene synthesis, in excellent accord with single-crystal measurements [4.12, 46]; again, this reflects the fact that (1 1 1) planes constitute a larger fraction of the surface area of large Pd particles than of small ones.

Rucker et al. [4.13] used single-crystal methods to study the dependence of tricyclisation rate on surface crystallography over the pressure range 0.1 – 1.0 atmospheres. They concluded that the (1 1 1) and (1 0 0) faces have equal activity with the (1 1 0) face one quarter as active. This contrasts with observations made in the same laboratory under UHV conditions, where the (1 1 1) face was reported 20 times more active than the (1 1 0) face and 6 times more active than the (1 0 0) face. It is also in apparent disagreement with our own findings. However, it seems probable that sample edge effects are of importance in such measurements. For any given specimen $\approx$1 mm thick, the edges ($\approx$20% of the total area) will expose a substantial area of (1 1 1) surface. If, as we believe, the (1 1 1) plane is overwhelmingly more efficient than (1 0 0), (1 1 0) etc., then erroneous conclusions may be drawn unless precautions are taken to exclude the effect of the sample edges on the observed catalytic rate. We shall return to this point in Sect. 4.6. *Rucker* et al. [4.13] used post-reaction CO titration to infer that in the steady state there were only 5 – 7% of Pd(1 1 1) surface atoms available for chemisorption and reaction with ethyne – the remainder of the surface having been blocked by carbon deposition resulting from ethyne decomposition. Corresponding values of 12 – 15% and $<$ 2% were obtained for the (1 0 0) and (1 1 0) faces. This appears to disagree with our results (see above, Fig. 4.15a and [4.27] and those of other workers [4.47] were it was found that ethyne blocks all CO chemisorption sites.

In summary, ethyne to benzene conversion over high-area supported Pd particles, can be operated as a genuine catalytic process under conditions of steady state turnover at elevated pressure. Furthermore, there is a direct correspondence between structural, mechanistic and reactivity data obtained with single crystal model systems under idealized conditions and the behaviour of practical catalysts under working conditions (10^{13} times higher pressure). The data obtained with dispersed catalysts provide compelling evidence in support of the single-crystal-derived hypothesis that a C_4H_4 surface intermediate plays an essential role in ethyne→benzene conversion. Conversely, the observed dependence of practical catalyst performance on metal loading and on pre-sintering is fully understandable in the light of the single crystal results.

4.4 The Effects of Promoters, Poisons and Other Coadsorbed Species

In principle, the use of Pd alloys presents one attractive way of investigating the effect of controlled variations in surface electronic properties on the tricyclisation reaction. Our current work in this area involves using Pd/Au (1 1 1) thin films of varying thickness and morphology and Au/Pd surface alloys in order to look at "geometric" and "electronic" effects on the catalytic chemistry. This will be described and discussed more fully in Sect 4.6.

A related approach is to make use of deliberately introduced non-metal foreign atoms which may act as either promoters or poisons. *Somorjai* and coworkers have investigated the effects of Si, P, S, Cl and K on the conversion of ethyne to benzene both under UHV conditions and at elevated pressures; they obtained data on all three low index faces of single crystal Pd [4.48]. A limited number of observations were made and the authors reported very different kinds of behaviour in the two different pressure regimes. Their results led them to conclude that at high pressures on the (1 1 1) and (1 0 0) surfaces, electron donating additives (K) enhance the rate of benzene formation whilst electron acceptors (Cl and S) reduce cyclisation activity; the reaction was essentially completely inhibited for coverages > 0.2. Phosphorus was also found to reduce the cyclisation rate on these two surfaces. However, on the (1 1 0) surface it was found that K actually *inhibits* the reaction whilst phosphorus *enhances* it. Taken as a whole, these high pressure data are not easy to rationalise.

Corresponding data obtained by the same research under UHV conditions are also interesting but it is not easy to devise a coherent picture which encompasses all the results obtained with all promoters/poisons on all surfaces both at high pressures and under UHV conditions. Thus under UHV conditions it was found that on the (1 1 1) surface K strongly suppressed the benzene desorption yield, S and Cl had a small detrimental effect, Si a small positive effect, whilst P actually *doubled* it; this latter point is in apparent complete opposition to the behaviour of phosphorus on the same surface at high pressures (see above). It was also found that S, Cl and P all suppress ethyne decomposition but whereas P leads to an increased yield of ethene, S and Cl suppress ethene formation; large variations in benzene desorption temperature were observed for the different additives but not discussed – this is a point that we will return to later. In common with their high pressure results, the authors found that the behaviour of the (1 0 0) and (1 1 0) surfaces under UHV conditions again differed from that of the (1 1 1) surface. Thus a dramatic *enhancement* in benzene formation was found for Si on the (1 0 0) face while Cl had the opposite effect to S on the (1 0 0) face: typically, S and Cl exert similar effects on a given reaction.

The inhibiting effect of K on Pd(1 1 1) in UHV may be rationalised in terms of the way alkali atoms are known to disperse on metal surfaces, namely there is no tendency towards island formation: such blocking of adsorption sites could simply prevent sufficiently close approach of adsorbed ethyne molecules: close proximity is known to be a crucial prerequisite for benzene synthesis, as demonstrated above by using nitric oxide [4.18] and atomic oxygen [4.49] as coadsorbates. The other additives investigated by *Somorjai* and coworkers have a greater tendency towards island formation; they might therefore by expected to be less effective inhibitors. However, at the large additive coverages employed ($\theta \approx 0.25$) one might still expect that the simple effects of site blocking would exert a negative effect on the rate of benzene formation in all cases. It was therefore proposed that electron withdrawing adatoms must promote benzene formation and inhibit ethyne decomposition; unfortunately, the most strongly electronegative adatom, Cl. does not fit this pattern. The results do serve to indicate that there are several factors which effect the efficiency of cyclisation and that these may interact in a complex way making it difficult to separate the different effects on the basis of only limited experimental information.

In passing, we should mention the work of *Marchon* [4.50] who used HREELS to study the behaviour of ethyne on clean and P-doped Pd(1 1 1). *Marchon* concluded that P enhances benzene formation at 150 K, based on the appearance of a loss at 690 cm^{-1} which he assigned to the γ_{CH} mode of benzene. However, in a similar HREELS investigation, *Timbrell* et al. [4.51] demonstrated that this loss can be attributed to negative ion resonance scattering from the chemisorbed ethyne molecule and that only ethyne itself is present on the surface in this temperature regime; this in agreement with the earlier HREELS work of *Kesmodel* and co-workers [4.52, 53].

Patterson investigated the effect of adsorbed sulfur on the chemisorption and reactivity of ethyne on Pd(1 1 1) [4.29] finding that decomposition of the reactant was effectively quenched for $\theta_S \geqslant 0.46$. The benzene yield increased significantly relative to that of the clean surface for $\theta_S = 0.35$ but decreased at higher coverages of sulfur. At the same time the desorption yield of unreacted ethyne increased monotonically with sulfur coverages becoming large for high values of θ_S. At this point all benzene desorption was found to occur in the low temperature peak, and it was also noted that the threshold for benzene formation occurred at lower ethyne coverages on the clean surface. We may understand both these latter points in terms of the effects of compression of the reactant overlayer by S in an analogous manner to that observed with coadsorbed NO (see above). In the present case the benzene yield is not enhanced to the same extent as it is by coadsorbed NO, possibly because, unlike NO, the S adatoms are not mobile at low temperatures; this would be consistent with LEED observations which indicate that NO on Pd(1 1 1) does indeed have a much greater tendency towards island formation than does sulfur. The foregoing interpretation of the effects of S on benzene formation is, of course, in terms of structural or geometric arguments and makes no particular appeal to the effects of electronic perturbation induced by sulfur atoms, although, of course, we cannot rule out a possibility that such effects are of significance.

Ormerod [4.23] recently studied the effect of oxygen coadsorption on the tricyclisation reaction. Not surprisingly this proved to be a complex system because in this case the additional possibility exists of reactions between the foreign atom and all the adsorbed hydrocarbon species, ethyne, C_4H_4 and benzene. (The nature of such oxidative processes will not be discussed here [4.54].) LEED observations showed that ethyne molecules chemisorb within the (2×2) oxygen structure to form a mixed phase, XPS pointing to a small electronic interaction between the coadsorbed species. Low oxygen precoverages resulted in increased benzene desorption from the high-temperature state and a reduction in the threshold acetylene coverage for benzene formation, reminiscent of the behaviour with coadsorbed NO and S [4.18, 29]. At the highest oxygen coverages ($\theta = 0.25$) benzene formation was strongly suppressed; interestingly, no changes in peak temperatures were observed. At all oxygen precoverages it was found that relative to the clean surface a larger fraction of benzene desorption occurred in the high-temperature peak. Figure 4.16 displays the benzene desorption yield from TPR spectra as a function of ethyne exposure for various oxygen precoverages. The decreasing benzene yield with increasing oxygen precoverage can readily be seen as can the lowering of the coverage threshold for benzene synthesis at low oxygen coverages.

Simple model calculations based on the formation of a mixed phase with ethyne adsorption occurring in three-fold hollow sites can quantitatively

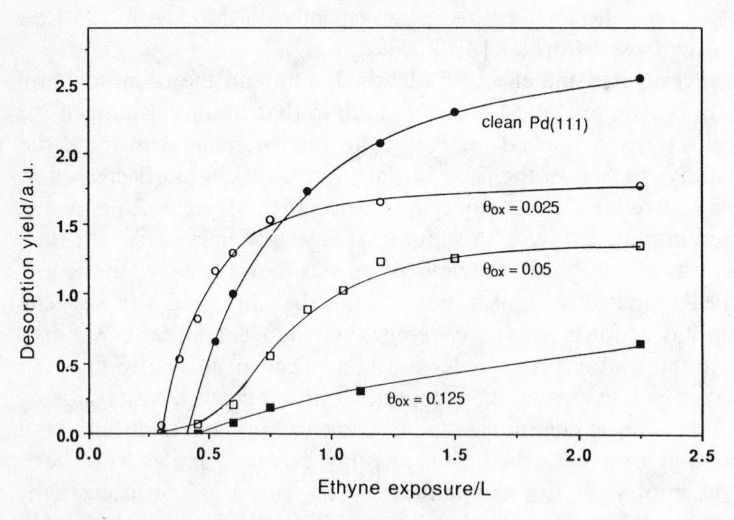

Fig. 4.16. Benzene desorption yields from Pd(1 1 1) as a function of C_2H_2 exposure at 170 K for a series of different oxygen precoverages

account for the dependence of benzene yield on the amount of preadsorbed oxygen [4.49]; these calculations demonstrate that oxygen induced suppression of benzene formation is *not* due to competing oxidative reactions but is the result of O atoms inhibiting the required close approach of reactant molecules necessary for cyclisation to occur. *The model assumes that benzene synthesis requires population by ethyne of all three 3-fold sites around a given Pd atom*, the metal surface acting as an efficient template. Agreement between experiment and calculation was found to be good, particularly in predicting a very small but non-zero conversion of ethyne to benzene on a fully oxygen-saturated Pd(1 1 1) surface ((2×2) structure). This suggests that we have identified the minimum ensemble necessary for tricyclisation of ethyne.

Precovering the surface with oxygen was found to completely inhibit the formation of ethene from ethyne; this may be attributed to efficient scavenging of surface hydrogen by the oxygen atoms, resulting in water formation. Ethyne decomposition was also suppressed on the oxygen covered surface while dosing an ethyne-precovered surface with oxygen led to essentially identical behaviour to that observed on the clean surface; this sequence of adsorption leads to the formation of separate domains of oxygen and ethyne, but without significant compression of the latter.

In Sect 4.2.3. we discussed the dramatic effects of NO coadsorption on the tricyclisation reaction [4.18]. This effect is independent of the order of adsorption of the two species and arises because the two are forced into mutually exclusive domains resulting in ethyne islands of high local coverage. These results and essentially all the other results obtained in the presence of a variety of foreign atoms suggest that the principle effect that such atoms or molecules have on benzene formation is a "geometric" one.

In other words, the degree to which a particular foreign particle acts as a promoter or poison is largely determined by the extent to which it enhances or inhibits adsorbed ethyne from achieving the critical density required for the two-step cyclisation reaction to occur.

Electronic factors appear to exert comparatively small influences which can be difficult to distinguish in the presence of large geometric effects. As noted earlier, one way of attempting to unravel the two is to examine the tricyclisation reaction on suitably chosen alloy surfaces. We therefore undertook a study of the Au(1 1 1)/Pd system and our findings are discussed below.

4.5 The structure and Bonding of Ethyne Chemisorbed on Transition Metal Surfaces

The need for a fundamental understanding of adsorbate bonding at metal surfaces has motivated extensive research on the chemisorption and reactivity of unsaturated hydrocarbon molecules, most notably ethene and ethyne, on transition-metal surfaces.

Although we are principally concerned with the low-temperature tricyclisation of ethyne to benzene on Pd(1 1 1), knowledge of the interaction of ethyne with other transition-metal surfaces is also of relevance. Many such systems have been studied, including those involving the following metals; Pd [4.10, 11, 50, 52, 53, 55–58], Pt [4.55, 59–63], Ni [4.55, 57, 64–69], Cu [4.57, 70–72], Ag [4.73], Rh [4.74, 75], Ir [4.76], Co [4.77], Fe [4.78–80], Ru [4.81, 82], W [4.83–85], Mo [86], Gd [4.87]. In addition, there have been a number of theoretical studies devoted to the ethyne-metal surface interaction – see, for example [4.88–91].

In the particular case of Pd(1 1 1) the chemisorption of ethyne has been investigated by High-Resolution Electron Energy Loss Spectroscopy (HREELS) [4.52, 53, 56], angle-resolved Ultraviolet Photoelectron Spectroscopy (ARUPS) [4.14, 55], Low-Energy Electron Diffraction (LEED) [4.14, 52], near-edge X-ray absorption fine structure spectroscopy (NEXAFS) [4.19]. X-ray Photoelectron Spectroscopy (XPS) [4.19, 23]. Metastable De-excitation Spectroscopy (MDS) [4.11], molecular beam measurements [4.14], work-function measurements [4.23], and Temperature-Programmed Reaction Spectroscopy (TPRS) [4.10, 12–14]. Recently, *ab initio* quantum calculations have also been performed for this system [4.91].

Demuth carried out an extensive photoemission study of ethyne and ethene chemisorbed on Ni(1 1 1), Pd(1 1 1) and Pt(1 1 1) [4.55], concluding that ethyne adopts a distorted geometry on all three metals, though the degree of distortion is less on Pd and Pt than on Ni. The C–C bond length was estimated to be 1.34–1.39 Å from the photoemission spectra, which compared to the observed gas phase separation of 1.20 Å [4.92] suggests significant rehybridisation of the molecule upon adsorption.

From their HREELS investigations on Pd(1 1 1) *Kesmodel* and coworkers [4.52, 53, 56] inferred that following low-temperature chemisorption, ethyne is extensively rehybridised to approximately $sp^{2.5}$; this conclusion was reached by comparing the ν_{C-H} and ν_{C-C} stretching frequencies with gas phase data [4.93]. The spectra of C_2H_2 and C_2D_2 together with the peak assignments are shown in Fig. 4.17. The ν_{CH} mode occurs at 2992 cm^{-1} and the ν_{CC} mode at 1402 cm^{-1}, from which the authors propose a C–C bond length of 1.42 $\pm$ 0.03 Å, somewhat greater than that suggested by *Demuth* [4.55]. The ν_{CH} and ν_{CC} modes are observed at 2920 and 1220 cm^{-1} on Ni(1 1 1) [4.67], confirming that a stronger molecular distortion does indeed occur in the case of Ni, whilst intermediate behaviour is observed on Pt(1 1 1) [4.59]. Comparison with the ν_{CH} and ν_{CC} modes of the dicobalt complex $(C_2H_2)Co_2(CO)_6$ [4.94] gives excellent agreement for the ν_{CC} frequencies, but the ν_{CH} modes are $\approx$100 cm^{-1} higher in the complex. This may be taken as evidence that chemisorbed ethyne is more distorted on Pd(1 1 1) than in the dicobalt complex, and is coordinated to more than two metal atoms.

Gates and *Kesmodel* [4.52] used vibrational assignments and exploited HREELS selection rules to discuss the surface geometry of C_2H_2 on Pd(1 1 1) at low temperature. Adsorbed ethyne may have C_{2v}, C_2 or C_s symmetry, depending on whether interactions with the second layer of metal atoms are considered

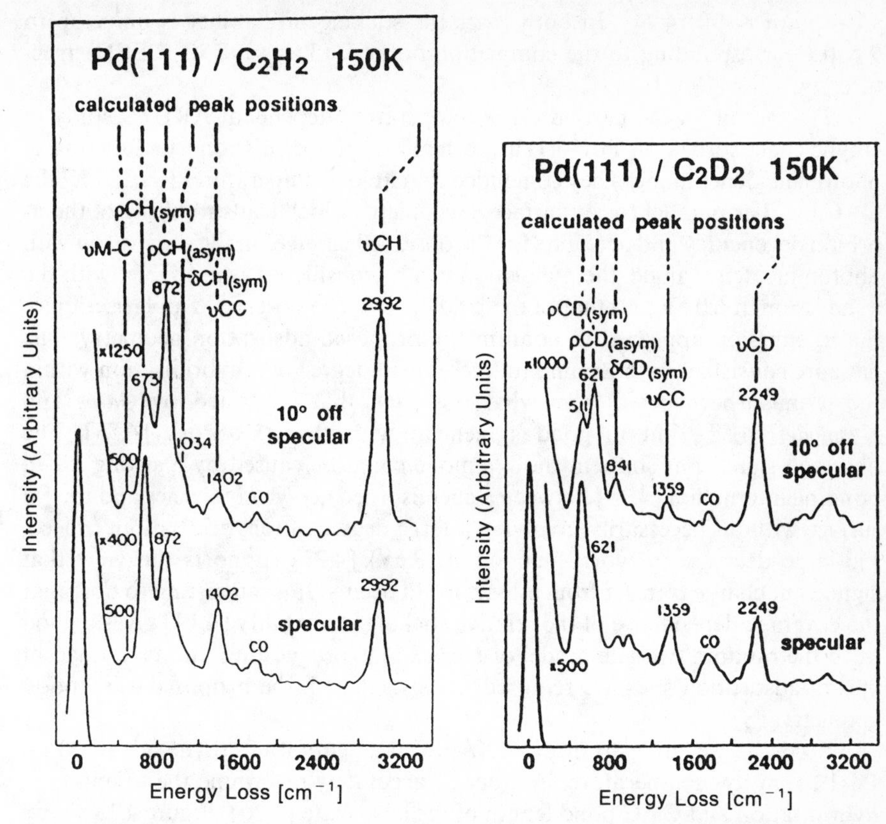

Fig. 4.17. Electron energy loss spectra for C_2H_2 and C_2D_2 on Pd(1 1 1) at 150 K, for specular and off-specular scattering [4.52]. Also shown are the peak assignments and the calculated peak positions for adsorption of acetylene in the three fold hollow site [4.91]

significant. However, the ρ_{CH} mode at 673 cm^{-1} was found to be dipole enhanced, which rules out C_{2V} symmetry, whilst the fact that the ρ_{CH} mode at 872 cm^{-1} was not dipole enhanced supports a geometry with C_2 or C_S symmetry. Although *Gates* and *Kesmodel* could not completely rule out a lower-symmetry (C_1) structure, C_S symmetry was favoured. The estimated C–C bond length and the strong molecular distortion resemble those observed in trinuclear transition metal complexes such as $Os_3(CO)_{10}(C_2Ph_2)$, where the C–C bond length is 1.44 Å and the molecule is symmetrically coordinated near the centre of the three metal atoms [4.95]. Bearing in mind the observed $(\sqrt{3} \times \sqrt{3})$ R30° diffraction pattern, a model was proposed in which the C–C axis lies parallel to the surface and the molecular plane is tilted slightly with respect to the surface normal, the molecule being *cis*-bent with a CCH angle of $\approx 122°$.

Tysoe et al. used molecular beam and C (1s) XPS measurements to show that the saturation coverage of the species formed by low temperature adsorption is 0.46, while the saturation adsorbate coverage after room temperature adsorption is 1.0 [4.14]. In both cases the sticking probability is unity up to $\theta \cong 0.3$ corresponding to the completion of the $(\sqrt{3} \times \sqrt{3})$ R30° LEED structure.

Tysoe et al. [4.14] carried out a temperature-dependent ARUPS study of ethyne chemisorbed on Pd(1 1 1) at normal exit photoemission as a function of photon incidence angle. They concluded that at low temperatures (< 220 K) the C–C axis lies parallel to the surface, a geometry which leads to lifting of the π-orbital degeneracy and accounts for the observed changes in the π-emission with photon incidence angle: the π-emission can be considered as a doublet with the π_z emission at 6.0 eV and that of the π_y at 4.8 eV. The observed z-polarisation of the π_z emission appeared to confirm the proposed adsorption geometry. The data are consistent with a comparatively small degree of rehybridisation within the chemisorbed species somewhat less than that suggested by *Gates* and *Kesmodel* [4.52], but in good agreement with *Demuth's* work [4.55]. The observed significant shifts in the ν_{CC} mode could be caused by a strong $\pi \to d$ bonding interaction [4.96] between a chemisorbed acetylenic species and the Pd surface without necessarily involving a large degree of rehybridisation. Indeed, the large decrease in work function (1.53 eV) [4.23] supports the view that significant charge transfer from ethyne to Pd occurs. It is interesting to note that the coverage dependence of the relative sticking probability [4.14] gives a good fit to the random two-site model of *Kisluik* (4.97) suggesting that two adjacent vacant adsorption sites are required, consistent with the proposed adsorption geometry.

Recently, we carried out a NEXAFS investigation of ethyne adsorbed on Pd(1 1 1) at low temperature, in order to accurately determine the orientation, hybridisation and C–C bond length of the adsorbate [4.20]. Figure 4.18 shows carbon K-edge NEXAFS spectra following a saturation dose of ethyne at 90 K,

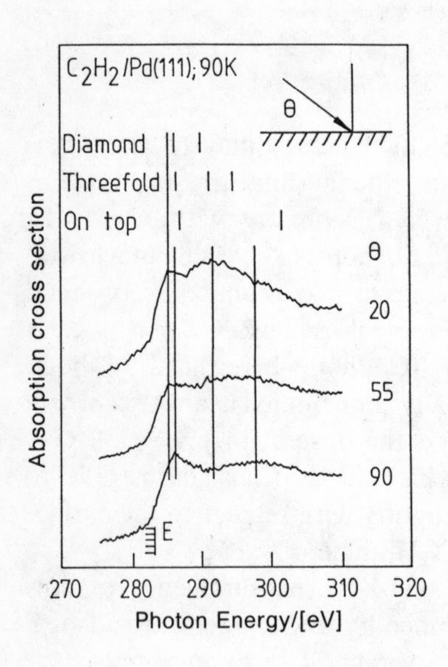

Fig. 4.18. Near-edge X-ray absorption fine structure spectra for ethyne adsorbed on Pd(1 1 1) at 90 K. Data is shown for three different photon incidence angles, θ. Also shown are the calculated peak positions for three adsorption geometries

recorded as a function of photon incidence angle (θ); θ is therefore the angle between the electric field vector and the surface normal. Intensities have been normalised so that the edge jump above 305 eV is identical for each spectrum. It can be seen that the spectrum exhibits a broad absorption at a photon energy of 297.9 ± 0.5 eV which increases in intensity with increasing θ, another absorption at 291.6 ± 0.5 eV which displays the opposite polarisation dependence, and a fairly sharp resonance centred at approximately 286 eV which reveals only a small variation in intensity with incidence angle. This latter peak does, however, show some dispersion with incidence angle, shifting from 285.2 ± 0.2 eV at $\theta = 20°$ to 286.3 ± 0.2 eV at $\theta = 90°$.

The 286 eV peak is assigned to a transition from the C (1s) level into π^* antibonding orbitals. For an undistorted molecule which maintains its cylindrical symmetry, the lack of angular variation would suggest that the carbon–carbon axis of the ethyne is tilted by an angle close to the magic angle (55°) with respect to the surface [4.98]. However, it is clear that this peak does reveal some dispersion as the incidence angle is varied, suggesting that the π orbitals lose their degeneracy upon adsorption, and thus the molecule can no longer be considered cylindrically symmetric.

A recent theoretical study of ethyne adsorption on Pd(1 1 1) has been carried out by *Sellers* [4.91], who determined the structures and vibrational frequencies for adsorption in the on-top position, the three-fold hollow site and the two-fold bridge site using clusters having up to ten atoms and two layers. Comparison of the computed structures and vibrational spectra for each of these configurations with the HREELS spectrum of ethyne adsorbed on Pd(1 1 1) [4.52, 53] is displayed in Fig. 4.17 and suggests that the molecule exists in a rehybridised conformation in the three-fold hollow, with a C–C bond length of 1.42 Å, a C–H bond length of 1.09 Å and a CCH angle of 117°. The molecular plane is calculated to be tilted by 22° from the surface normal.

We have calculated [4.20] the theoretical NEXAFS spectra for ethyne adsorbed at the three-fold hollow, two-fold bridge and on-top site using the equilibrium geometries calculated for these sites by *Sellers* [4.91]. A one-electron calculation of the X-ray absorption cross-section was carried out, explicitly including the effects of multiple scattering of the photoelectron by atoms immediately surrounding the absorbing atom. The exchange contribution to the energy was included by means of an appropriate local-density approximation. The resulting peak positions for the three geometries are shown in Fig. 4.18. The best agreement is observed for ethyne adsorbed in the three-fold site. In particular, excellent agreement is obtained for the resonances at $\approx$286 and 291.6 eV, whilst there is a discrepancy of $\approx$4 eV for the resonance at 297.9 eV. However, the resonance at 297.9 eV is assigned to a 1s$\rightarrow\sigma^*$ transition, and it has been proved experimentally that this resonance is very sensitive to small changes in the C–C bound length [4.99]. The experimental value for the energy difference between the σ^* and π^* resonance yields a C–C bond length of 1.31 ± 0.02 Å for ethyne adsorbed on Pd(1 1 1) at 100 K – 0.11 Å shorter than the value predicted theoretically by *Sellers*. It is noteworthy that the calculated

NEXAFS spectra for ethyne adsorbed at the three-fold hollow site with a C–C bond length of 1.31 Å show extremely good agreement with the experimental spectra [4.20]. Additionally, an X-ray structure determination of the Pd complex, $Pd(PPh_3)_2(C_2(COOMe)_2)$ [4.34] indicates that the Pd atom and both unsaturated carbons are virtually coplanar, with the ethyne bent into a cis-configuration (R–C–C angle = 149°) and a C–C bond length of 1.28 Å in good agreement with our results for an extended metal surface.

When the low-temperature ethyne overlayer is annealed, transformation to a different surface species occurs between 220 and 270 K. This surface rearrangement has been observed by HREELS [4.56], ARUPS [4.14] and XPS [4.20, 23]. *Gates* and *Kesmodel* [4.56] suggested that both vinylidene ($=C=CH_2$) and ethylidyne ($\equiv C-CH_3$) are present, along with some unreacted ethyne (Fig. 4.19). In the presence of preadsorbed hydrogen, ethylidyne is exclusively formed. The formation of vinylidene was deduced on the basis of a ν_{CH_2} mode at 2986 cm^{-1}, analogous to the organometallic compound

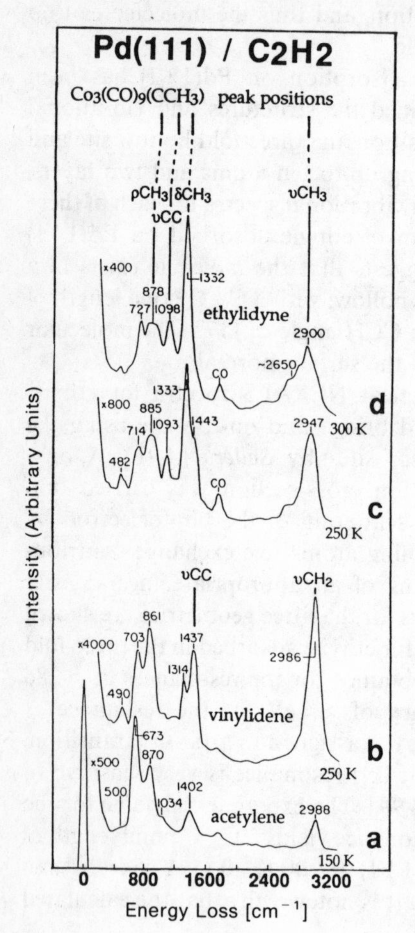

Fig. 4.19. Electron energy loss spectra for chemisorbed acetylene on Pd(1 1 1) at 150 K and subsequent annealed. (a) 150 K (b) annealed to 250 K. (c) and (d) annealed to 250 and 300 K in 10^{-7} Torr of H_2 [4.53]. Also shown are the peak assignments for ethylidyne and vinylidene. and the peak positions for $Co_3(CO)_9(CCH_3)$ [4.101]

$H_2Os_3(CH_2C)(CO)_9$, which contains a vinylidene species [4.100], whilst ethylidyne was deduced by analogy with the organometallic compound $(CO)_9Co_3(CCH_3)$ [4.101], being characterised by a ν_{CH_3} mode at 2900 cm^{-1}, a ν_{CC} mode at 1098 cm^{-1}, a δ_{CH_3} mode at 1332 cm^{-1} and a ρ_{CH_3} mode at 923 cm^{-1}. This latter species had also been proposed in the cases of Pt(1 1 1) [4.59, 102, 103] and Rh(1 1 1) [4.74] in the presence of adsorbed hydrogen, and there is some evidence for vinylidene formation in the absence of hydrogen on Pt(1 1 1) [4.59, 102], though *Avery* has concluded that there is no evidence for even an intermediate vinylidene species [4.63], instead proposing that ethyne disproportionates between 330 and 420 K to ethylidyne and a C_2H residue. In contrast, on Ni(1 1 1) chemisorbed ethyne is stable up to 400 K, decomposing above this temperature to form a CH species [4.66]. This species, along with C_2H, also forms on Pd(1 1 1) above 400 K, following decomposition of the vinylidene/ethylidyne phase [4.53, 56]; at elevated temperatures progressive decomposition of the C_2H and CH species occurs to form a carbonaceous overlayer.

From an analysis of their ARUPS data, *Tysoe* et al. [4.14] also concluded that ethyne transforms into vinylidene between 220 and 270 K. The thermal

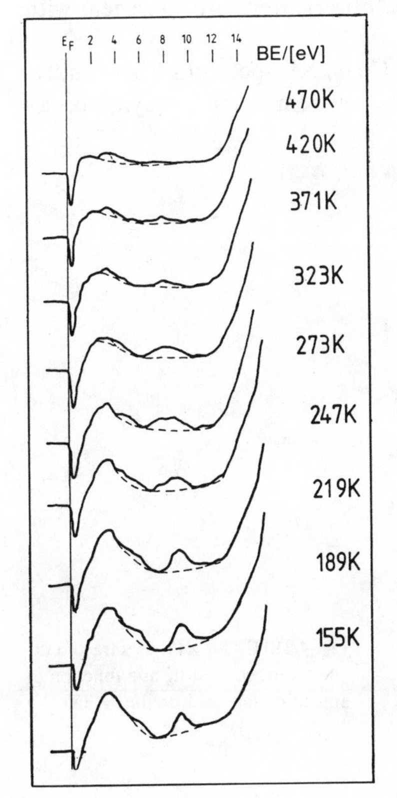

Fig. 4.20. HeI Pd(1 1 1) + 2.3L C_2H_2, $\psi = 50°$, $\theta = 40°$. Temperature dependent ARUPS of ethyne on Pd(1 1 1)

evolution of the He I UP spectrum over the range 155 to 470 K following ethyne adsorption at 155 K is exhibited in Fig 4.20 and the change in the adsorbate-induced emission is clearly apparent between 219 and 273 K. A spectrum essentially identical to the 273 K spectrum is obtained when ethyne is adsorbed on Pd(1 1 1) at 300 K, and this correlates well with the spectrum of gaseous ethene, in accord with the findings of *Demuth* [4.96]. Analysis of the variation in intensity of the emission features with photon incidence angle [4.14] led *Tysoe* et al. to conclude that vinylidene is formed with the C–C axis perpendicular to the Pd surface. Figure 4.20 shows that vinylidene starts to decompose at 371 K, in agreement with the HREELS observations of *Gates* and *Kesmodel* [4.56]; notice that additional small features are present in the 420 K spectrum. By 470 K two broad features at 3.6 and 7.2 eV remain: these may be assigned to $C_{[ad]}$ and $CH_{[ad]}$, respectively, in accord with the conclusions of *Demuth* based on observed and calculated orbital ionisation energies ([4.65] and references therein).

We have recently also carried out a NEXAFS investigation of the species which is formed by adsorption of ethyne on Pd(1 1 1) at *room temperature*. Preliminary analysis reveals two bound state resonances at ≈285 and ≈289 eV, suggesting the formation of an unsaturated species. Furthermore, the 285 eV resonance is most intense at normal photon incidence indicating that the C–C axis is perpendicular to the surface. Both these observations are consistent with the presence of vinylidene.

Figure 4.21 displays the variation in the C (1s) XP spectrum with temperature, following adsorption at 90 K [4.20]. The low-temperature ethyne species

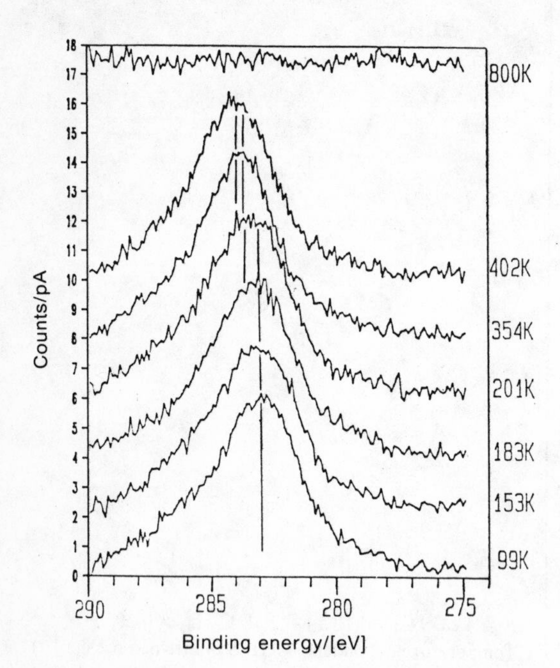

Fig. 4.21. C (1s) XPS spectra of ethyne adsorbed at 99 K as a function of annealing temperature ($h\nu = 350$ eV)

is characterised by a C (1s) binding energy of 283.0 eV, which remains constant with heating up to a temperature of 201 K. By 350 K, C (1s) binding energy has increased to 284.0 eV, corresponding to vinylidene formation; exactly the same binding energy is observed following room-temperature ethyne adsorption. Some reduction in the intensity of C (1s) emission occurs below 600 K, and complete removal of surface carbon has occurred by 800 K, indicating very efficient diffusion of residual carbon into the bulk metal. This contrasts strongly with the behaviour of ethyne on Pt(1 1 1) and Ni(1 1 1), where successive adsorption – desorption cycles lead to a progressive build-up of graphitic carbon and eventual complete carbiding of the surface [4.104, 105]. No such cumulative carbiding occurs for Pd(1 1 1): the initial sticking probability of ethyne remaining constant even after many consecutive adsorption – desorption cycles [4.14]. This is of considerable practical utility when studying ethyne cyclisation with single-crystal Pd specimens.

In summary, we conclude that ethyne chemisorbs on Pd(1 1 1) at 100 K in the three-fold hollow site, with the C – C bond parallel to the surface and with charge transfer from ethyne to Pd. The C – C bond length is 1.31 Å and the CCH angle is 117°, indicating some molecular distortion but to a lesser extent than that which occurs on the congener metals, Ni(1 1 1), Pt(1 1 1) and Rh(1 1 1). The HCCH plane is titled by 22° with respect to the surface normal. Between 220 and 270 K an intramolecular transformation to a perpendicularly oriented vinylidene species occurs, which itself decomposes above 400 K to form smaller hydrocarbon fragments, before eventually forming surface carbon which is removed by very efficient diffusion into the Pd bulk above 650 K, thus restoring the clean surface. The proposed adsorption geometry of ethyne on Pd(1 1 1) at low temperature is shown schematically in Fig 4.22.

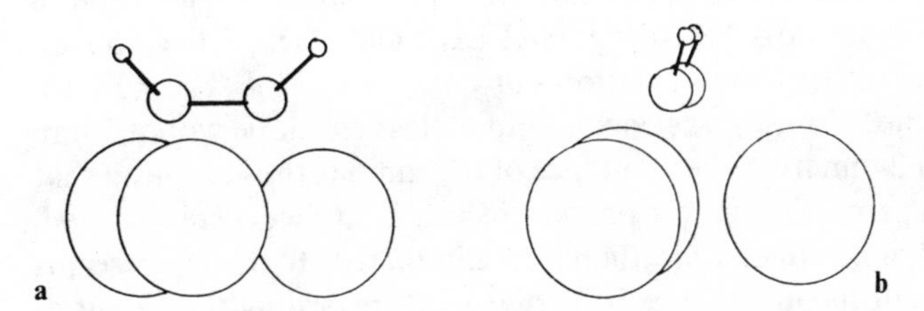

Fig. 4.22.a,b Schematic diagram showing the proposed adsorption geometry of acetylene in the three-fold hollow site on Pd(1 1 1)

4.6 Why is Tricyclisation so Specific to Palladium and Why is the (1 1 1) Plane so Strongly Favoured?

We may now consider specifically the cyclotrimerisation of ethyne. Two very interesting aspects of this reaction are (i) why is it almost unique to Pd? and (ii) why is the (1 1 1) surface so strongly preferred?

We may address these two questions by investigating the reactive properties of thin Pd overlayers of varying degrees of perfection on an Au(1 1 1) substrate. Additional information may be obtained by performing similar measurements with Au/Pd surface alloy phases derived from such overlayer systems. We shall discuss our findings and conclusions in this section, but we begin with a brief summary of previous work which suggests that Pd and in particular the (1 1 1) face is strongly favoured for benzene formation.

Gentle and *Muetterties* studied ethyne chemisorption on the three low Miller index planes of Pd [4.12] and the results of a similar investigation have been reported by *Rucker* et al. [4.13]. In both instances the (1 1 1) face was found to be considerably more active than either the (1 1 0) or the (1 0 0) surfaces. However, in the latter case *Rucker* et al. proposed that the (1 1 0) face is more than three times more active than the (1 0 0) plane; the relative activities of the (1 1 1) : (1 1 0) : (1 0 0) faces being 100 : 18 : 5, whereas in the former case the (1 1 0) surface is found to be the least active, with only a very small amount of benzene being formed (detected by displacement with trimethylphosphine). The two sets of results for the (1 1 0) and (1 0 0) surfaces are clearly inconsistent, though both support the view that the (1 1 1) face is the most active towards benzene formation. Significant differences were also observed in the respective H_2 desorption spectra, arising from ethyne decomposition.

More recently *Yoshinobu* et al. [4.58] carried out a detailed investigation of ethyne chemisorption on Pd(1 1 0) using HREELS and TDS. They looked carefully for evidence of benzene formation *but none was found*. They used a mass spectrometer whose ioniser was enclosed in a glass envelope with a 2 mm exit diameter located 2 mm from the front face of the sample, thus ensuring that only gases desorbing from the central portion of the front face were detected. Effects due to the sample edges were therefore eliminated. It therefore seems likely that the apparent inconsistencies in earlier work regarding the extent of benzene formation on the three low Miller index faces of Pd may be explained by cyclisation occurring on the sample edges and at (1 1 1) facets.

Rucker et al. [4.13] noted that benzene evolution from Pd(1 1 0) and Pd(1 0 0) could only be observed after much larger exposures to ethyne than were necessary in the case of Pd(1 1 1). Commercially available research-grade ethyne contains trace amounts of benzene impurity which can only be removed by repeated bulb-to-bulb distillation. Unless this is done there is a real danger of adsorption/desorption of impurity benzene being mistaken for reactive formation of benzene from ethyne. Such effects are particularly serious at high gas exposures, as might be expected.

Given these possibilities for experimental artefacts due to specimen edges and gas contamination, we infer that the (1 1 0) face of Pd is inactive towards benzene formation and consider it likely that the (1 0 0) face is also essentially inactive, a point which we will discuss later. The conclusion is tricyclisation of ethyne to benzene is very sensitive to surface crystallography, with the (1 1 1) face being the only active surface. This is in harmony with the TPD studies carried out at high pressure on high-area supported Pd particles which were discussed in Sect 4.3. The efficient tricyclisation of ethyne is also rather specific to palladium surfaces; with the exception of Cu(1 1 0) [4.72] and possibly Ni(1 1 1) [4.106], no other metal surface has been reported as being active towards benzene formation. In particular on both Pt(1 1 1) [4.63, 107] and Rh(1 1 1) [4.75] ethyne adsorbs irreversibly, with H_2 being the only desorption product resulting from the transformation of ethyne and the subsequent dehydrogenation of various C_xH_y species. Ethyne adsorption on Ni(1 1 1) is also essentially irreversible [4.67], though *Bertolini* et al. [4.106] suggested that after high ethyne exposures and suitable thermal treatment, a vibrational spectrum resembling that of benzene is observable. However, no benzene desorption was detected, suggesting the extent of conversion was small. We have recently shown that both Au(1 1 1) [4.46] and Ag(1 1 1) [4.108] are totally inactive towards benzene formation from ethyne.

Taken together these observations suggest an additional reason for the almost unique efficiency of palladium in the tricyclisation of ethyne to benzene. The molecule interacts sufficiently strongly with the Pd surface so that it is still adsorbed at the temperatures necessary for C_4H_4 formation (unlike Ag and Au); it is not too strongly hybridised (unlike Pt, Rh and Ni) so that cyclisation is favoured over fragmentation. A measure of the extent of rehybridisation of ethyne which occurs on adsorption can be gained from the stretching frequency of the carbon – carbon bond; the lower the frequency, the lower the bond order and hence greater the degree of rehybridisation. The ν_{CC} stretching frequencies of ethyne adsorbed on low Miller index transition-metal surfaces are particularly revealing: the representative values are listed in Tables 4.2, together with the appropriate references.

Although these measurements were recorded with different spectrometers it can clearly be seen that on Pd(1 1 1) the ν_{CC} energy loss occurs at a significantly higher frequency than on any other transition-metal surface [with the exception of Ag(1 1 0) where the ν_{CC} mode is not observed] indicating a low degree of rehybridisation. It is apparent that ν_{CC} on Pd(1 1 0) [4.58] and Pd(1 0 0) [4.109] is significantly lower than on Pd(1 1 1) [4.52] (1230 and 1210 cm^{-1} compared to 1402 cm^{-1}) indicating that ethyne is considerably more strongly distorted on Pd(1 0 0) and Pd(1 1 0) than on Pd(1 1 1). This may explain, at least partly, the extreme structure sensitivity of ethyne tricyclisation to the (1 1 1) face of Pd. Similarly, the observed ν_{CC} values for ethyne on transition-metal single-crystal surfaces provide a possible explanation as to why Pd(1 1 1) catalyses the tricyclisation reaction whilst the (1 1 1) faces of the congener metals do not: the reactant molecule/surface interaction is now so strong that disruption of the

Table 2

Metal surface	ν_{CC} Stretching frequency/cm^{-1}	Reference
Pd(1 1 1)	1402	52,53
Pd(1 0 0)	1210	109
Pd(1 1 0)	~ 1230	58
Pt(1 1 1)	1310	59,63
Ni(1 1 1)	1220	65,67
Rh(1 1 1)	1260	75
Cu(1 1 1)	1307	57
Ir (1 1 1)	1160	76
Ru(0 0 0 1)	1135	81
	1110	82
Ag(1 1 0)	not observed	73
Cu(1 1 0)	1305	72
Ni(1 1 0)	1305	110
Ni(1 0 0)	1345	111
Ni[5(1 1 1) × (1 1 0)	1200	66
W(1 1 0)	1130	84
Fe(1 1 0)	1240	79
Fe(1 1 1)	1145	80

molecular framework occurs in preference to C_4H_4 formation. This argument focusses on the reactant species; it seems possible that similar considerations could apply to the intermediate (C_4H_4) and product (C_6H_6). That is, even if these higher hydrocarbons were formed in significant yield on the (1 1 1) faces of other metals, the higher adsorption energy would favour decomposition over cycloaddition and desorption, respectively.

Such effects are not restricted to the tricyclisation of ethyne; a highly selective partial hydrogenation of ethyne to ethene occurs on Pd if H_2 is present [4.112, 113]. We have studied the interaction of ethene with Pd(1 1 1) by NEXAFS, photoelectron spectroscopy and TDS, demonstrating that at low temperatures ethene lies flat on the metal surface and is π-bonded to Pd [4.114, 115]. Both the valence-level spectrum and NEXAFS indicate that the molecule is essentially undistorted following adsorption; furthermore, the observed weakening of the C–C bond is less than that which occurs on the congener metal surfaces, Pt(1 1 1) [4.116] and Ni(1 1 1) [4.117]. In addition, both XPS and UPS indicate that upon annealing Pd(1 1 1) to 300 K, molecularly adsorbed ethene undergoes essentially completely reversible desorption, without detectable rearrangement or decomposition. This contrasts sharply with the thermal behaviour of ethene on other group-VIII metal surfaces; Pt(1 1 1) (4.60, 116], Rh(1 1 1) [4.74], Ni(1 1 1) [4.117], Pd(1 0 0) [4.118] and Pd(1 1 0) [4.119], where large hydrogen desorption yields due to ethene decomposition are observed in all cases, and ethylidyne, or other related species, are formed on thermal treatment. The very low sticking probability of ethene on Pd(1 1 1) at

300 K ($S \approx 0.01$) [4.115] compared to Pt(1 1 1) and Rh(1 1 1) [4.61, 74, 120], is consistent with this picture: efficient adsorption to form ethylidyne occurs on Pt(1 1 1), Rh(1 1 1), but not on Pd(1 1 1).

Thus the ability of the Pd(1 1 1) surface to conserve the C=C bond in ethene appears to be an important factor in the outstanding selective hydrogenation properties of Pd catalysts. Similarly, the bonding of benzene at high surface coverages in a weakly-bonded tilted geometry appears to be unique to Pd(1 1 1) and is crucial to efficient catalytic synthesis of the molecule [4.19]. *Wong and Hoffman* have very recently performed calculations on the bonding of ethene to Ni(1 1 1), Pd(1 1 1) and Pt(1 1 1) [4.121], and their results agree with the experimental finding that ethene is substantially less distorted on Pd(1 1 1) compared with its congeners.

Avery's observation of low-temperature ethyne cyclisation on Cu(1 1 0) is worthy of comment [4.72]. Benzene evolution was observed at 325 K and the HREEL spectrum of ethyne on Cu(1 1 0) was reported to resemble that observed by *Bandy* et al. on the Cu(1 1 1) surface [4.57]. In turn, the latter scientists have pointed out that the HREEL spectra of ethyne on Cu(1 1 1) and Ni(1 1 1) are similar; they are also very different to those observed on Pt(1 1 1), Pd(1 1 1), Rh(1 1 1), Pd(1 1 0) and Ni(1 1 0) which are themselves generally similar. This was interpreted in terms of the smaller metal-metal distances in Cu and Ni leading to a different adsorption site: both carbon atoms of the ethyne molecule were thought to be located above three-fold hollow sites – rather than the di-$\sigma + \pi$ conformation above a three-fold site which is adopted in the case of adsorption on second- and third-row transition metals. It is therefore possible that the extent of rehybridisation of ethyne on Cu(1 1 0) may be comparable to that on Pd(1 1 1). thus accounting for the activity of Cu(1 1 0) towards benzene formation. However, this explanation does not account for the weak interaction of ethyne with Cu as compared to Ni. Thus, although it is possible that the extent of rehybridisation on Cu(1 1 0) may be comparable to that on Pd(1 1 1), the unfortunate absence of direct evidence about the cyclisation activity (or lack of it) of Cu(1 1 1) precludes us from drawing any firm conclusions about the correlation between the degree of hybridisation and catalytic behaviour.

As mentioned earlier in this section, the properties of the Au/Pd system may be exploited in order to shed light on the relative importance of structural and electronic effects in the tricyclisation of ethyne to benzene. Here we summarise our recent results using Pd overlayers and Pd/Au surface alloys on a Au(1 1 1) substrate in order to investigate these effects. In complete contrast with Pd(1 1 1), Au(1 1 1) is, of course, totally inert towards cyclisation and all other reactions of ethyne.

Figure 4.23 exhibits the desorption of reactively formed benzene from unannealed Pd films deposited on Au(1 1 1) at < 170 K, as a function of the overlayer film thickness. The bottom spectrum confirms the inertness of the bare Au(1 1 1) surface while the second spectrum demonstrates that even a 1 ML unannealed film of Pd is active for tricyclisation, though the yield of benzene is

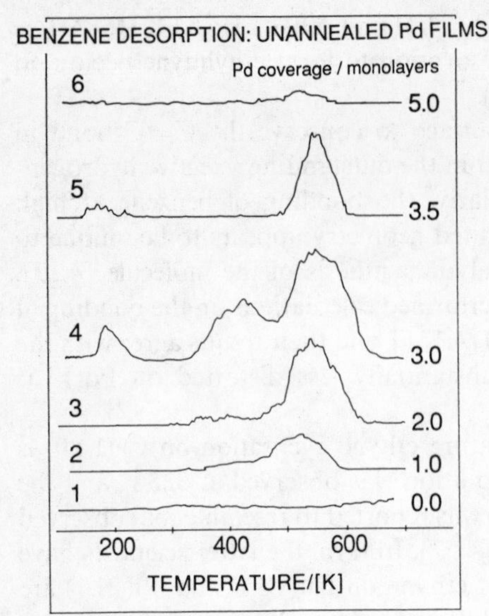

Fig. 4.23. Benzene desorption from unannealed Pd films of increasing thickness deposited on Au(1 1 1) at > 170 K, following a saturation exposure of ethyne (6 L/170 K)

quite small. 2 ML and 3 ML Pd films give rise to increased benzene formation and a new desorption state appears at $\approx$410 K. Note that this new state is intermediate in temperature between the desorption maxima observed with clean Pd(1 1 1); these are associated with weakly bound tilted molecules ($\approx$230 K) and strongly bound flat-lying molecules ($\approx$530 K) respectively [4.14]. We attributed this feature to the effects of surface roughness of the Pd film which arises from the high metal deposition rate and low substrate temperature. Finally, it can be seen from Fig. 4.23 that for 5 ML *unannealed* Pd films, benzene formation is almost completely suppressed. By this stage, electronic modification of the surface Pd atoms due to the underlying Au(1 1 1) must be negligible. This effect must be morphological in origin, reflecting the very rough surface produced by a large dose of Pd rapidly delivered to a cold Au(1 1 1) substrate (this was confirmed by the extremely diffuse (1×1) LEED pattern observed compared with that of the clean Au(1 1 1) surface). It demonstrates the extreme structure sensitivity of tricyclisation to the presence of (1 1 1)-like ensembles of Pd atoms.

If the Pd films are annealed prior to ethyne adsorption and reaction, the results are quite dramatically different; Figure 4.24 illustrates our findings for 1 ML *annealed* Pd films as a function of increasing pre-annealing temperature. Pretreatment at 300 K leads to a very substantial increase in the subsequent benzene yield; also, the low-temperature desorption peak appears. This behaviour may be understood in terms of a progressively increasing degree of perfection of the Pd(1 1 1) film (which has been observed by LEED). Following 400 K annealing, essentially *all* benzene desorption occurs within the low-temperature peak, a process which goes to completion for films preannealed to

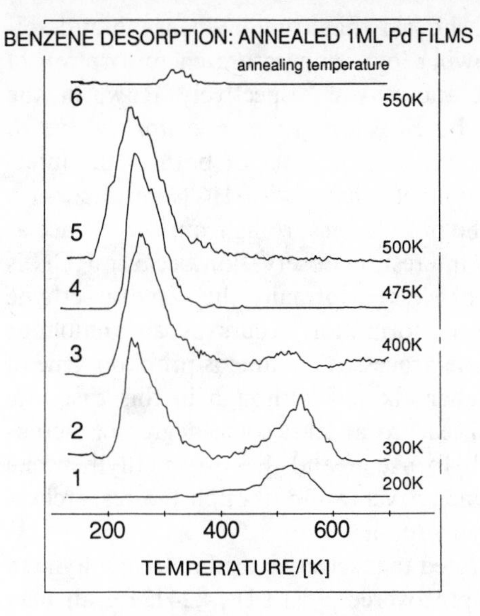

Fig. 4.24. Benzene desorption following saturation ethyne exposure (6 L/170 K) from 1 ML Pd films on Au (1 1 1), pre-annealed to the indicated temperatures

475 K. The observed trend towards increasing desorption from the low-temperature state reflects some mechanism which causes flat-lying benzene molecules to be displaced into a titled configuration. It seems plausible that this is the result of the progressive break-up of Pd ensembles at the surface resulting from increasing interdiffusion of Au and Pd with increasing preannealing temperature. In some respects this is analogous to the NO-induced behaviour described in Sec. 4.2.3 [4.18]. For temperature above 525 K metal interdiffusion is rapid so that for films preannealed to 550 K benzene formation is almost completely suppressed.

It thus appears that the extent of ethyne conversion and the binding energy of the resulting benzene depend on the atomic composition, morphology and electronic properties of the surface. We have found that thin, annealed (1 1 1)-oriented Pd films are less efficient at catalysing the reaction than is bulk Pd(1 1 1); bulk Pd behaviour is reached by $\approx$3 ML thickness. Rough Pd films are less efficient than smooth ones and appear to give rise to a new binding site for the reactively formed benzene. For a given benzene yield, Au/Pd alloy formation has a major effect on the desorption temperature of reactively-formed benzene.

4.7 Other Cyclisation Reactions

Since the original reports of ethyne tricyclisation on Pd(1 1 1), a number of other cyclisation reaction have been observed on single crystal metal surfaces, these are discussed briefly below.

Gentle and coworkers [4.12, 122] showed that trimethylbenzene and toluene can be formed on Pd(1 1 1) following low temperature chemisorption of propyne and coadsorption of propyne and ethyne, respectively. However, the yields are appreciably lower than for benzene formation, presumably due to increased steric hindrance inhibiting formation of either or both of the intermediate and final product. Coadsorption of ethyne with HCN resulted in a small amount of pyridine formation, the product desorbing from the surface at ≈370 K [4.122]. This is a particularly interesting observation because HCN is isoelectronic with C_2H_2; the overall process is formally the same as ethyne tricyclisation so it is likely that pyridine formation occurs by an analogous pathway to that for the ethyne→benzene process. The same is probably true of cyclisation reactions involving the higher alkynes, although in this case one might expect that steric factors would lead to at least some degree of stereospecific synthesis. Thus 1, 2, 4-trimethylbenzene and 1, 3, 5-trimethylbenzene could both be formed from the propyne tricyclisation, though the researchers did not identify which was the favoured product.

As noted earlier, we have demonstrated that selective oxidation of ethyne to furan (C_4H_4O) can occur on oxygen-precovered Pd(1 1 1) [4.24]. Furan may also be synthesised by partial oxidation of the C_4H_4 species formed by dissociative chemisorption of cis-3, 4-dichlorocyclobutene (DCB) (Fig. 4.25). Similarly, in an earlier report, *Gentle* et al. observed the formation of thiophene following chemisorption of both ethyne and DCB on a sulphided Pd(1 1 1) surface [4.25]. This strongly suggests that heterocyclisation reactions proceed by a pathway analagous to that for ethyne tricyclisation, relying on the

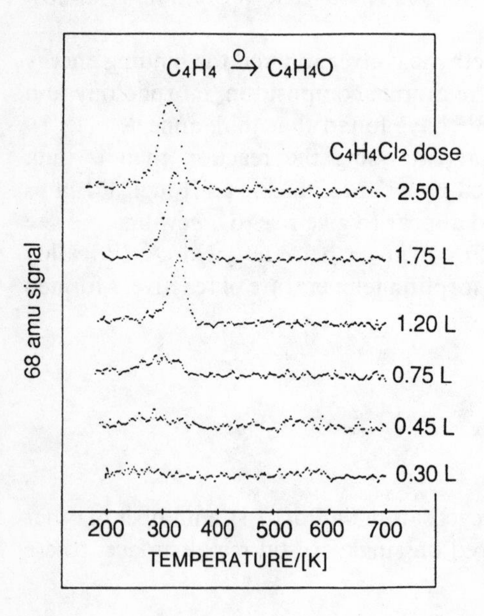

Fig. 4.25. Partial oxidation of C_4H_4 intermediate to furan. TPR spectra from oxygen precovered Pd(1 1 1) obtained after DCB adsorption at 270 K. $\theta_{ox} = 0.025$

formation of a C_4H_4 intermediate. It is interesting to note that, as in the case of benzene formation, evolution of the reactively-formed furan and thiophene is limited by the rate of product desorption.

The fact that both furan and thiophene can be reactively formed by incorporation of chemisorbed oxygen and sulphur at low temperature seems rather surprising, given the strong interaction of atomic oxygen and sulphur with the Pd(1 1 1) surface [4.123, 124]. In particular, the fact that any furan at all is formed in such a highly oxidising environment is both interesting and significant. Furan formation is favoured at lower oxygen precoverages; this may be attributed to two factors. First, and most importantly, at higher oxygen coverages the total oxidation of hydrocarbons to CO_2, CO and H_2O is favoured. Secondly, at lower oxygen coverages a higher surface concentration of C_4H_4 is attained (shown by UPS), due to the increased availability of chemisorption sites. Figure 4.26 illustrates the effect of varying oxygen precoverage on furan formation, for a fixed 1.75 DCB exposure [$\theta_{oxygen}(max) = 0.25$]. This behaviour stands in marked contrast to the observations of *Gentle* et al. [4.25] who found that thiophene was formed only at the highest sulphur coverages employed ($\theta_s = 0.33$), suggesting that furan formation is considerably favoured relative to thiophene formation. At low oxygen precoverages, the overall selectivity towards furan formation is as high as 80%.

Recently, *Gellman* has studied the thiophene yield from an extensively sulphided Pd(1 1 1) surface as a function of surface pretreatment. CO uptake was used to titrate the number of bridge sites available for chemisorption, and the

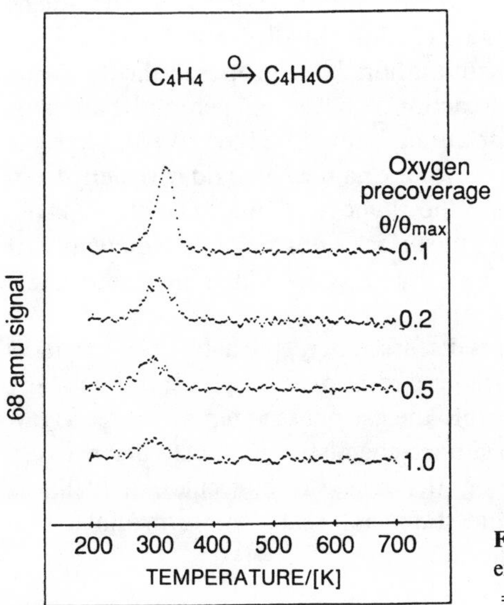

Fig. 4.26. Partial oxidation of C_4H_4 showing effect of oxygen precoverage. (DCB exposure = 1.75)

correlation was found between this quantity and the thiophene yield [4.26]. These bridge sites are created in the sulphur overlayer either by annealing > 800 K or by using initial sulphur coverages just below the saturation value – though no thiophene synthesis was observed until the sulphur coverage approached saturation. The thiophene yield was small (< 0.1% of a monolayer).

Madix and coworkers have recently demonstrated the formation of furan and 2, 5-dihydrofuran following butadiene chemisorption at 100 K on oxygen-precovered Ag(1 1 0) [4.125]. As with furan synthesis on oxygen-covered Pd(1 1 1), CO_2 and H_2O were formed in competitive reactions though CO was not. In addition, maleic anhydride, 2(5H)-furanone and 4-vinylcyclohexene were formed, the latter by dimerisation of butadiene and the first two by oxidation of 2, 5-dihydrofuran.

The fact that ethyne tricyclisation does not occur on Ag single-crystal surfaces, whilst the heterocyclisation reaction does, appears to confirm our proposal that it is the formation of the C_4H_4 metallocycle intermediate which is the crucial rate-determining step in benzene formation. In Sect. 4.2.2 we commented on the similarities between the tilted C_4H_4 intermediate and a cis-bonded butadiene [4.28], and hence the cycloaddition of oxygen and butadiene on Ag(1 1 0) is a very similar process to the reaction between oxygen and C_4H_4 on Pd(1 1 1), though there are, of course, two more hydrogen atoms present in butadiene. The similarities between the two processes are confirmed by the formation of 2, 5-dihydrofuran (C_4H_6O) from butadiene (C_4H_6), exactly analogous to synthesising furan (C_4H_4O) from C_4H_4.

However, the behaviour on Ag(1 1 0) is somewhat more complex; furan is also evolved with identical kinetics to 2, 5-dihydrofuran peak temperature (465 K). However, when 2, 5-dihydrofuran is adsorbed on oxygen-precovered Ag(1 1 0) desorption occurs at 250 K and no furan is observed. Butadiene desorption is complete by 300 K so that reactions involving this species must occur at lower temperatures. The present authors have proposed the following scheme to account for their results. Reaction between oxygen and butadiene occurs below 300 K to form 2, 5-dihydrofuran, but before the product molecule can desorb, further reaction with oxygen occurs, namely an oxidative dehydrogenation of an allylic C – H bond, to form the allylic furan intermediate, C_4H_5O. The allyl species is stable on the Ag(1 1 0) surface until ≈465 K, whereupon it disproportionates to form furan and 2, 5-dihydrofuran, which are immediately evolved into the gas phase.

For most of the cyclisation reactions discussed here, the activation barrier is considerably lower than that of the gas phase reaction, being most dramatically manifested by the evolution of benzene into the gas phase at temperatures as low as 200 K in the case of Pd(1 1 1). The surface chemistry obviously plays a very critical role in appropriately orienting the reactant molecules, inducing a suitable amount of rehybridisation, and stabilising reactive intermediates in a way which cannot occur in the gas phase.

4.8 Conclusions

Single-crystal studies of the chemistry and physics of metal surfaces are capable of contributing to our understanding of heterogeneously catalysed reactions in a genuinely fundamental way. This has often been claimed, and indeed there have been some notable successes; however, the reader will be aware that these achievements are heavily outnumbered by those cases in which much is claimed by way of relevance but little is delivered by way of insight. The relevance or otherwise of single-crystal observations to the interpretation of catalytic data of practical importance depends crucially on the extent to which experimental information may be realiably transferred between the two fields. Successful examples of this approach include studies on methanation, ammonia and methanol synthesis, CO oxidation by O_2 and NO, and ethene epoxidation; the list is not comprehensive but it could not be lengthened substantially. Ethyne tricyclisation, the principal subject of the preceding review, may be counted as another success. In certain respects it stands apart. Here, we have been able to unambiguously identify the key intermediate, characterise its bonding, and also the bonding geometry of the reactant and product species. This information appears to be of direct relevance to the successful rationalisation of the behaviour of practical catalytic systems operating at elevated pressures. In addition, it points to new chemistry. The discovery of chemisorbed C_4H_4 provides vital insight into the nature of heterocyclisation reactions undergone by ethyne. Last but not least, partial oxidation of ethyne to furan over a transition metal surface is not something which would have been readily predicted; nor would it have been understandable in the absence of the detailed information obtained from the single-crystal measurements.

Acknowledgement. RMO holds a University of Cambridge Oppenheimer Research Fellowship.

References

4.1 D.P. Curran: (ed) *Advances in Cycloaddition*; (Jai Press, Greenwich, Conn. 1990) Vol. 2
4.2 A. Wassermann: *Diels-Alder Reactions* (Elsevier, Amsterdam 1965)
4.3 M. Berthelot, Ann. Chem. **141,** 173 (1867)
4.4 W. Reppe, O. Schlichting, K. Klager, T. Toepal: Justus Liebigs Ann. Chem. **560,** 1 (1948)
4.5 G.W. Parshall: *Homogeneous Catalysis* (Wiley, New York 1980)
4.6 K.P.C. Vollhardt: Acc. Chem. Res. **10,** 1 (1977)
4.7 R. Ugo: Catalysis Rev. **11,** 225 (1975)
4.8 E.L. Muetterties, T.N. Rhodin, E. Band, C.F. Brucker, W.R. Pretzer: Chemical Rev. **79,** 91 1979
4.9 Y. Inoue, I. Kojima, S Moriki, I. Yasumori: Proc. Int'l. Congr. Catal. 6th Chem. Soc. Letchworth, UK (1976)
4.10 W.T. Tysoe, G.L. Nyberg, R.M. Lambert: J. Chem. Soc. Chem. Commun., 623 (1983).

4.11 W. Sesselman, B. Woratschek, G. Ertl, J. Küppers, H. Haberland: Surf. Sci. **130,** 245 (1983)
4.12 T.M. Gentle, E.L. Muetterties: J. Phys. Chem. **87,** 2469 (1983)
4.13 T.G. Rucker, M.A. Logan, T.M. Gentle, E.L. Muetterties, G.A. Somorjai: J. Phys. Chem. **90,** 2703 (1986)
4.14 W.T. Tysoe, G.L. Nyberg, R.M. Lambert: Surf. Sci. **135,** 128 (1983)
4.15 C.H. Patterson, R.M. Lambert: J. Phys. Chem. **92,** 1266 (1988)
4.16 C.H. Patterson, R.M. Lambert: J. Am. Chem. Soc. **110,** 6871 (1988)
4.17 C.H. Patterson, J.M. Mundenar, P.Y. Timbrell, A.J. Gellman, R.M. Lambert: Surf. Sci. **208,** 93 (1989)
4.18 R.M. Ormerod, R.M. Lambert: Surf. Sci. **225,** L20 (1990)
4.19 H. Hoffmann, F. Zaera, R.M. Ormerod, R.M. Lambert, L.P. Wang, W.T. Tysoe: Surface Sci. **232,** 259 (1990)
4.20 H. Hoffmann, F. Zaera, R.M. Ormerod, R.M. Lambert, D.K. Saldin, J.M. Yao, L.P. Wang, D.W. Bennet, W.T. Tysoe Surf. Sci. **268,** 1 (1992)
4.21 R.M. Ormerod. R.M. Lambert: in preparation.
4.22 W.T. Tysoe, R.M. Lambert: Surface Sci. **199,** 1 (1988)
4.23 R.M. Ormerod: Structural and electronic effects in the catalytic chemistry of palladium: Ph.D. Thesis, Cambridge (1989)
4.24 R.M. Ormerod, R.M. Lambert: Catalysis Lett. **6,** 121 (1990)
4.25 T.M. Gentle, K.P. Walley, C.T. Tsai, A.J. Gellman: Catalysis Lett. **2,** 19 (1989)
4.26 A.J. Gellmann: Langmuir **7,** 827 (1991).
4.27 R.M. Ormerod, R.M. Lambert: J. Chem. Soc. Chem. Commun. 1421 (1990).
4.28 R.M. Ormerod, R.M. Lambert, H. Hoffmann. F. Zaera, D.K. Saldin, J.M. Yao, L.P. Wang, D.W. Bennet, W.T. Tysoe: Surf. Sci.
4.29 C.H. Patterson: Reactions of sulphur and acetylene on a palladium (111) surface Ph.D. Thesis, Cambridge (1985)
4.30 G. Bieri, F. Burger, E. Heilbronner, J.P. Maier: Helvetica Chim. Acta. **60,** 2213 (1977)
4.31 W. Hubel: Chem. Ber. **95,** 1155 (1962)
4.32 M.A. Bennett, R.N. Johnson, T.W. Turney: Inorg. Chem. **15,** 107 (1976)
4.33 H. Yamazaki, K. Yasutuku, Y. Wakatsuki: Organometallics **2,** 726 (1983)
4.34 P.M. Maitlis: Pure and App. Chem. **33,** 489 (1973) and references therein
4.35 M.A. Bennett, P.B. Donaldson: Inorg. Chem. **17,** 1995 (1978)
4.36 B.L. Booth, R.N. Haszeldine, I. Perkins: J. Chem. Soc. Dalton Trans. 2593 (1981)
4.37 M.H. Chisholm, K. Folting, J.C. Huffmann, I. P. Rothwell: J. Am. Chem. Soc. **104,** 4389 (1981)
4.38 J.P. Collman, J.W. Kang. W.F. Little, M.F. Sullivan: Inorg. Chem. **7,** 1298 (1968)
4.39 G.M. Whitesides, W.J. Ehmann: J. Am. Chem. Soc. **91,** 3800 (1969)
4.40 A. Nakamura, N. Hagihara: Bull. Chem. Soc. Jpn. **34,** 452 (1961)
4.41 F.P. Netzer, J.U. Mack: J. Chem. Phys. **79,** 1017 (1983)
4.42 D. Osella, S. Aime, D. Boccardo, M. Castiglioni, L. Milane: Inorg. Chim. Acta. **100,** 97 (1985)
4.43 J.H. Sinfelt: J. Catalysis **29,** 308 (1973)
4.44 J.P.S. Badyal, R.M. Lambert: in preparation
4.45 R.M. Ormerod: unpublished results
4.46 R.M. Ormerod, C.J. Baddeley, R.M. Lambert: Surf. Sci. **259,** L709 (1991)
4.47 W.G. Durrer, H. Poppa, J. Dickinson: J. Catalysis **115,** 310 (1989)
4.48 M.A. Logan, T.G. Rucker, T.M. Gentle, E.L. Muetterties, G.A. Somorjai: J. Phys. Chem. **90,** 2709 (1986)
4.49 R.M. Ormerod, R.M. Lambert: J. Phys. Chem. **96,** 8111 (1992)
4.50 B. Marchon: Surf. Sci. **162,** 382 (1985).
4.51 P.Y. Timbrell, A.J. Gellman, R.M. Lambert, R.F. Willis: Surf. Sci. **206,** 339 (1988)
4.52 J.A. Gates, L.L. Kesmodel: J. Chem. Phys. **76,** 4218 (1982)
4.53 L.L. Kesmodel, G.D. Waddill, J.A. Gates: Surf. Sci. **138,** 464 (1984)
4.54 R.M. Ormerod, R.M. Lambert: in preparation.
4.55 J.E. Demuth: Surf. Sci. **84,** 315 (1979)
4.56 J.A. Gate, L.L. Kesmodel: Surf. Sci. **124,** 68 (1983)

4.57 B.J. Bandy, M.A. Chesters, M.E. Pemble, G.S. McDougall, N. Sheppard: Surf. Sci. **139,** 87 (1984)
4.58 J. Yoshinobu, T. Sekitani, M. Onchi, M. Nishijima: J. Phys. Chem. **94,** 4269 (1990)
4.59 H. Ibach, S. Lehwald: J. Vac. Sci. Technol. **15,** 407 (1978)
4.60 H. Steininger, H. Ibach, S. Lehwald: Surf. Sci. **117,** 685 (1982)
4.61 M.R. Albert, L.G. Sneddon, W. Eberhardt, F. Greuter, T. Gustafsson, E.W. Plummer: Surf. Sci. **120,** 19 (1982)
4.62 N. Freyer, G. Pirug, H.P. Bonzel: Surf. Sci. **126,** 487 (1983)
4.63 N.R. Avery: Langmuir **4,** 445 (1988)
4.64 J.E. Demuth, D.E. Eastman: Phys. Rev. Lett. **32,** 1123 (1974)
4.65 J.E. Demuth: Surf. Sci. **69,** 365 (1977)
4.66 S. Lehwald, H. Ibach: Surf. Sci. **89,** 425 (1979)
4.67 H. Ibach, S. Lehwald: J. Vac. Sci. Technol. **18,** 625 (1981)
4.68 G. Gasalone, M.G. Ciattania, F. Merati, M. Simonetta: Surf. Sci. **120,** 171 (1982)
4.69 L. Hammer, T. Hertein, K. Muller: Surf. Sci. **178,** 693 (1986)
4.70 K.Y. Yu, W.E. Spicer, I. Lindau, P. Pianetta, S.F. Lin: Surf. Sci. **57,** 157 (1976)
4.71 D. Arvantis, U. Dobler, L. Wenzel, K. Baberschke, J. Stohr, Surf. Sci. **178,** 696 (1986)
4.72 N.R. Avery: J. Am. Chem. Soc. **107,** 6711 (1985)
4.73 E.M. Stuve, R.J. Madix, B.A. Sexton: Surf. Sci. **123,** (1982) 491.
4.74 L.H. Dubois, D.G. Castner, G.A. Somorjai: J. Chem. Phys. **72,** 5234 (1980)
4.75 C.M. Mate, C.T. Kao, B.E. Bent, G.A. Somorjai: Surface Sci **197,** 183 (1988)
4.76 Ts. S. Marinova, K.L. Kostov: Surf. Sci. **181,** 573 (1987)
4.77 M.R. Albert, L.G. Sneddon, E.W. Plummer: Surf. Sci. **147,** 127 (1984)
4.78 C. Brucker, T.N. Rhodin: J. Catalysis **47,** 214 (1977)
4.79 W. Erley, A.M. Baro, H. Ibach: Surf. Sci. **120,** 273 (1982)
4.80 U. Seip, M.C. Tsai, J. Küppers, G. Ertl: Surf. Sci. **147,** 65 (1984)
4.81 J.E. Parmeter, M.M. Hills, W.H. Weinberg: J. Am. Chem. Soc. **108,** 3563 (1986)
4.82 P. Jakob, A. Cassuto, D. Menzel: Surf. Sci. **187,** 407 (1987)
4.83 C. Backx, B. Feuerbacher. F. Fitton, R.F. Willis: Surf. Sci. **63,** 193 (1977)
4.84 C. Backx, R.F. Willis: Chem. Phys. Lett. **53,** 471 (1978)
4.85 J.C. Hamilton, N. Swanson, B.J. Waclawski, R.J. Celotta: J. Chem. Phys. **74,** 4156 (1981)
4.86 L. Wang, W.T. Tysoe: Surf. Sci. **230,** 74 (1990)
4.87 R.J. Simonson, J.R. Wang, S.T. Ceyer: J. Phys. Chem. **91,** 5681 (1987)
4.88 A.B. Anderson: J. Am. Chem. Soc. **100,** 1153 (1978)
4.89 D.B. Kang, A.B. Anderson: Surf. Sci. **155,** 639 (1985)
4.90 J. Silvestre, R. Hoffmann: Langmuir **1,** 621 (1985) and references therein
4.91 H. Sellers: J. Phys. Chem. **94,** 8329 (1990)
4.92 Tables of Interatomic Distances and Configurations in Molecules and Ions, Chem. Soc. Special Publ. 11 (1958)
4.93 G. Herzberg: *Molecular Spectra and Molecular Structure* (Van Nostrand, New York 1945), Vol. II.
4.94 Y. Iwashita, F. Tamura, A. Nakamura: Inorg. Chem. **8,** 1179 (1969)
4.95 M. Tachikawa, J.R. Shapley, C.G. Pierpont: J. Am. Chem. Soc. **97,** 7172 (1975)
4.96 J.E. Demuth: Chem. Phys. Letts. **45,** 12 (1977)
4.97 P. Kisluik: J. Phys. Chem. Solids **3,** 95 (1957)
4.98 J. Stohr, R. Jaeger: Phys. Rev. B **26,** 411 (1982)
4.99 J. Stohr, F. Sette, A.L. Johnson: Phys. Rev. Lett. **53,** 1684 (1984)
4.100 A.J. Deeming, M. Underhill: J. Chem. Soc. Chem. Commun., **277,** (1973); J. Chem. Soc. Dalton Trans, 1415 (1974)
4.101 P. Skinner, M.W. Howard, I.A. Oxton, S.F.A. Kettle, D.B. Powell, N. Sheppard: J. Chem. Soc. Faraday Trans. II **77,** 1203 (1981)
4.102 L.L. Kesmodel, L.H. Dubois, G.A. Somorjai: J. Chem. Phys. **70,** 2180 (1979)
4.103 H. Ibach, D.L. Mills: *Electron Energy Loss Spectroscopy and Surface Vibrations* (Academic, New York 1982) p. 326

4.104 P.C. Stair, G.A. Somorjai: J. Chem. Phys. **66,** 2036 (1977)
4.105 J.E. Demuth: Surf. Sci. **69,** 365 (1977)
4.106 J.C. Bertolini, J. Massardier, G. Dalmai-Imelik: J. Chem. Soc. Faraday Trans. I **74,** 1720 (1978)
4.107 C.E. Megiris, P. Berlowitz, J.B. Butt, H.H. Kung: Surf. Sci. **159,** 184 (1985)
4.108 S. Hawker: unpublished results
4.109 L.L. Kesmodel: J. Chem. Phys. **79,** 4646 (1983)
4.110 J.A. Stroscio, S.R. Bare, W. Ho: Surf. Sci. **148,** 499 (1984)
4.111 F. Zaera, R.B. Hall: J. Phys. Chem. **91,** 4318 (1987)
4.112 W.T. Tysoe, G.L. Nyberg, R.M. Lambert: J. Phys. Chem. **90,** 3188 (1986)
4.113 J.E. Germain: *Catalytic Conversion of Hydrocarbons* (Academic, London 1969) Chap. 3
4.114 W.T. Tysoe, G.L. Nyberg, R.M. Lambert: J. Phys. Chem. **88,** 1960 (1984)
4.115 L.P. Wang, W.T. Tysoe, R.M. Ormerod, R.M. Lambert, H. Hoffmann, F. Zaera: J. Phys. Chem. **94,** 4236 (1990)
4.116 P. Berlowitz, C.E. Megiris, J.B. Butt, H.H. Kung: Langmuir 1 206 (1985)
4.117 J.E. Demuth: Surf. Sci. **76,** L603 (1978)
4.118 E.M. Stuve, R.J. Madix: J. Phys. Chem. **89,** 105 (1985)
4.119 M. Nishijima, J. Yoshindou, T. Sekitani, M. Onchi: J. Chem. Phys. **90,** 5114 (1989)
4.120 L.L. Kesmodel, L.H. Dubois, G.A. Somorjai: Chem. Phys. Lett. **56,** 267 (1978)
4.121 Y.T. Wong, R. Hoffmann: J. Chem. Soc. Faraday Trans. **86,** 4083 (1990)
4.122 T.M. Gentle, V.H. Grassian, D.G. Klarup, E.L. Muetterties: J. Am. Chem. Soc. **105,** 6766 (1983)
4.123 H. Conrad, G. Ertl. J. Kuppers, E.E. Latta: Surf. Sci. **65,** 245 (1977)
4.124 C.H. Patterson, R.M. Lambert: Surf. Sci. **187,** 339 (1987)
4.125 J.T. Roberts, A.J. Capote, R.J. Madix: J. Am. Chem. Soc. **113,** 9848 (1991)

5. Model Organic Rearrangements on Aluminum Surfaces

L.H. Dubois, *B.E. Bent*, and *R.G. Nuzzo*

To what extent are the reactivity patterns of discrete organometallic complexes predictive of reaction pathways on surfaces? This question is a central focus of much current research, and the general understanding which is beginning to emerge suggests that a close similarity may often exist. The best established correlations in these so-called cluster-surface analogies, are ones which are structural in nature [5.1]. Considerable advances have been made in recent years and it is now clear that the bonding patterns of hydrocarbon moieties on surfaces find many analogies in corresponding discrete transition and main group organometallic complexes. This progress notwithstanding, structural determinations of these often transient surface species remain far from routine, require the application of multiple techniques, and frequently are beset by controversy.

It is thus not unexpected that even less is known about the surface reaction mechanisms of all but the simplest transformations. The approach taken by us and others to explore the mechanisms of complex chemical processes mediated by a surface is to examine the structure-reactivity profiles exhibited in closely related model systems. This chapter summarizes studies which explore such correlations in both the formation and the thermal decomposition of metal-alkyl intermediates on aluminum surfaces [5.2, 3]. Our interest in these intermediates derives from their intermediacy in the growth of aluminum thin films by MetallOrganic Chemical Vapor Deposition (MOCVD) [5.4, 5].

5.1 Background

Alkyl groups bound to metal surfaces are normally transient intermediates in surface reactions and thus little is known about the mechanisms of their reactive chemistry. Several approaches for isolating stable alkyl fragments on metal surfaces have been demonstrated recently, however. These include (1) exposing the surface to a flux of alkyl radicals [5.6], (2) adsorbing a labile precursor which upon thermal, photochemical, or electron-beam treatment dissociates to form the alkyl of interest [5.7, 8], (3) impinging a supersonic molecular beam of alkanes with sufficient kinetic energy to form the alkyl and adsorbed hydrogen upon collision with the surface [5.9], and (4) adsorbing metal alkyl compounds

Springer Series in Surface Sciences, Vol. 34
Surface Reactions Ed.: R.J. Madix

onto a surface of the same metal [5.4, 10, 11]. A recent elegant study has also demonstrated the possibility of trapping alkyl groups formed by the insertion of an olefin into a metal-hydrogen bond [5.12]. For the case of alkyls on aluminum, we have compared two of these approaches, the first being the thermal dissociation of alkyl halides and the second being the activated dissociative adsorption of trialkyl aluminum compounds [5.2–5]. These approaches are shown schematically below:

$$\mathrm{Al(iBu)_3} \xrightarrow{\mathrm{Al}} \mathrm{Al(iBu)_3/Al} \xrightarrow{T > 450\ \mathrm{K}} \text{isobutylene} + H_2\,, \qquad (5.1)$$

$$\mathrm{iBu\text{-}I} \xrightarrow{\mathrm{Al}} \mathrm{I + iBu/Al} \xrightarrow{T > 450\ \mathrm{K}} \text{isobutylene} + H_2\,. \qquad (5.2)$$

The advantage of the alkyl halide approach, if the alkyl can be formed, is the ease of handling and synthesizing isotopically labeled precursors. The disadvantage is the presence of the halogen atom which remains coadsorbed on the surface throughout the temperature range where the alkyl group reacts [5.2]. With aluminum alkyls, a pure alkyl monolayer can be formed, but these pyrophoric, low-vapor pressure compounds are more difficult to handle, and the added aluminum adatom could, in principle, effect the chemistry as well. We discuss here first the alkyl halide approach, returning to aluminum alkyls and a comparison of these methods in a later section.

By way of background, it is useful to consider the range of mechanistically distinct processes which might take place if the analogy to organometallic chemistry in solution were to hold. For discrete transition [5.13] and main group metal [5.14] compounds in solution, the reaction pathways of alkyl ligands have been extensively investigated. Figure 5.1 shows examples of the more important pathways, as they might be extrapolated to a surface alkyl intermediate. The nomenclature used in this figure, which is derived from organometallic chemistry [5.13–15], will be used throughout this chapter.

The focus here is on the thermal chemistry of the isobutyl groups as a representative example. The reactive chemistries of concern in Fig. 5.1 are dominated by transformations of both carbon–hydrogen and carbon–carbon bonds. These processes are frequently coupled to metal–carbon bond rupture. These pathways might seem peculiar if one considers that some of the intermediates we discuss below are derived from the dissociative chemisorption of the corresponding alkyl halide. Indeed, one might expect that the chemistry of the organic groups would be significantly perturbed by the coadsorbed halogen atom. For example, it is well-known that the reaction of aluminum with alkyl

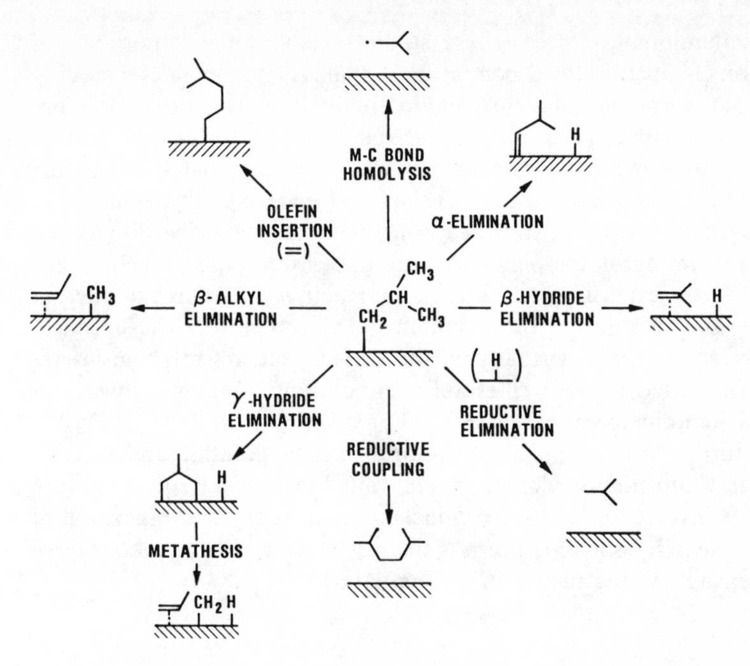

Fig. 5.1. Potential thermal decomposition pathways for an alkyl group (isobutyl) bound to a metal surface. The pathways shown are based upon the established chemistry of discrete organometallic complexes in solution

halides provides a powerful and industrially important process for the synthesis of aluminum sesquihalides in bulk quantities [5.16].

$$3RX + 2Al \rightarrow R_2AlX + RAlX_2. \tag{5.3}$$

This aluminum etching chemistry is quite facile in solution; under UltraHigh Vacuum (UHV) conditions, however, and for reasons not fully understood, alkyl decomposition predominates. The halide species is lost as an inorganic metal halide product (AlX) at a significantly higher temperature than that where hydrocarbon decomposition and product evolution are noted [5.2]. For example.

$$C_3H_7I + Al \rightarrow C_3H_6 + \tfrac{1}{2}H_2 + AlI. \tag{5.4}$$

It is quite fortuitous that the interactions of the hydrocarbon decomposition products with the aluminum surface are weak [5.2–4]. It is therefore possible to monitor the desorbing flux of species by mass spectrometry and directly determine the *kinetics of the underlying surface chemical process* by Temperature-Programmed Reaction Spectrometry (TPRS) (for example, the evolution of olefin derived from a rate determining β-hydride elimination [5.2, 4, 5]).

We have utilized this alkyl-halide adsorption chemistry to prepare and study the thermal chemistry of a wide range of surface hydrocarbon species, including aliphatic [5.2], alkenyl [5.3], and metallacycle [5.3] moieties, on single-crystal aluminum surfaces. These studies establish the importance of several reaction channels in the decomposition of alkyl intermediates including most notably, the formation of olefins and hydrogen by β-hydride elimination. Of particular significance in our observation that C–H and C–C bond-forming reactions can also occur on aluminum surfaces with adsorbates of suitable structure. In addition, we find evidence for the occurrence of β-alkyl elimination reactions as well as homolytic metal-carbon bond cleavage. The discussions which follow present our developing understanding of this important chemistry in the context of the extensive literature on the reactivity patterns exhibited in the gas phase and in solution by aluminum alkyl complexes. The chemical similarities are striking, and virtually all the surface-reaction mechanisms we observe can be related to the well-established solution reaction pathways of mononuclear aluminum compounds (Fig. 5.1).

We now turn to a systematic description of the preparation and reactive chemistry of alkyl moieties on aluminum. Our initial focus will be on the nature of the reactions involved in and the products formed by the decomposition of alkyl halides. The first issue of concern is the mechanism and the energetics of C–X bond cleavage on the surface of this material.

5.2 Carbon-Halogen Bond Cleavage

5.2.1 Reactive Sticking Probability

It is well known that the structure of an alkyl group can exert an enormous influence on the rates and product yields of preparative reactions involving carbon–halogen bond cleavage in an alkyl halide by an active metal (for example, as in the Grignard reaction [5.7]). Similar structural effects are also obtained in the rates evidenced in UHV where the reactive sticking probability (S_r) is found to be sensitive to the structure of the adsorbate. The most dramatic effect observable is that due to the length of the alkyl chain with the most significant changes in reactive sticking probability occurring between C_1 and C_3 [5.2]. Methyliodide, as has been reported previously [5.18] and confirmed by our studies [5.2], has an extremely small reactive sticking probability of $< 10^{-4}$ at 300 K on aluminum surfaces. Ethyliodide is more reactive, but even with doses of > 100 l we were not able to produce a significant concentration of adsorbed alkyl species. The apparent lack of reactivity for this molecule, as evidenced by the absence of products in temperature-programmed reaction spectrometry, does not reflect the selective loss of the ethyl fragment from the surface during a dissociative adsorption since very little iodine was found to be present by Auger Electron Spectroscopy (AES) subsequent to this prolonged gas exposure. When the chain length reaches C_3 (or longer) we are able to achieve orders of magnitude larger surface coverages with doses of only 5–10 l (see below).

These observations establish in a qualitative way that a simple extension of the alkyl chain length, for example from methyl to *n*-butyl, is sufficient to change the reactivity of a homologous series of *n*-alkyl iodides towards Al(1 0 0) by approximately 3–4 orders of magnitude. Thus $S_r(C_1) \lesssim 10^{-4} \simeq S_r(C_2) \ll S_r(C_3) \lesssim S_r(C_4) \simeq S_r(C_6) \gtrsim 0.1$. This is, needless to say, a striking change in reactivity, given that the C–I bond strength changes by only ≈ 1.2 kcal/mol across this series and therefore is *not* the dominant contribution to this effect. We have found as well that the differences in reactivity observed in such experiments are relatively insensitive to surface temperature over a rather wide range (≈ 130–400 K). Studies using either propyl or *n*-butyliodide show that dosing a multilayer (≈ 15 l) at the lower surface temperature followed by warming to room temperature results in the deposition of only slightly smaller quantities of dissociatively adsorbed material as does directly dosing at room temperature.

In solution, the measured reaction rates and/or product yields accompanying reactions of alkyl halides tend to sensitively reflect the energetics associated with reactive *intermediates*. Consider as an example, the above-mentioned formation of the Grignard reagent

$$\mathrm{RX} + \mathrm{Mg_{(s)}} \xrightarrow{\text{slow}} [\mathrm{R}\cdot + \text{“}\mathrm{Mg_{(s)}} - \mathrm{X}\text{”}] \longrightarrow \mathrm{RMgX}. \quad (5.5)$$

$$\mathbf{1} \qquad \mathbf{2} \qquad \mathbf{3}$$

This system has been studied in considerable detail and it is known unequivocally that the product forming pathway leading to the reagent **1** involves the intermediacy of a free radical trapped in a hydrodynamic boundary layer [5.17]. The mechanism is less well defined as to the nature of the metal–halogen species which may be present or the manner in which the species indicated by **2** are removed from the lattice as product. The rate of product formation (in the absence of transport limitations) and/or its yield can vary sensitively with the choice of alkyl halide. Alkyl substituents influencing the stability of **1** are very important in this regard and provide one piece of *prima facia* evidence supporting the intermediacy of such radical species. In UHV the point of concern is somewhat different. These particular rate effects – presented above as a chain-length dependence of the reactive sticking probability – can be explained by a kinetic model which takes into account the differing heats of adsorption of the molecular precursors.

Figure 5.2 shows a schematic, one-dimensional potential–energy diagram which can be used to help visualize the effect of molecular weight on S_r. The very low value of S_r for methyliodide suggests that the dissociative adsorption is activated by at least 4–5 kcal/mol. The heat of adsorption of this molecule is also low, of the order of ≈ 10 kcal/mol [5.18]. This number provides a reference point for the potentials indicated in the figure. It is also clear that the attractive potentials for the molecular precursors must differ with chain length. The addition of each methylene unit increases the number of attractive dispersive interactions occurring between the adsorbate and surface. We can estimate the

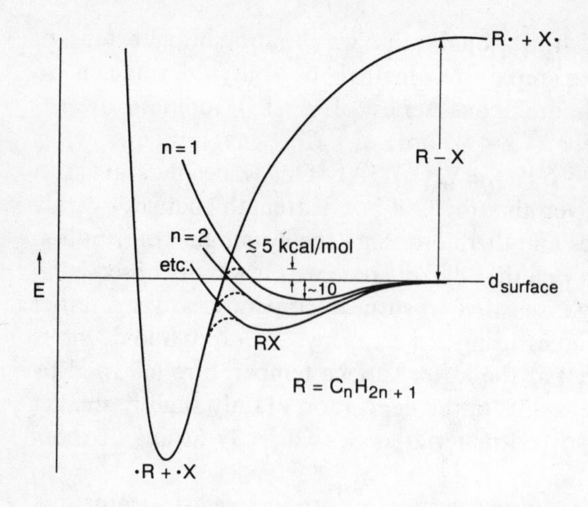

Fig. 5.2. One-dimensional potential energy diagram for the adsorption of an alkyl halide, RX ($R = C_nH_{2n+1}$), on a surface. The shallow well represents the weakly bonded intact molecule, while the deep well is for the dissociated species. As the chain length increases, the depth of the physisorption well increases until there is no barrier to dissociation

magnitude of this effect as being *at least* as large as the group constituent heat of vaporization for *n*-alkanes, ≈ 0.7 kcal/mol per CH_2 group. Indeed, *Gellman* and coworkers have found that the molecular heats of adsorption of C_1–C_5 *n*-alcohols on silver surfaces increase by a somewhat larger value, $\approx 1.15 \pm 0.1$ kcal/mol per CH_2 group [5.19]. This effect can easily explain the branching seen between molecular desorption and dissociative adsorption for C_1–C_4 *n*-alkyl iodides on aluminum.

The key structure–reactivity correlation, as embodied schematically in Fig. 5.2, is that the increased heat of adsorption seen in the physisorbed precursor state is also manifested in the transition state for C–X bond cleavage: i.e., the transition state is stabilized by attractive dispersion forces to a degree comparable to that reflected by the increased heats of adsorption of the molecular precursors (both the physisorption well *and* the transition state energies are lowered comparably). The subtle but important point that emerges from this picture is that it is not the *barrier* to dissociation which is decreasing with increasing chain length, rather it is the branching between dissociation and molecular desorption that is changing. In this view of the mechanism, the activation barrier estimated for methyliodide ($\approx$4–5 kcal/mol [5.2]) should disappear after the addition of ≈ 3 methylene groups to the chain. This assumes a 0.8 kcal/mol reduction in the C–I homolytic bond strength upon substitution and an increased heat of adsorption of 1.2 kcal/mol per CH_2 group.

We have demonstrated that both the size of the halogen and the strength of the C–X bond are relevant to understanding S_r. The reactive sticking probabilities of alkyl iodides, for example, tend to be slightly higher than those of the structurally analogous bromides ($\approx 2\times$ for Bu–I vs. Bu–Br) despite a large difference in the depth of the physisorption well [5.18]. It is most striking that the size of this effect is not as large as the ≈ 16 kcal/mol difference in bond strengths suggest it should be. We feel this points to a very clear mechanistic conclusion: the process by which the C–X bond is cleaved *must* involve significant metal–halogen bonding in the transition state, bonding which must compensate the differing bond energies of the C–X groups.

5.2.2 High-Resolution EELS and TPRS Observations of C–X Bond Cleavage

The identification of adsorbed alkyl intermediates derived from the reactions discussed above follows from several lines of evidence, both spectroscopic and chemical. By far the most useful spectroscopic method has proven to be high-resolution Electron Energy Loss Spectroscopy (EELS). In the case of the more reactive alkyl iodides, we have been able to monitor the surface-mediated dissociation of the carbon-halogen bond directly using this technique [5.2]. A suitable example is provided by the high-resolution electron energy loss spectra shown in Fig. 5.3 for *n*-propyliodide adsorbed on Al(1 0 0). Temperature-programmed studies presented later imply that a similar chemistry is operative for the larger alkyls whose vibrational spectra are substantially more complex.

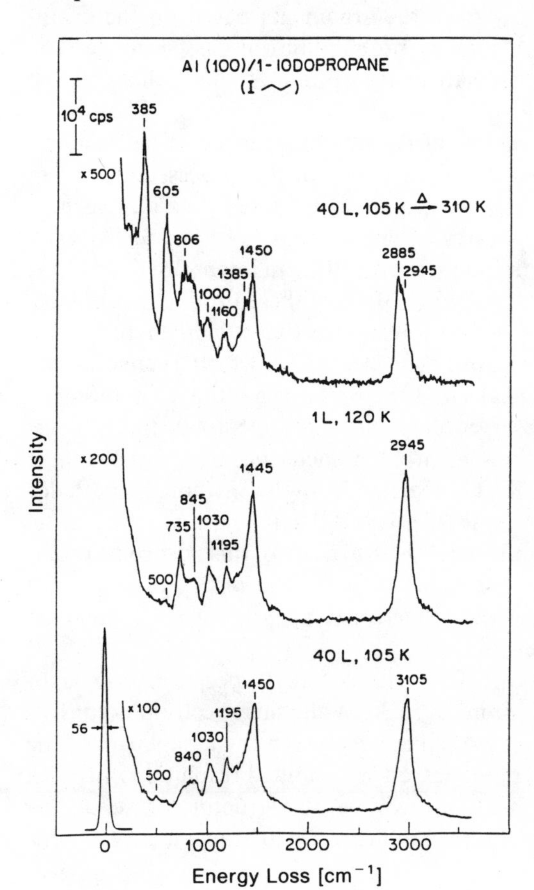

Fig. 5.3. Specular high resolution electron energy loss spectra for 1-iodopropane adsorbed on an Al(1 0 0) surface: (A) 40 l at 105 K to form a multilayer, (B) 1 l at 120 K, and (C) 40 l adsorbed at 105 K followed by briefly annealing to 310 K to induce carbon–iodine bond-breaking. All spectra were recorded at 105 K. Peak assignments are summarized in Table 5.1 and [5.2]

The spectrum depicted in the lower portion of Fig. 5.3 corresponds to a multiplayer (40 l exposure) of *n*-propyliodide condensed on an Al(1 0 0) surface at 105 K. The loss features seen can *all* be attributed to modes of the *molecular* species. The same is also true for spectrum (B) which corresponds to a monolayer (1 l) exposure at 120 K. Not only is the C–I stretch detected in this spectrum (weak peak at 500 cm^{-1}), but also there are no features seen which can be assigned to either Al–I or Al–C stretching vibrations. On warming, dissociation occurs (dissociative adsorption is facile above $\approx$125 K [5.2]), yielding species whose spectra can be ascribed as those of surface bound iodide and propyl fragments. A representative spectrum is exhibited Fig. 5.3c. The features seen can be assigned as modes corresponding to those expected for a propyl fragment. Specific mode assignments are summarized in Table 5.1 and are discussed in more detail elsewhere [5.2]. The two most diagnostic features, located at $\approx$385 and 605 cm^{-1}, are due to the Al–I and Al–C stretching vibrations, respectively [5.2, 3]. Based on the temperature dependence of the appearance of these two modes, we estimate the barrier for C–I bond cleavage to be $\approx$7–8 kcal/mol [5.2]. This is a fairly low value, one which is well within the range predicted to follow plausibly from the arguments based on the chain-length dependence of the reactive sticking probabilities discussed above. The data thus establish the facile dissociative adsorption of this halide on the Al(1 0 0) surface.

Additional insight into the nature of carbon–halogen bond cleavage on aluminum as well as confirmation that the barriers to this process are low are obtained by studies of the reactivity of vicinally substituted dihalides such as 1,2-dibromopropane. At low exposures of either an Al(1 1 1) or an Al(1 0 0) surface to this molecule, two features are seen in TPRS: an intense peak at 135 K and a very much weaker feature at $\approx$235 K. Both of these product peaks have been identified as propylene; we note for reference that adsorbed propylene desorbs molecularly from clean Al(1 0 0) at $\approx$110 K (see below). Increasing the exposure further saturates these peaks and the onset of multilayer (molecular) desorption of the dibromide is observed. It would thus appear that the cleavage of both carbon-halogen bonds in this adsorbate is facile, occurring for the most part at temperatures below 135 K. The barrier is low even though the halogen cleaved is a bromide and not an iodide. It is difficult to assess whether the 1,2-dihalide subsitution pattern influences the reactivity of this adsorbate or if it is indeed representative of all disubstituted halides. The results of this study provide an additional example of the reductive cleavage of a C–X bond on aluminum by a facile, low-barrier process.

We return to a consideration of the stable alkyl species whose high-resolution EELS spectrum is shown in Fig. 5.3c. While this spectrum is consistent with the formation of a surface propyl moiety, given the complexity of the adsorbate, the attendant number of modes, and the limited resolution of EELS, we cannot independently establish either the molecular structure or the orientation of this species definitively [5.2]. Confirmation of the identity of this

Table 5.1. Comparison of the vibrational frequencies for 1-iodopropane in solution and adsorbed on an Al(1 0 0) surface.[a]

Al(1 0 0)		Liquid		
1.0 l at 120 K[b]	40 l at 105 K, warm to 310 K[c]	trans conformation[d]	*gauche* conformation[d]	Approximate mode assignments[d]
2945	2945	2967	2967	CH_3 asym stretch, CH_2 sym stretch
	2885	2876	2876	CH_3 sym stretch
1445	1450	1458	1458	CH_3 asym bend
		1435	1435	CH_2 bend
~ 1350, sh[e]	1385	1380	1382	CH_3 sym bend
~ 1300, sh	~ 1300, sh	1325	1340	CH_2 wag[f]
		1278	1276	CH_2 twist
1195	1160	1185	1194, 1148	CH_2 wag, twist
1030	1000	1020	1020	C–C stretch, CH_3 rock
		1012	1012	CH_2 twist
845, br[e]	806, br	819	816	CH_3, CH_2 rock
735		728	764	CH_2 rock
500		594	505	C–I stretch

[a] All frequencies in cm^{-1}.
[b] Observed frequencies from Fig. 5.3 for a submonolayer coverage of molecularly adsorbed 1-iodopropane.
[c] Observed frequencies from Fig. 5.3 for a monolayer of chemisorbed propyl groups.
[d] Frequencies and approximate mode assignments from [5.20].
[e] sh: shoulder; br: broad.
[f] All modes below $\approx 1350\,cm^{-1}$ are highly coupled, see [5.20].

intermediate can be obtained, however, from other lines of evidence, ones largely based on the reactivity profiles of these and related materials.

One simple example of such a profile comes from temperature-programmed reaction studies. If a monolayer of dissociatively adsorbed alkyl halide is deposited on either a (1 0 0) or (1 1 1) surface of aluminum and the temperature is ramped, product evolution is noted. As shown for the specific example of propyliodide adsorbed on Al(1 0 0) (Fig. 5.4), ions diagnostic of the evolution of hydrocarbon (m/e = 41) *and* H_2 (m/e = 2) are detected coincidently (at a temperature far above where adsorbed propylene is seen to desorb, Fig. 5.4b). The figure shows as well a representative example of the desorption profile of H_2

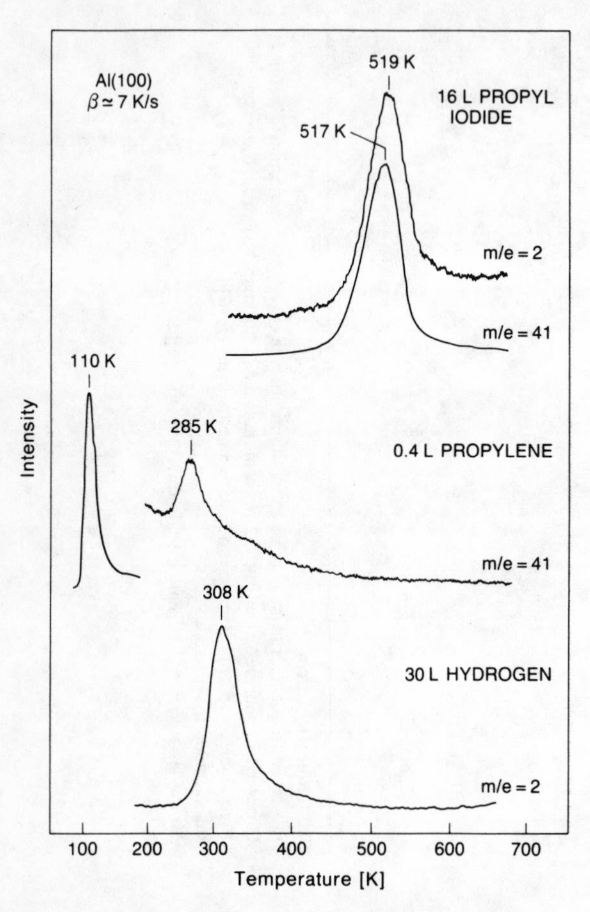

Fig. 5.4. Temperature programmed reaction traces of (A) 161 of propyliodide adsorbed on an Al(1 0 0) surface at 300 K. The data shown represents both propylene (*m/e* = 41) and hydrogen (*m/e* = 2) desorption, the expected products form a surface β-hydride elimination reaction. The desorption traces are a measure of the surface reaction kinetics since the desorption of either propylene (0.4 l, B) or hydrogen (adsorbed as hydrogen atoms, C) occurs at significantly lower temperatures

(Fig. 5.4c), adsorbed independently as hydrogen atoms [5.21]. It is clear that the hydrogen evolution seen with propyliodide occurs at a much higher temperature, strongly suggesting that its formation is kinetically coupled to a more highly activated decomposition process involving the alkyl moiety. We also note that H_2 is not evolved at lower temperatures and that the amount of H_2 desorbing in this peak is far greater than the amount observed from the cracking of the product olefin by the mass spectrometer. This reasoning implicity establishes the existence of an alkyl moiety whose C–H bonding framework has not been changed by reactions with the surface at temperatures below at least 450 K [5.4].

This latter deduction has been proven explicitly using reflection-adsorption infrared spectroscopy [5.22]. Figure 5.5. shows a series of infrared spectra recorded after exposing a clean Al(1 0 0) surface to a saturation dose ($\approx$60 l) of triisobutylaluminum (TIBA) at 345 K and flashing the sample to the indicated temperatures. Peak assignments are summarized in Table 5.2. The observed

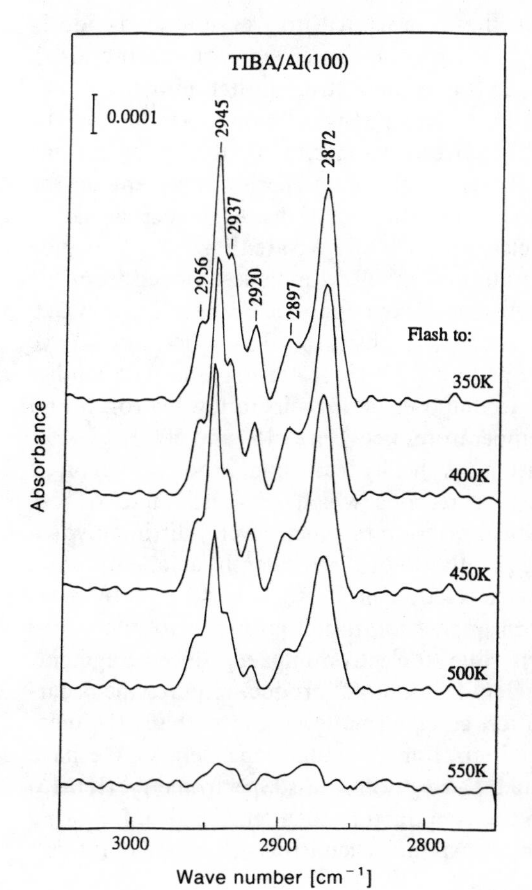

Fig. 5.5. Reflection–absorption infrared spectra of isobutyl groups derived form the dissociative adsorption of triisobutylaluminum on Al(1 0 0). The clean aluminum surface was exposed to $\approx$ 60 l of TIBA at 345 K and then cooled to 135 K (upper trace). Mode assignments are summarized in Table 5.2 and [5.22]. All subsequent spectra were recorded at 135 K after flashing the sample to the indicated temperatures. Desorption begins to occur above $\approx$ 450 K

Table 5.2. Vibrational mode assignments for TIBA adsorbed on Al(100).[a]

TIBA/Al(100)[b]	1-iodo-2-methylpropane$^{c}_{(g)}$	Assignment[c]
2956	2959, 2955	CH_3 asym stretch
2945	2943	CH_3 asym stretch
2937	2934[d]	
2920	2922	CH_2 sym stretch
2897	2904[e]	CH stretch
2872	2876	CH_3 sym stretch

[a] All frequencies in cm^{-1}.
[b] 60 l of TIBA adsorbed at 345 K, spectra recorded at 135 K.
[c] From [5.23].
[d] Solid phase, no mode assignment given; band at 2942 cm^{-1} for matrix isolated species assigned to CH_2 sym stretch [5.24].
[e] Solid phase, peak at 2896 cm^{-1} in Raman spectrum [5.23].

frequencies are very similar to those measured for gas-phase 1-iodo-2-methylpropane (isobutyliodide) [5.23], in good agreement with the proposed, terminally bonded structure [5.4, 25]. It is most striking that *no* new features and/or shifts in peak frequencies or *relative* intensity changes are seen in the spectra even after flashing the TIBA covered surface to 500 K. This is but one argument [5.22] that only a *single* type of isobutyl species is present on the surface *at all temperatures*, further suggesting that the Al–C bonds of the incident TIBA molecule must be cleaved by 300 K. We are thus led to conclude that both the structure *and* orientation of the alkyl moieties derived from the dissociative chemisorption of TIBA remain fixed up to the temperature at which they decompose thermally. The spectral intensity begins to decrease above $\approx$450 K, in good agreement with previous TPRS, scattering, and AES studies which revealed that this species decomposes cleanly liberating hydrogen and isobutylene to the gas phase at temperatures between $\approx$450 and 500 K [5.4, 5].

The unambiguous identification of the hydrocarbon product(s) evolved requires more complex analyses for reasons which are easily understood. Consider, for example, a hypothetical reaction in which several distinct hydrocarbon products form and subsequently desorb coincidently as a result of a reaction occurring on the surface. Detection in TPRS is performed by mass spectrometry in which the electron impact ionization gives rise to many ions (some molecular, with the majority due to electron impact-induced fragmentation). Ideally, the unambiguous identification of a product requires the accurate interpretation of these ionization cracking patterns and perhaps the measurement of such factors as their ionization potential dependences. We have found the technique of Integrated Desorption Mass Spectrometry (IDMS) [5.26] to be a convenient method for conducting such studies. The following section describes this method using the specific chemistries which are of interest to this chapter as an example.

5.3 Integrated Desorption Mass Spectrometry

Integrated Desorption Mass Spectrometry (IDMS) is a straightforward, high signal-to-noise technique for obtaining a *complete* mass spectrum during a *single* TPRS and/or temperature-programmed desorption (TPD) experiment. In its simplest implementation, IDMS is a two step process. In the first step, a molecule is adsorbed on a surface and a conventional temperature-programmed experiment is performed following a single characteristic mass (i.e., the m/e = 2 cracking fragment for the desorption of a hydrocarbon). One then notes the beginning and ending temperatures (and/or times) of the desorption peak.

A conventional Quadrupole Mass Spectrometer (QMS) run in a pulse-counting mode and a Personal Computer (PC) based Multichannel Scalar (MCS) card are then used in the second step to sum a series of mass spectra while the molecules are desorbing from the surface (Fig. 5.6). The start and stop times

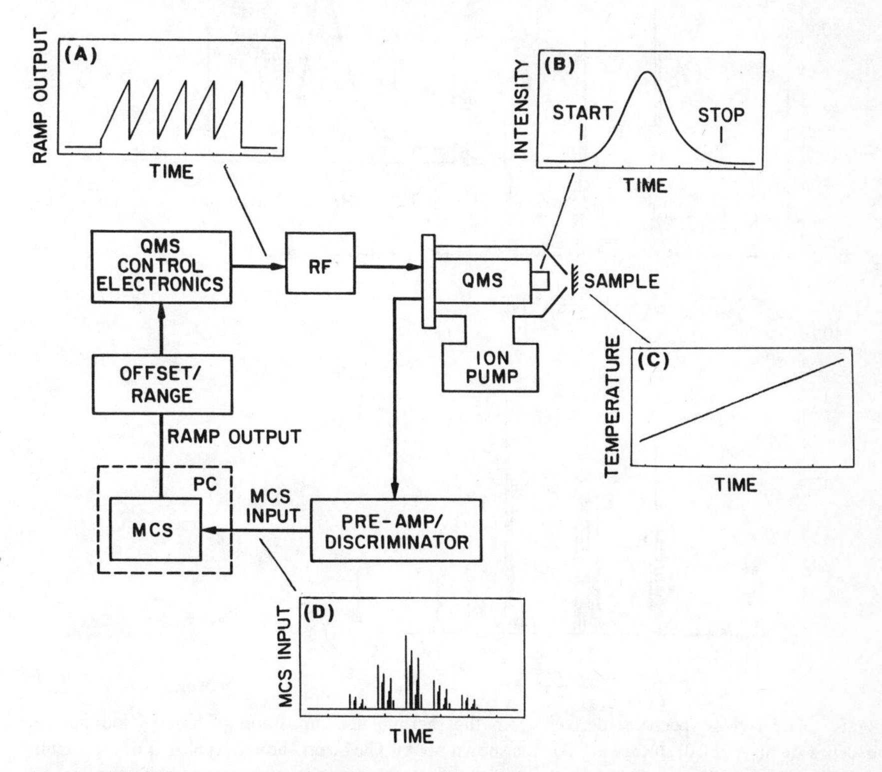

Fig. 5.6. The IDMS experimental set-up [5.26]. As the sample temperature is increased linearly with time (insert C), the desorbing molecules are ionized by the QMS (insert B). The mass spectrometer control electronics are still used to drive the RF box and to control the ionizer and lenses, but the output of the electron multiplier is sent to a preamplifier/discriminator and then to the PC based MCS card (insert D). The built-in ramp generator controls the mass axis of the QMS (insert A). A typical choice of start and stop times (temperature) are shown in insert B. In our case the quadrupole mass spectrometer is differentially pumped using a 40 l/s ion pump

are determined by the heating rate and the temperature measured in the previous TPD/TPRS experiment while the ramp generator built into the MCS card controls the mass scale of the QMS. With heating rates between 1 and 20 K/s, more than 100 mass spectra can be acquired and summed during the desorption process (we record *at least* 10 data points per amu ensuring that peak shapes, and therefore maximum intensities are accurately determined). The signal levels measured are typically between 10 and 10^4 counts with background levels less than 2 counts. The limiting dynamic range is $> 10^6$ [5.26].

Figure 5.7a illustrates an example of a typical IDMS spectrum for the species derived from the adsorption of 8 l of 1-iodohexane onto a clean Al(1 0 0) surface at 300 K. The sample temperature was ramped at $\approx$8 K/s; the data

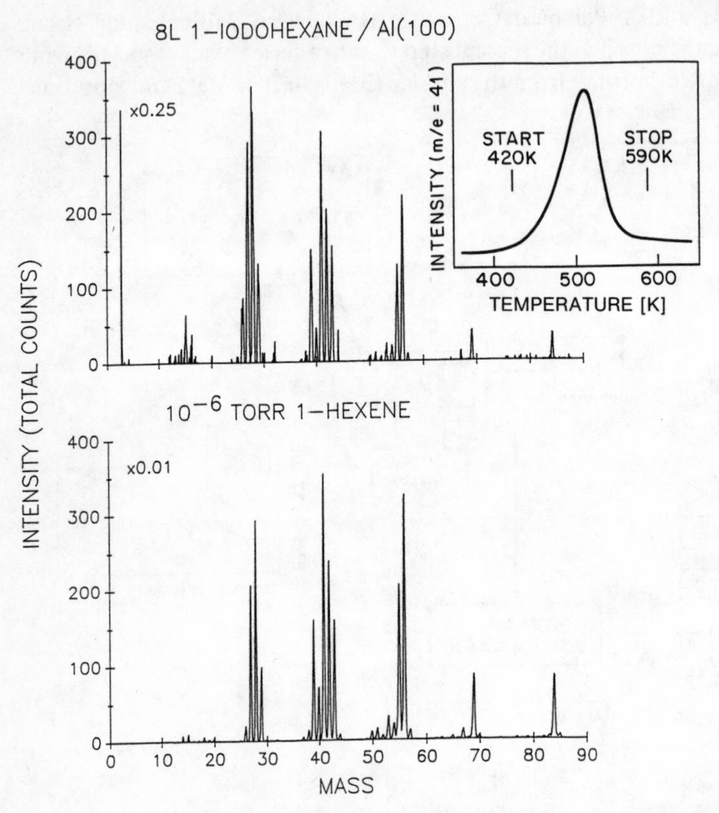

Fig. 5.7. The IDMS spectrum derived from the thermal decomposition of 8 l of 1-iodohexane adsorbed on an Al(1 0 0) surface at 300 K is shown in (A). The insert shows a typical TPD spectrum recorded at $m/e = 41$ amu from which an activation energy and pre-exponential factor can be derived. Typical start and stop temperatures are also shown. By comparison with the mass spectrum of 10^{-6} Torr of 1-hexene (B), we conclude that the desorbing product in (A) is also 1-hexene. Note that the excess intensity at $m/e = 2$ in (A) is due to the desorption of hydrogen, the other β-hydride elimination product of 1-iodohexane. Care must be taken in setting-up the mass spectrometer *identically* for these two types of experiments since the mass dependent transmission of the instrument is *very* sensitive to tuning [5.27]

acquisition was begun at 420 K and was stopped at 590 K (see insert). In this example 1024 channels of data from 0 to 88 amu were recorded (≈ 12 channels/amu). The dwell time was 100 μs/channel and 150 mass spectra was summed over a 15 s period. The signal-to-noise is excellent throughout the spectrum with the diagnostic peaks all containing between 100 and 200 counts and the background only 1–2 counts. Despite the relatively low signal level, an accurate cracking pattern is obtained. Due both to the number of important masses and to the low count rate, recording data of this kind using a more conventional multiplexed mass spectrometer system would be nearly impossible. Comparison of the data in Fig. 5.7a to standard mass spectra *recorded using identical mass spectrometer settings* reveals that the desorbing species is $> 95\%$ 1-hexane, the β-hydride elimination product of associatively adsorbed 1-iodohexane. The particularly diagnostic masses are 56, 69, and 84 amu, although *every* peak between 24 and 84 amu can be assigned to the cracking of 1-hexene [5.27].

It is also possible to use IDMS to measure the ionization energy dependence of the mass spectra, a procedure which allows more sophisticated product identification but generally is complicated by very low signal levels at low electron kinetic energies ($\leqslant 20$ eV) [5.26]. The utility of this latter type of measurement is illustrated in the sections which follow.

5.4 Alkyl Surface Chemistry

The reactions of aluminum with a large number of alkyl iodides and bromides are reported in this section. A brief overview of the general results obtained is given in Table 5.3. As suggested above, we find that in almost every case the alkyl halide adsorbs dissociatively on aluminum surfaces at temperatures between 130 and 450 K by carbon-halogen bond cleavage to give coadsorbed alkyl and halide species. The focus here, as was mentioned in Sect. 5.1., is on the thermal chemistry of the adsorbed alkyl group; the coadsorbed halogen in most instances does not significantly perturb the reaction pathways *under the specific conditions employed in these studies* [5.2]. In general, the adsorbed alkyl ligands formed by the dissociation of the C–X bond on the surface are thermally stable up to ≈ 450 K [5.22]. The products of the alkyl thermolysis which occurs above this temperature are volatile, and the reaction channels can be identified by monitoring the species evolved into the gas phase with a mass spectrometer (TPRS and IDMS). The product desorption rates are generally much faster than the rates of the surface reaction and, as such, can be used to determine the surface reaction kinetics. The dominant surface reaction for most adsorbed alkyl fragments is β-hydride elimination to evolve olefin and hydrogen. Other chemistries observed on a much more limited scale include the reductive elimination of carbon–hydrogen bonds (hydrogenation), intramolecular olefin insertion into Al–C bonds, β-alkyl elimination, and homolytic Al–C bond cleavage to release alkyl radicals (see below). We discuss each reactivity class in turn.

Table 5.3. Thermal decomposition of surface alkyl groups on Al(1 0 0) summarized

Surface Alkyl Group	Adsorbate	Major Hydrocarbon product[a]	Peak Temperature [K][b]	Mechanism	Reference
Methyl	Trimethylaluminum[c]	Methyl Radical,[d] Methane[e]	210	Bond homolysis, hydrogenation	5.10
	Iodomethane	—[f]	—	Dehydrogenation	5.2, 18
Methylene	1,2 diiodomethane[c]	Ethylene, methylene radical	110, 105	Coupling, bond homolysis	5.28
Ethyl	Ethyliodide	Ethylene/ethane	580 → 480[g]	β-Hydride elimination	5.2
	1,2-Diiodoethane	Ethylene	105, 325 → 270[g]	Olefin desorption	This work
	1,2-Dibromopropane	Propylene	135, 265 → 235[g]	Olefin desorption	5.3
Propyl	1-Iodopropane	Propylene	520	β-Hydride elimination	5.2
	1-Iodopropane-2,2-d_2	Propylene-d_1	540	β-Hydride elimination	5.2
C_3 Metallacycle/ allyl	1,3-Diiodopropane	Propylene	510	Formal C–H bond rearrangement	5.3
	1,3-Dibromopropane	Propylene	515	Formal C–H bond rearrangement	5.3
	1,3-Dibromopropane-d_6	Propylene-d_6	525	Formal C–H bond rearrangement	5.3
	3-Iodo-1-propene	Propylene[h]	505	Formal C–H bond rearrangement	5.3
	3-Bromo-1-propene	Propylene[h]	520	Formal C–H bond rearrangement	5.3
	1,3-Dibromobutane	Butene[i]	(465)[j]	Formal C–H bond rearrangement	5.3
n-butyl	1-Iodobutane	Butene[i]	535	β-Hydride elimination	5.2
	1-Iodobutane-d_9	Butene[i]	545	β-Hydride elimination	5.2
Isobutyl	1-iodo-2-methyl propane	Butene[i]	505[k]	β-Hydride elimination	5.4
	Triisobutylaluminum	Butene[i]	520[e]	β-Hydride elimination	5.4

C_4 metallacycle	1,4-Diiodobutane	1,3-Butadiene	490	β-Hydride elimination	5.3
	4-Bromo-1-butene	1,3-Butadiene[h]	495[m]	β-Hydride elimination	5.3
Neopentyl	1-Iodo-2,2-dimethyl propane	Neopentyl radical[h]	→ 645[g,m]	Bond homolysis	This work
n-hexyl	1-Iodohexane	1-hexene	520–535	β-Hydride elimination	5.2, 26
C_6 metallacycle	1,6-Diiodohexane	Methylene cyclopentane[h]	535	β-Hydride elimination and cyclization	This work
	6-Bromo-1-hexene	Methylene cyclopentane[h]	550	β-Hydride elimination and cyclization	This work
	Iodobenzene	Benzene[h]	525		This work

[a] Products identified by integrated desorption mass spectrometry over a temperature range of ≈ 150 K symmetric about the thermal desorption peak maximum. Both the mass spectral cracking patterns [5.29] and ionization cross-sections [5.30] for other potential products are well-known, and their presence can therefore be ruled out.
[b] Saturation doses of the indicated alkyl halide; typical temperature ramp = 7 K/s. When other ramp rates were used, peak temperatures were corrected for the differing heating rates.
[c] Polycrystalline aluminum substrate.
[d] Methyl radicals were detected by laser induced fluorescence at surface temperatures above 600 K in a scattering experiment.
[e] Methane production is enhanced in the presence of co-adsorbed hydrogen form the thermal decomposition of dimethylaluminum hydride [5.10].
[f] Molecular desorption form both Al(1 1 1] [5.18] and Al(1 0 0) [5.2]; reactive sticking probability $< 10^{-4}$ on both surfaces.
[g] Peak shift with increasing coverage.
[h] Lesser amounts of other products are also produced.
[i] Isomer not readily determined by mass spectrometry.
[j] Temperature believed to be in considerable error due to a misplacement of the thermocouple during these experiments.
[k] Peak shifts to 490 K on Al(1 1 1) [5.2, 4,5].
[l] peak shifts to 505 K on Al(1 1 1) [5.4, 5].
[m] Unresolved multipeak desorption.

5.4.1 Iodoalkanes with β-Hydrogens

Alkyl Decomposition Products. Having established that carbon–iodine bond scission is quite facile on Al(1 0 0), at least for adsorbates of the form $CH_3(CH_2)_nI$ and $n \geqslant 2$, we now concentrate on the subsequent thermal chemistry of alkyl groups containing β-hydrogens. The *major* gas phase products and the temperatures at which they are formed are summarized in Table 5.3. It is evident from an inspection of this data that the decomposition of all surface alkyl groups with β-hydrogens involves predominantly the loss of one hydrogen to evolve alkene. This class of reaction is one with considerable precedent in the literature for organometallic compounds in either the gas phase [5.31] or solution [5.32]. A schematic description of the mechanism is shown in (5.6).

$$\text{M–CH}_2\text{–CH}_2\text{R} \underset{k_1}{\overset{k_{-1}}{\rightleftharpoons}} \text{M(H)}\cdots(\text{CH}_2\text{=CHR}) \underset{k_2}{\overset{k_{-2}}{\rightleftharpoons}} \text{M–H} + \text{CH}_2\text{=CHR}, \quad (5.6)$$

$$2\text{M-H} \rightarrow H_2 + 2\text{M}$$

Decomposition of the related surface species proceeds cleanly in UHV; after heating the surface to > 700 K, neither carbon nor iodine are detected by Auger Electron Spectroscopy (AES) which did show that the iodine atom remained on the surface *throughout the temperature range of the alkyl decomposition* (450–550 K), however [5.2]. The desorption of AlI (m/e = 154) was observed at the end of the TPD ramp, generally at temperatures above ≈ 650 K [5.2]. The only other desorption product detected is H_2, which desorbs at the same temperature as the alkene (Fig. 5.4 and [5.2–4]).

The formation of olefin as a product of the thermolysis of the alkyl group is expected to involve the abstraction of a β-hydrogen atom. To test explicitly for the relevance of this pathway to reactions occurring on an aluminum surface, we studied the decomposition of isotopically labeled adsorbates. An example is provided by a comparison of the thermolyses of 1-idodopropane and 1-iodopropane-2,2-d_2 on Al(1 0 0). Figure 5.8 compares the temperature programmed desorption and integrated desorption mass spectra obtained for saturation coverages of these two related adsorbates. Analysis of the IDMS data (taken with a 15 eV ionizer energy to accentuate the molecular ion) indicates that the iodopropane product has a molecular ion with m/e = 42; while the cracking pattern from a similar IDMS spectrum recorded with an ionizer energy of 70 eV identifies the product as propylene (as opposed to cylcopropane [5.2]). Since the IDMS data for the labeled product is virtually identical above 30 amu, but shifted by *one* amu, we conclude that the dideuteroiodopropane product is *exclusively* propylene-d_1. This product requires loss of one deuterium from the β-carbon in the starting material, as shown in (5.6), and provides conclusive evidence for the involvement of a β-hydride elimination pathway.

It is also evident in the data that the product peak maximum shifts ≈ 2 K to higher temperature when the deuterated isomer is used, indicating that the

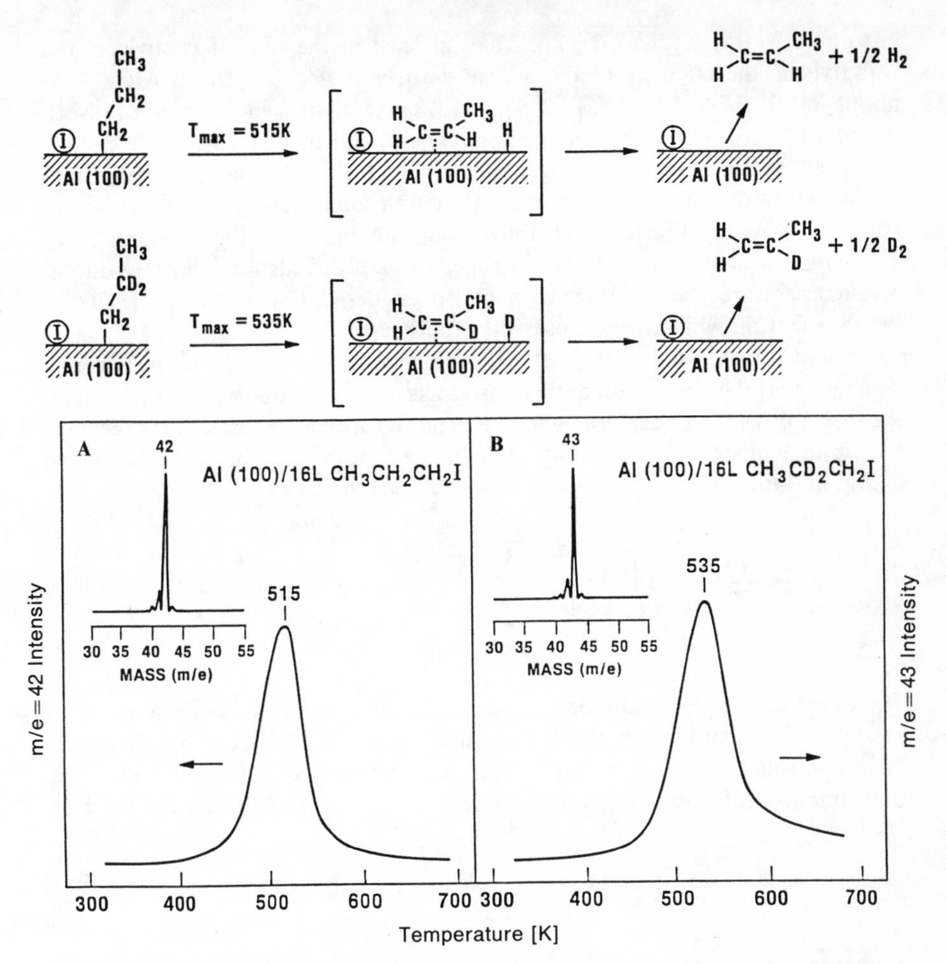

Fig. 5.8. TPD and IDMS data showing propylene desorption derived form the adsorption and thermal decomposition of propyliodide on an Al(100) surface. In (A), the adsorbate used was 1-iodopropane while in (B) 1-iodopropane-2,2-d_2 was studied. An exposure of 16 l was used in each with a heating rate of 10 K/s. The schemes at top of the figure show that the hydrocarbon products resulting form a single β-hydride elimination on the surface should be exclusively propylene and propylene-d_1 (process *b*)

desorption process shows a pronounced kinetic isotope effect: the perhydro compound decomposes at a faster rate (as judged by the lower peak temperature) than the perdeutero compound. These results suggest a value that is in the lower end of the range expected for classical kinetic isotope effects in rate-determining C–H (C–D) bond-breaking processes such as occur in β-hydride elimination reactions (see below) [5.33].

The question arises as to whether desorption of hydrogen and/or the alkene product is the rate-determining process in the thermal desorption experiments. We have addressed this concern by adsorbing hydrogen and alkene separately onto an Al(1 0 0) surface. We find that, at least in the case of propylene and isobutylene, the majority of the alkene desorbs molecularly from Al(1 0 0) at about 110 K (Fig. 5.4 and [5.2, 4]). Molecular hydrogen does not adsorb dissociatively on aluminum surfaces, but experiments in which atomic hydrogen was evaporated from a hot tungsten filament onto an aluminum surface at 100 K showed that hydrogen recombination and desorption from either Al(1 0 0) or Al(1 1 1) occurs just above room temperature (Fig. 5.4) [5.21, 34]. These results suggest that any dehydrogenation of the adsorbed alkyl group at low temperature should produce a hydrogen desorption peak in the 300 K range, a feature which is *not* observed experimentally.

Based on the facility of both H_2 and olefin desorption, one might also conclude that the desorption of these products is *not* the rate-determining step in the thermolysis. To understand why this is not rigorously required it is necessary to consider again the elementary steps which are or may be involved in the β-elimination process. A summary of these is given in (5.7)

$$\underset{\mathbf{A}}{\text{R}} \; \underset{k_1}{\overset{k_{-1}}{\rightleftharpoons}} \; \underset{\mathbf{B}}{\text{R}} \; \underset{\mathbf{C}}{\text{H}} \quad \begin{matrix} \xrightarrow{k_3} 1/2\,H_2 \\ \xrightarrow{k_2} \text{R} \end{matrix} \tag{5.7}$$

The desorption of H_2 and olefin are clearly irreversible in UHV. The point of concern rests simply with the kinetic importance k_{-1}. Figure 5.9 depicts a potential, schematic energy diagram for a closely related model process. The most important feature of this plot is its illustration of the fact that the reaction

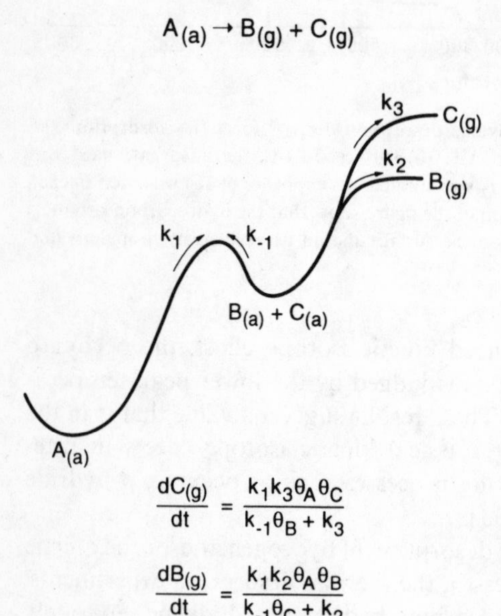

Fig. 5.9. One-dimensional potential energy diagram and kinetic expressions for the reaction of an adsorbed A molecule to gas phase products B and C. The endothermicity of the reaction can be overcome by the increased entropy of the gas phase products

depicted is *endothermic*. Thus, the elementary rate constant k_{-1} can be significant and the rates of olefin and/or hydrogen desorption, although very fast when adsorbed *separately* onto the surface, *may* affect the rate of product evolution from β-hydride elimination. It is therefore crucial to test for the reversibility of the elementary steps to arrive at a proper understanding of the rate behavior.

We conducted two independent studies to test for the reversibility of the β-elimination step in a more direct manner. First, the coadsorption of olefins such as propylene or isobutylene with varying coverages of hydrogen atoms on Al(1 0 0) or Al(1 1 1) surfaces does *not* yield surface alkyl groups. Similarly, thermolysis of butyliodide on an Al(1 0 0) surface bearing D atoms yields olefin containing little or no isotopic label. There is, however, a concern that a surface ensemble effect may operate here (that is, the coadsorbates *may* phase separate); the kinetic descriptions given in Fig. 5.9 will only pertain if adsorbates can interdiffuse on the surface. Thus, the most definitive evidence comes from coadsorption studies: butyliodides d_0 and d_9 when coadsorbed on Al(1 0 0) give olefin products on thermolysis showing no isotopic exchange. In the absence of reversibility, the forward reaction described by the rate constant, k_1, *must be* rate limiting.

The conclusion that follows naturally from these observations is that carbon-iodine bond cleavage at low temperature produces a surface alkyl intermediate which remains intact until alkene is evolved at temperatures $\gtrsim$450 K. This has been shown explicity for the case of isobutyl (Fig. 5.5), *n*-butyl, and *n*-hexyl species using high resolution EELS [5.4] and reflection-absorption infrared [5.22] spectroscopies. Perhaps most significantly, it also obtains (at least for the simple *n*-alkyl groups) that the alkene temperature programmed reaction spectrum *directly reflects* the decomposition kinetics of the adsorbed alkyl group. The combined weight of both the kinetic and spectroscopic studies support the formation of the alkyl surface species posited above.

Alkyl Reaction Kinetics. A rate-determining *unimolecular decomposition* of the adsorbed alkyl intermediates would be expected to show first-order kinetics. This presumption is supported experimentally by temperature programmed reaction studies. As a representative example, Fig. 5.10a shows the butene product spectra obtained for various doses of 1-iodo-2-methylpropane (isobutyliodide) adsorbed on an Al(1 1 1) surface at < 150 K. The asymmetric peak shape and the lack of a significant peak-temperature shift with varying coverage are characteristic of a first-order desorption process with coverage *independent* kinetic parameters [5.35]. Isobutyl species can also be formed on aluminum surfaces from the decomposition of triisobutylaluminum (Fig. 5.5). When TIBA is condensed in submonolayer quantities on an aluminum surface held at 150 K and the substrate is subsequently heated, decomposition and *not* molecular desorption of the aluminum compound is observed [5.4, 22]. The isobutylene and hydrogen gas phase products are identical to those observed for 1-iodo-2-methylpropane and the temperature for product evolution is $\approx$500 K

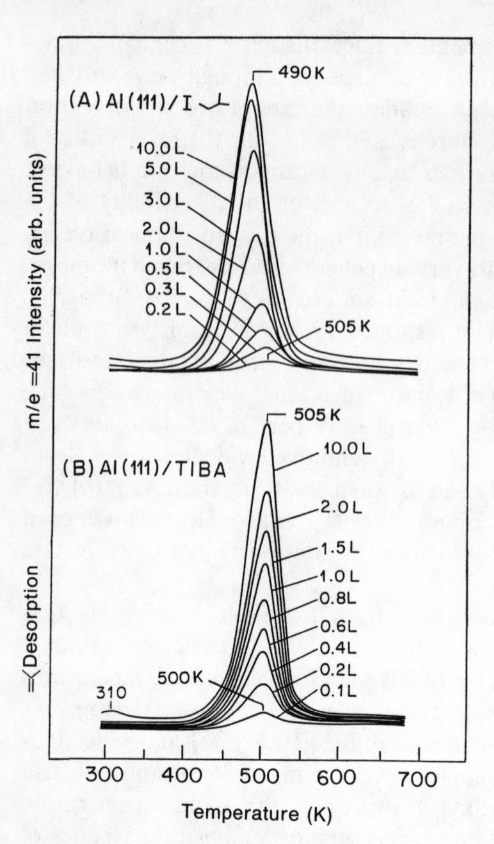

Fig. 5.10. Temperature programmed desorption data monitoring $m/e = 41$ as a function of (A) 1-iodo-2-methylpropane (isobutyliodide) and (B) triisobutylaluminum (TIBA) dose on Al(1 1 1). The single peak at *ca.* 500 K has been identified by IDMS and is due to butene (presumably isobutene [5.2, 4]) desorption. The asymmetric peak shapes and lack of a temperature shift with increasing coverage are indicative of a first-order process whose kinetic parameters are independent of coverage. These parameters are determined from the plot in Fig. 5.12 and are summarized in the text. The heating rate was ≈ 7 K/s in both (A) and (B)

(Fig. 5.10b). The observation of a single, narrow product peak in the temperature programmed reaction spectra of Fig. 5.10b establishes that the decomposition reactions of the three isobutyl groups derived from TIBA are kinetically indistinguishable [5.4]. The data shown in Fig. 5.5 and discussed in [5.22] indicate that they are indistinguishable *spectroscopically* over a temperature range which encompasses the decomposition process.

As was noted before for the corresponding iodide adsorbate, the data suggest a first-order desorption process with, to a first approximation, coverage independent kinetic parameters. This conclusion as to the order of the alkyl desorption is supported by the somewhat more detailed kinetic analysis whose results are summarized in Fig. 5.11 for the specific case of the decomposition of propyliodide on an Al(1 0 0) surface. Order plots ($[\ln(\text{rate}) - n \ln(\Theta)]$ vs. $1/T$ where Θ indicates a normalized surface coverage and n is the reaction order) [5.36] are shown in panel (Fig. 5.11b]). The parameters for this plot were extracted from the TPRS curves of Fig. 5.8a. The best fit value found for n is 1.0, a result confirming the assignment of a first-order rate process.

Given that the β-hydrogen and β-deuterium elimination reactions yielding the data shown in Fig. 5.8 are first order, we can quantify the kinetic isotope

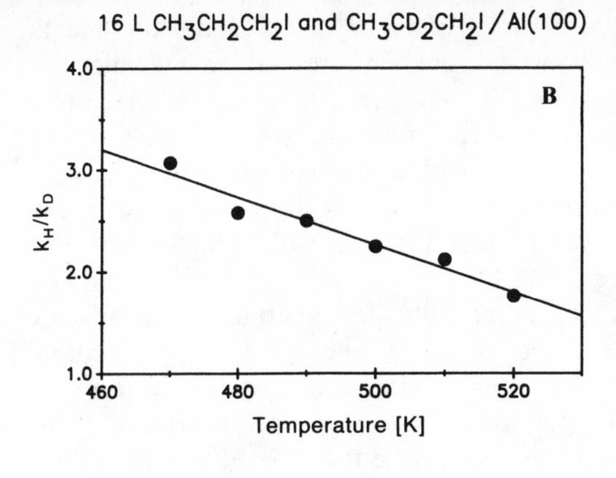

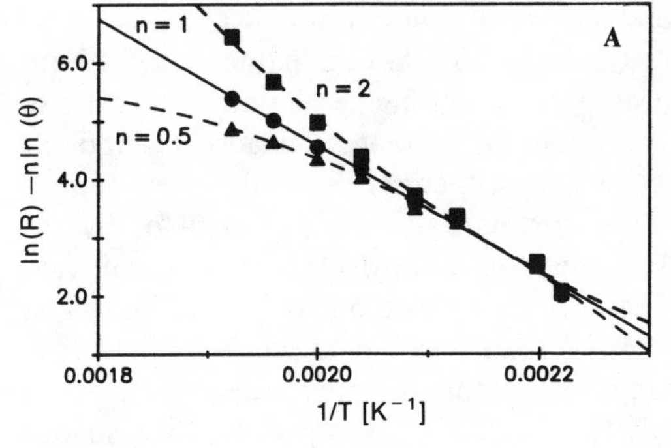

Fig. 5.11. (A) Order plots [5.20] for the thermal decomposition of propyliodide on Al(100) (the data is taken from Fig. 5.7 A). Clearly the best fit to the data is for $n = 1.0$, a first-order reaction. The activation energy for this process is R (the gas constant) times the slope of this curve. (B) The kinetic isotope effect (k_H/k_D) for the thermal decomposition of propyliodide is plotted as a function of surface temperature (the data is taken from Fig. 5.7 A and B). The measured values are in the range expected for a β-hydride elimination reaction [5.31–33]

effect (KIE, k_H/k_D) for this reaction using a simple model. Variants of these analyses can also be used to profile the temperature dependent kinetic isotope effects exhibited in this system. The plot shown in Fig. 5.11a was derived from TPRS data obtained using propyliodide and iodopropane-2, 2,-d_2 adsorbates as analyzed by the protocol described by *Madix* and coworkers [5.37]. The measured values of k_H/k_D (2.0–3.2 for $460 < T < 540$ K) conform to the range of classical kinetic isotope effects and are very similar to those measured in discrete organoaluminum alkyl compounds over the same temperature range (2.5–3.7) [5.31–33].

It is also possible to use the data of Fig. 5.10, as well as other analyses of temperature-programmed reaction data, to establish the Arrhenius kinetic parameters of the surface β-hydride elimination process. We have done this for many adsorbates on both Al(100) and Al(111) but will, in the context of the present discussion, restrict our attention to the isobutyl group (as this allows most convenient cross comparisons between adsorbate classes – alkyl halides and aluminum alkyls). Consider then, the values of these parameters as determined by varying the surface heating rate and monitoring the shift in the peak temperature [5.35]. An analysis of the results for 10 l exposures of 1-iodo-2-methylpropane on Al(111) at room temperature is exhibited in Fig. 5.11 [5.4, 5]. The measured activation energy (E_a) for β-hydride elimination is

23 kcal/mol with a pre-exponential factor (A) of 8×10^9 s^{-1}. Similar experiments on Al(1 0 0) gave thermal desorption peak temperatures that were uniformly 5–15 K higher in temperature than on Al(1 1 1) and with kinetic parameters of $E_a = 28$ kcal/mol and $A = 4 \times 10^{11}$ s^{-1}. The corresponding activation energy and pre-exponential factor for the decomposition of adsorbed isobutyl groups on Al(1 1 1) and Al(1 0 0) derived from TIBA, that is those deposited *in the absence of* coadsorbed halogen, are 27.7 and 32.6 kcal/mol and 3.8×10^{11} and 1.4×10^{13} s^{-1}, respectively.

The data depicted in Fig. 5.12 graphically demonstrate the significant surface structure sensitivity exhibited by β-hydride elimination reactions occurring on aluminum surfaces. The alkyl groups, whether derived from an organometallic reagent (i.e., TIBA) or from the dissociative adsorption of an alkyl iodide, decompose faster on the (1 1 1) surface than they do on the (1 0 0). This result is unaltered by the presence of a halide atom as a coadsorbate, although the halogen does serve to systematically shift the parameters for both surfaces. (The differences in the measured activation energies are ≈ 5 kcal/mol with both being lowered by 4–5 kcal/mol in the presence of coadsorbed iodine.) These activation energy differences, which of themselves would give rise to a very large difference in relative rates, are partially compensated for by the different pre-exponential factors. The values on Al (1 0 0) for both adsorbates are consistently 2 orders of magnitude larger than those on Al (1 1 1). The largest pre-exponentials occur in the absence of coadsorbed halogen.

The importance of site blocking as a factor underlying the effects of the coadsorbed halogen atom on reactivity is unclear to us at present. Even so, our belief is that the bulk of the noted structure/composition effects must be electronic in origin. Recent theoretical calculations attribute the faster reaction rate on the (1 1 1) surface to greater electron transfer in the transition state

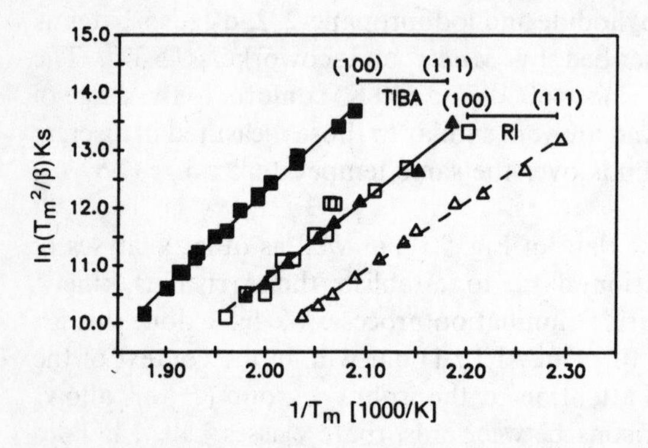

Fig. 5.12. The natural log of the temperature of the peak maximum (T_m) squared divided by the heating rate (β) is plotted vs. the reciprocal of T_m for both triisobutylaluminum (solid symbols) and 1-iodo-2-methylpropane (open symbols). ■, □ are data from Al(1 0 0) while ▲, △ are for adsorption on Al(1 1 1). Saturation exposures of the two adsorbates were used in all cases. The slope of these plots yields the activation energy for desorption [5.29]. The heating rate was varied from < 1 to 20 K/s

[5.25]. In addition, we find that the differences in E_a on (1 0 0) and (1 1 1) are exactly the same as the differences in the clean surface work functions, suggesting an important correlation between the rate of the β-elimination and the "electron affinity" of the surface [5.4]. Halogen atoms on either surface enhance the rate of the process, an effect in general accord with the reactivity profiles seen in discrete organoaluminum complexes [5.31, 32].

β-Elimination in the Gas Phase and in Solution. The asserted correspondence between gas-phase, solution, and surface aluminum alkyl decomposition chemistry is all the more plausible when one notes that the energetics for these various processes are very similar. For example, the activation energy for the elimination of one equivalent of isobutylene from triisobutylaluminum in the gas phase is 27.6 kcal/mol [5.31], which compares favorably to values measured on aluminum surfaces (28–33 kcal/mol) [5.2]. Additional kinetic similarities are also evident. The β-hydride elimination rate both at aluminum surfaces and in aluminum compounds increases with increasing alkyl substitution at the β-carbon. Specifically, the β-hydride elimination rate decreases (the olefin TPRS peaks move to higher temperatures, Table 5.3) in the order:

$$\text{isobutyl} > \text{butyl} \simeq \text{hexyl} \simeq \text{propyl} > \text{ethyl}.$$

As discussed in more detail in [5.2], this is the same sequence observed for similar groups in aluminum alkyl compounds [5.32].

This effect can be rationalized by the 4-center transition state geometry proposed for a β-hydride elimination reaction. Extensive calculations [5.25, 38] and studies of substituent effects support a structure for this transition state similar to that shown below.

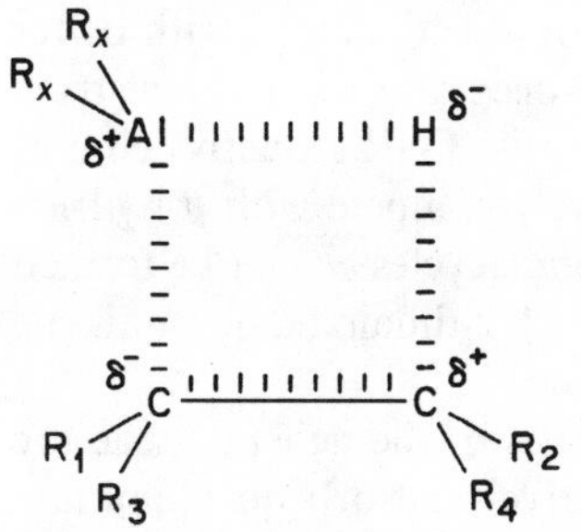

The perported polarization of the bonds (indicated by the partial charges) suggests that factors influencing the stability of such charge flow (e.g., alkyl substitution at the β-carbon) should lower the energy of the transition state significantly. This picture is intriguing in that it suggests that a single metal atom may be all that is required to effect these decompositions [5.25, 38]. The surface does not seem to play a major role as regards the need for an alkyl group to bond to multiple sites in order to undergo this decomposition reaction. In particular, the activation energy for β-hydride elimination on the surface is very similar to that observed in the gas phase [5.31]. It is clear, however, that the surface must, in other respects, play a major role for reasons that can easily be described.

Consider the decomposition of TIBA as a example. In the gas phase, TIBA undergoes a reversible β-hydride elimination to give one equivalent each of olefin (isobutylene) and diisobutylaluminum hydride (DIBAH) [5.31]. In the absence of a metal surface, the reaction stops here even at temperatures as high as 500 K. The bonding of TIBA to the surface *must* facilitate the facile conversion of this precursor to the products of the exhaustive thermolysis (3 equivalents of isobutylene and hydrogen). We feel this reflects the importance of bonding the central metal atom to a substrate, a mechanistic feature which allows the decomposition of the three alkyl groups of TIBA to proceed with uncoupled kinetics [5.4]. The very *large endothermicity* of the TIBA decomposition reaction ($\approx$63 kcal/mol) is overcome by several independent and less activated steps (see below) in addition to the increased entropy of the system [5.4]. the endothermicity and indeed the overall rate–structure profile of each of these steps closely resembles that of the TIBA–DIBAH equilibrium.

5.4.2 Dihaloalkanes

Formation of Metallacycles by the Adsorption of a Dihaloalkane. It was shown above that idodoalkanes could be used to generate surface alkyl groups on Al(1 0 0) and Al(1 1 1) and that the thermal decomposition of adsorbed alkyls containing β-hydrogens occurs almost exclusively by β-hydride elimination. In studying dihaloalkanes our hope was to generate metallacycles by the dissociation of two C–X bonds [5.3]. Non-bonded interactions (for example, steric and strain energies) can be exhibited very differently in transition states involving cyclic and acyclic intermediates. We might expect, in direct analogy with better understood organometallic complexes, that other decomposition chemistries, for example cyclization, hydrogenation (hydrogenolysis), C–H reductive elimination, and reversible H, D exchange, might effectively compete with β-hydride elimination. As was noted above, stable surface metallacycles *cannot* be formed on aluminum via the dissociative adsorption of 1, 2-dihaloalkanes (alkene desorption is observed well below room temperature).

We have found in our studies that dihaloalkanes of the general structure $RCHX(CH_2)_nCH_2X$ ($n \geqslant 1$ and X = Br, I) dissociatively adsorb on aluminum and yield what we believe to be stable metallacycle surface species [5.3]. Spectroscopic evidence for the formation of these intermediates comes from high-resolution EELS studies of various 1,3- and 1,4-dihaloalkanes which show that both C–X bonds are broken at temperatures below 300 K. Typical vibrational spectra from such studies are displayed in Fig. 5.13. Figure 5.13a is for an adsorbed layer of 1, 3-diiodopropane while Fig. 5.13b is the low-frequency region for the corresponding dibromide. Both species were dosed at 300 K. The difference to note in these spectra are the bands at 380 cm^{-1} (ν(Al–I)] and 590 cm^{-1} [ν(Al–C)] in (A) and 440 cm^{-1} [ν(Al–Br)] and 590 cm^{-1} [ν(Al–C)] in

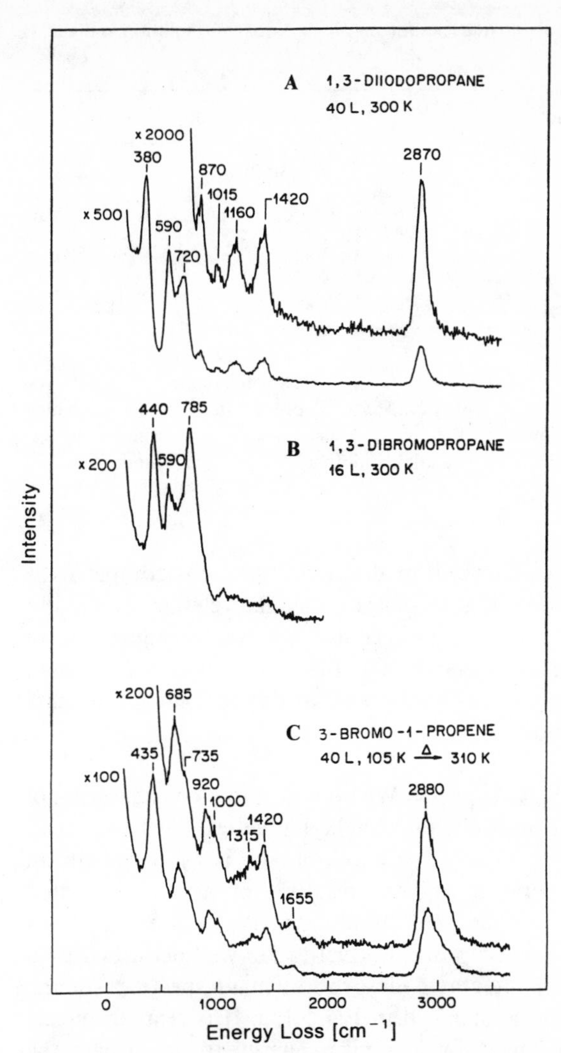

Fig. 5.13. Specular high resolution electron loss spectra of (A) 40 l of 1,3-diiodopropane and (B) 16 l of 1,3-dibromopropane adsorbed on an Al(100) surface at 300 K to form a monolayer of a 3-carbon metallacycle species. In (C), a condensed multilayer of allyl bromide was warmed to 300 K to form a surface allyl species. The important vibrational modes are discussed in the text and specific assignments are summarized in Table 4 and [5.3]

(B). The peak assignments for the former are consistent with those given earlier for propyliodide (Table 5.1 and [5.2]). An important point to note in these spectra is that no detectable peaks for either the C–I or C–Br stretches remains (≈ 530 and 610 cm^{-1}, respectively). In addition, the EELS peaks in the fingerprint region (700–1500 cm^{-1}) are reasonably similar to the infrared frequencies of liquid *gauche*, *gauche* 1,3-diiodopropane [5.3, 20], the expected chain conformation for a surface C_3 metallacycle derived from either adsorbate (Table 5.4). Taken together with the TPRS results described below (which slow that the alkyl chain remains intact after C–X bond cleavage) these studies comprise our "proof" of the formation of a C_3 surface metallacycle. Similar lines of argument can be used to support the formation of longer chain homologs as well [5.3].

Table 5.4. Comparison of the vibrational frequencies for 1,3-diiodopropane in solution and adsorbed on an Al(1 0 0) surface[a]

40 l at 300 K[b]	*gauche, gauche* conformation[c]	Approximation mode assignments[c]
2870	2837	CH_2 sym stretch
1420	1423	CH_2 bend
1160	1208	CH_3 wag
1015	1019	CH_2 wag
870	914	CH_2 rock
720	732	CH_2 rock

[a] All frequencies in cm^{-1}.
[b] Observed frequencies from Fig. 5.13 for a monolayer of C_3 surface metallacycles.
[c] Frequencies and approximate mode assignments from [5.20]. The structure of the surface bound C_3 metallacycle is expected to be similar to that of the *gauche*, *gauche* conformation of 1,3-diiodopropane [3]. The structure of the other potential conformations would not model a surface metallacycle.

The evidence for such metallacycles in discrete aluminum complexes is sparse [5.32], which might indicate that the surface metallacycles are bonded to more than one metal atom, a relatively inaccessible bonding configuration in most known organoaluminum compounds. On the other hand, the thermal chemistry of the surface metallacycles shows similarities to that for alkenyl ligands in aluminum compounds.

Thermal Decomposition of C_4 Metallacylces. We now turn to a consideration of the thermal chemistries of these multiply bonded alkyl moieties. The first species we discuss is a C_4 metallacycle. When, 1,4-diiodobutane is adsorbed on an Al(1 0 0) surface at room temperature and the resulting monolayer is heated while monitoring m/e = 54, only one peak at $\approx$490 K is observed in TPRS (Fig. 5.14a and Table 5.3). The decomposition kinetics are well described by a simple first-order rate law. An integrated desorption mass spectrum of the species responsible for this peak is depicted in Fig. 5.15a. It is clear from this spectrum, which was taken using a 15 eV ionizer energy to accentuate the molecular ions, that the predominant product has m/e = 54, but that some (about 20%) m/e = 56 product is also produced. The 72 eV IDMS spectrum (not shown) conclusively identifies these products as 1, 3-butadiene and a butene (presumably 1-butene), respectively. In particular, the intensity ratio of the ions m/e = 54 and 39 (0.65) in this spectrum mitigates against 1, 2-butadiene and the butynes as sources of the m/e = 54 peak [5.29]. There is also no evidence in either the 15 or 72 eV IDMS data for ethylene, a potential product of β carbon–carbon bond scission.

The simplest conclusion which can be drawn from these results is that the dominant thermal decomposition pathway of a 4-carbon metallacycle yields, 1, 3-butadiene by β-hydride eliminations at either end of the chain. This conclusion is further supported by studies using 4-bromo-1-butene as the

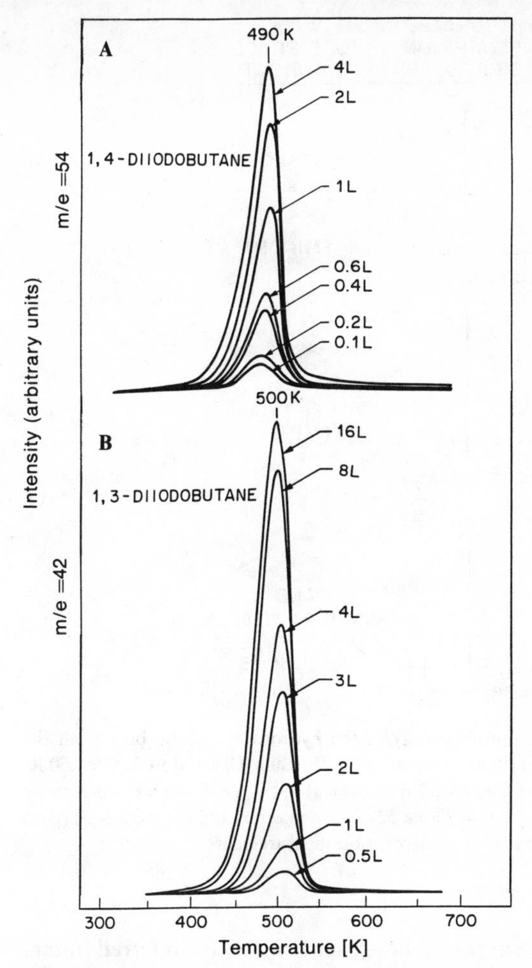

Fig. 5.14. (A) Butene evolution form an Al(1 0 0) surface after adsorbing between 0.1 and 4 l of 1,4-diiodobutane at room temperature and heating the surface at 5 K/s. (B) Propylene evolution from an Al(1 0 0) surface after adsorbing between 0.5 and 16 l of 1,3-diiodopropane at room temperature and heating the surface at 10 K/s. The asymmetric peak shapes in both (A) and (B) and the nearly constant value of the peak temperature with increasing gas exposure are characteristic of a first-order desorption process with coverage independent kinetic parameters

adsorbate. As shown in (5.8), the 3-butenyl species presumably generated by adsorbing this compound on Al (1 0 0) is the same as the surface species expected after one β-hydride elimination from the 4-carbon metallacycle.

Al ⇌ [Al, H] ⟶ (minor) + , H_2 (major) + Al (5.8)

An IDMS spectrum (obtained using a 15 eV ionizer energy) for 4-bromo-1-butene is shown in Fig. 5.15b. This spectrum is quite similar to that displayed in

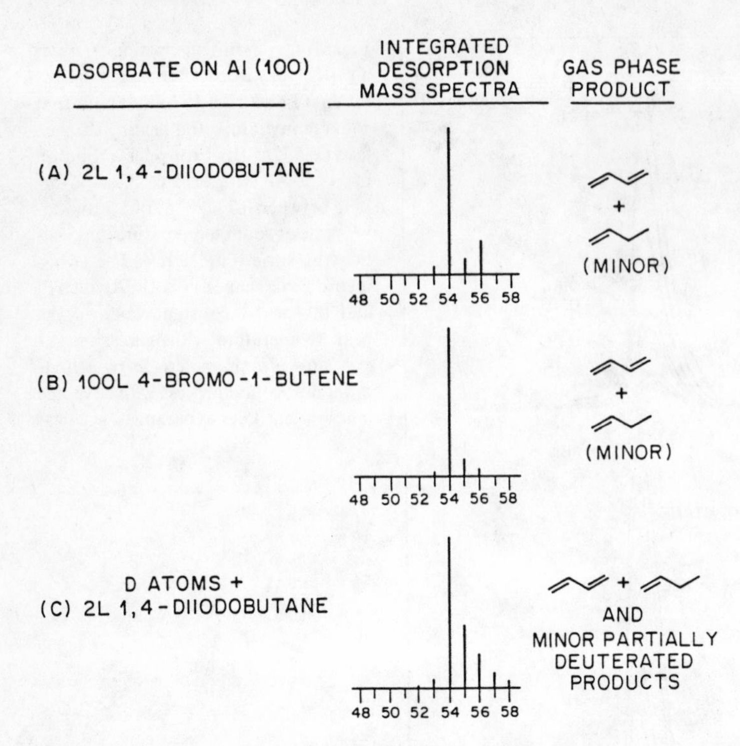

Fig. 5.15. Partial mass spectra at 15 eV ionizer energy of the hydrocarbons desorbing from the indicated monolayers on Al(1 0 0). (A) and (B) were dosed at 300 K, while (C) was dosed at < 150 K. Product desorption in these experiments was monitored over about a 150 K temperature range centered around the peak temperatures given in Table 5.3. The deuterium atom coverage in (C) is about half of saturation as determined by thermal desorption measurements

Fig. 5.15a for 1,4-diiodobutane, a result consistent with the inferred intermediacy of the 3-butenyl species shown in (5.8). The differences between the butenyl and metallacycle derived products (less formation of the hydrogenation product, $m/e = 56$, is observed for the butenyl moiety) are likely to reflect the fact that fewer surface hydrogen atoms from β-hydride elimination reactions are present in the latter case. One assumes that this feature operates to shift the pre-equilibrium β-elimination-olefin reduction steps toward the alkenyl intermediate. From these data alone, however, it is unclear how a low relative yield of the alkene would be observed.

The competitive production of butene in these butenyl/metallacycle experiments implicitly establishes that C–H reductive elimination (i.e., hydrogenation) processes can be kinetically competitive with β-hydride elimination. It is also found that, unlike the irreversible β-hydride elimination chemistry observed for adsorbed propyl and isobutyl species, the 4-carbon metallacycle does undergo some H,D-exchange. This is evident from the IDMS data obtained from an experiment in which the dihalide adsorbate was coadsorbed with deuterium atoms (Fig. 5.15c). It is clear from this spectrum that, while the major product of this particular exchange experiment is perhydro 1,3-butadiene (m/e = 54), there are minor products formed with masses up to m/e = 58. Masses 56, 57 and 58 could be hydrogenation products formed by addition of various combinations of H and D, but m/e = 55 (butadiene-d_1) must be the result of H,D exchange. This important observation definitively establishes the reversibility of the first β-elimination step in the adsorbed C_4 metallacycle.

While the extent of H,D exchange is small, its occurrence suggests that the double bond in the butenyl intermediate may weakly coordinate to the surface as shown in (5.8). A similar π-bonding interaction has been detected in aluminum alkenyl compounds by infrared spectroscopy for ligands of at least five carbon atoms in length [5.39]. The presence of adjacent, unsaturated metal atoms on the surface may facilitate such an interaction at shorter chain lengths.

In addition to exchange reactions, the 4-carbon metallacycle illustrated in (5.8) also disproportionates to a slight degree to produce measurable, though lesser, quantities of butene. It has been found in solution that butenyl species analogous to the proposed intermediate in this process will react with diisobutylaluminium hydride to produce not only butadiene, but also *cis*- and *trans*-2-butene, cyclobutane, isobutane, and butane [5.32]. These chemistries are believed to be the result of radical pathways, however (see below). The complex product mixture obtained in solution thus suggests that equilibria other than those shown in (5.8) need to be considered. The formation of a 1,3-bonded metallacycle by the reverse of the first β-hydride elimination with inverted regiospecificity seems to be the most plausible candidate to consider. We now explore the thermal chemistries of a C_3 metallacycle directly, the simplest homolog of a 1,3-bonded species.

Thermal Chemistry of C_3 Metallacycles. Figure 5.14 displays temperature-programmed reaction spectra derived from the adsorption of 1,3-diiodopropane on Al(1 0 0) at 300 K. As for the C_4 moiety, the kinetics reflect simple first-order behavior. The underlying reactions are more complicated than what this might suggest, however. The corresponding IDMS spectra of the products (temperature range: 450–550 K) for a saturation coverage of 1,3-diiodopropane and for a lower coverage coadsorbed with deuterium atoms on an Al(1 0 0) surface are shown in Fig. 5.16A and B, respectively. The 15 eV spectrum in Fig. 5.16 clearly shows that the product for the saturation layer has m/e = 42. The 72 eV cracking pattern identifies this product as propylene (cyclopropane, a potential product of intramolecular reductive coupling, has a m/e 42:41 ratio at 72 eV of 1.26 as compared with the ratio of 0.66 found here). Most interestingly, and consistent with these observations, the thermal decomposition of 1,3-diiodobutane yields *exclusively* butene [5.3]. Analogous studies with 1,3-dibromopropane showed that propylene is also the thermolysis product obtained for this adsorbate, and its peak temperature for desorption is to within experimental error the same as for the diiodo compound (Table 5.1).These results support the assertion that the atomic identity of the halide has little, if any, effect on either the desorption products or the surface reaction kinetics for this class of adsorbate. The data also suggest a general reaction scheme in which a 1, 3-bonded metallacycle partitions efficiently into an olefin forming reaction channel.

The IDMS results for the coadsorption of deuterium atoms (about 1/2 monolayer) with about 20% of a monolayer of 1, 3-diiodopropane on Al(1 0 0) are presented in Fig. 5.16. The 15 eV spectrumclearly shows that three different propylenes (d_0, d_1, and d_2) are produced. Further, the relative intensities ($\sim$ 1:2:1) are consistent with a statistical incorporation of H and D from a 50:50 mixture into two positions in the propylene. We were unable to determine from the mass spectral cracking patterns which two positions these were, however. Since *only* two hydrogen atoms readily exchange, we postulate that there is facile incorporation of deuterium into the β-position of the surface metallacycle yielding propylene-2, 3-d_2 through the intermediacy of an allyl moiety.

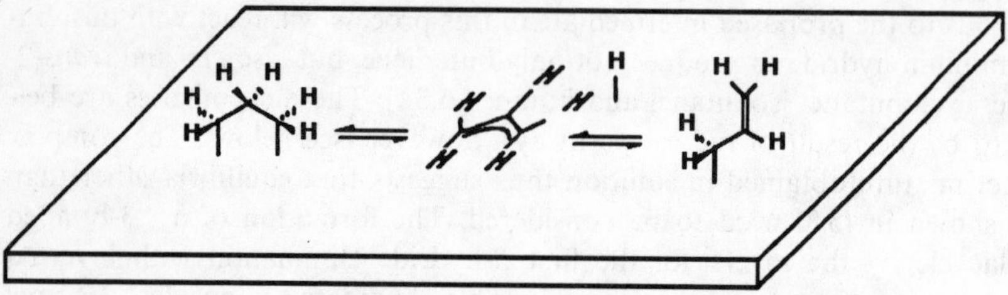

It is particularly significant that similar results were obtained by adsorbing *mixed monolayers* of 1, 3-dibromopropane and 1, 3-dibromopropane-d_6 [5.26]. The equilibration of H and D in these metallacycles must occur at relatively high temperatures, however, since heating a mixed monolayer (dosed at 300 K) to 400 K for 20 min caused no new peaks to appear in the high-resolution EELS spectrum which might be characteristic of a new surface species [5.3]. Separate thermal desorption experiments conducted with these perhydro and perdeutero compounds showed a 10 K peak shift in the desorption-peak temperature (Table 5.1), consistent with a rate-determining step involving the transformation of either C–H (C–D) or M–H (M–D) bonds. In any event, the facility of the H, D-exchange in the β-position and the formal hydrogen transfer to form the product propylene are unique among the surface alkyl chemistries we have observed on aluminum.

The results above establish the existence of distinct and efficient thermal reaction pathways for C_3 metallacycles. The data also establish the occurrence of reversible equilibria which appear to generate a surface allyl moiety. We have studied explicitly the role of such surface allyl species in these H, D-exchange processes by adsorbing 3-bromo-1-propene and 3-iodo-1-propene as the allyl precursors [5.3]. Figure 5.17 exhibits the IDMS results of these studies for the iodo compound. The 15 eV mass spectrum in Fig. 5.17 shows that adsorbed allyl moieties do indeed react in the presence of hydrogen atoms to produce

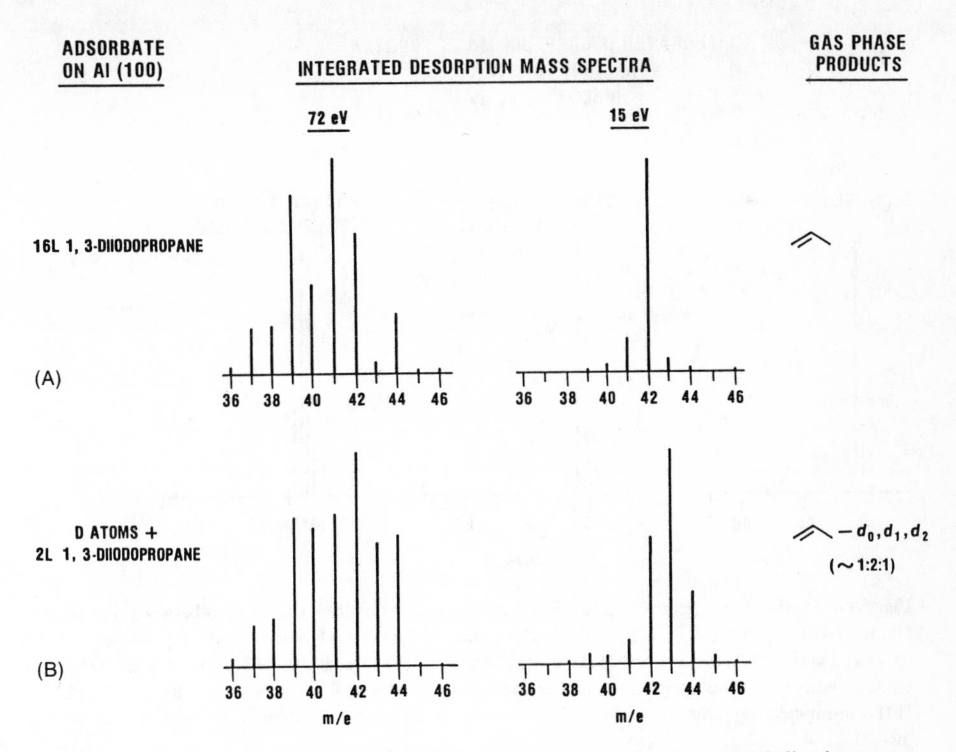

Fig. 5.16. Partial mass spectra at 72 and 15 eV ionizer energies showing that 1,3-dioodopropane decomposes on Al(1 0 0) to evolve propylene (A) which incorporates up to 2 deuterium atoms when deuterium is coadsorbed on the surface (B). The coadsorbed deuterium coverage in (B) was about half of saturation. The adsorptiun temperatures in (A) and (B) were 310 and < 150 K, respectively

propylene (m/e = 42) as one would infer from the H, D-exchange results discussed above for the 3-carbon metallacycles. Replacing the surface hydrogens with deuterium atoms (Fig. 5.17) results in the formation of d_0, d_1, and d_2 propylene consistent with rapid equilibration between a surface allyl and a 1,3-disigma-bound 3-carbon metallacycle, as shown in the scheme above.

The data in Fig. 5.17 reveal that even in the *absence* of additional hydrogen atoms, adsorbed allyl species produce a significant amount of propylene and essentially no propadiene (the β-hydride elimination product). The hydrogen necessary to form propylene presumably comes from the dehydrogenation of some of the surface allyl groups. Note that the high mass data in Fig. 5.17 and demonstrate that oligimerization of adsorbed allyl groups accompanied by dehydrogenation can occur. We have not been able to identify these products as yet. Little residual carbon is detected by AES, however.

Based on the fact that hydrogen recombines and desorbs from Al(1 0 0) with a peak temperature of about 320 K (Fig. 5.4 and [5.21]), one might conclude that H, D exchange via the metallacycle/allyl equilibrium occurs at 320 K or below. By this same reasoning, one would also infer that, in the temperature

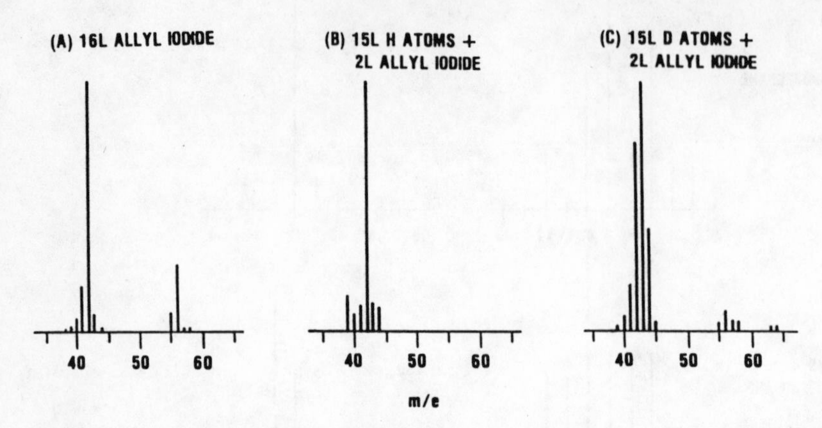

Fig. 5.17. Partial mass spectra at 15 eV ionizer energy of the species that desorb between 460 and 590 K upon heat an Al(1 0 0) surface covered with monolayers of allyl iodide (3-iodo-1-propene) with (B,C) and without (A) coadsorbed H (D) atoms. The Al(1 0 0) surface was held at room temperature while forming the monolayer in (A), but was cooled to < 150 K to form monolayers (B) and (C). The H (D) atom coverages are about half of saturation (as calibrated by thermal desorption measurements)

range between 400 and 500 K (above the H_2 desorption temperature), the adsorbed surface species *should be* the allyl group. These conclusions are not supported by the data, however, We have found that, in the presence of adsorbed alkyl iodides, the hydrogen desorption peak broadens and often overlaps the hydrocarbon desorption peaks around 500 K. The facility with which the 1, 3-dihalopropanes generate propylen (with no other products being detected) suggests that the predominant surface species present prior to decomposition must be the 3-carbon metallacycle.

Regardless of the position of the allyl/metallacycle equilibrium, this equilibration process *must* be rapid. The question arises, however, as to what is special about the surface allyl species (or the metallacycle) that promotes the facile olefin reduction (hydrogenation) chemistry shown above. One possibility is that the allyl moieties exist as flat-lying, conjugated species on the surface. We have probed this question with EELS studies of the structure obtained through the adsorption of 3-bromo-1-propene on Al(1 0 0) (Fig. 5.13 and [5.3]). It is clear that such spectra are substantially different from those for 1, 3-dihalopropane shown in Figs. 5.13a and b. A small, but particularly significant difference between the metallacycle and allyl spectra is that, in the latter, one observes a peak at 1655 cm^{-1}. This mode, which is absent in the metallacycle spectra, is outside the region in which C–H bending and stretching frequencies normally occur [5.40]. We assign this mode to the stretching vibration of a C–C double bond. It also *can* be inferred given the presence of this band, that the adsorbed species is not a flat-lying, conjugated allyl moiety, since such species do not have vibrational modes in the 1500–1700 cm^{-1} region [5.40]. The presence of a carbon–carbon double bond is also supported by the broadening of the C–H stretching peak to higher frequency. We therefore conclude that the adsorbed allyl group on Al(1 0 0) bounds formally as a propenyl intermediate.

The data described above establish the utility of model systems and structure-reactivity profiles in defining complex surface reaction mechanisms. The combined weight of the data suggest strongly that the 3-carbon metallacycle undergoes facile H, D exchange at the β-carbon position. This exchange occurs at temperatures near where irreversible decomposition sets-in and is well explained by a facile pre-equilibrium dehydrogenation to a surface allyl/propenyl species. Finally, the formation of propylene (which conserves the number of C–H bonds in the initial adsorbate) occurs preferentially to the type of double β-hydride elimination observed for the 4-carbon metallacycle.

In comparing this 3-carbon metallacycle surface chemistry with the thermolysis of aluminum alkyl compounds, we are struck by the absence of cyclopropane as a product. Cyclopropane is formed efficiently from aluminum alkyls in solution by α-, γ-elimination from 3-haloakyl ligands [5.14], potential transient precursors to these metallacycle intermediates [5.32].

$$Et_2AlCH_2X \xrightarrow{C_2H_4} Et_2Al(CH_2)_3X \longrightarrow Et_2AlX + \Delta \tag{5.9}$$

$$\phi_2C{=}C{=}CH_2 \xrightarrow{(i-Bu)_2AlH} \begin{array}{l} \phi_2C = CHCH_2Al(i-Bu)_2 \\ \phi_2C - CH = CH_2 \\ Al(i-Bu) \end{array} \tag{5.10}$$

It is also interesting that even when 3-alkenyl fragments are generated on the surface, no detectable propadiene, the β-hydride elimination product, is evolved during thermolysis. While this chemistry is also not observed in discrete complexes, the reverse reaction (5.10) does occur readily. For the adsorbed 3-alkenyl intermediate, some surface hydrogen-producing reaction(s) must occur, given that a major decomposition product is again found to be propylene. It would appear that the source of this hydrogen is *not* a β-hydride elimination in an allyl moiety, since propadiene (when adsorbed separately) desorbs from Al(1 0 0) at about 100 K rather than coupling to form the high-mass species we observed [5.3]. We can envision several factors which may inhibit β-hydride elimination from this alkenyl intermediate. Among these we include: (1) the olefinic character of the second β-hydrogen atom in the propenyl intermediate; (2) the orientation of the β-hydrogen atom with respect to the surface (aluminum atom) in the propenyl intermediate; (3) the nonplanar geometry of the product propadiene; and (4) the destabilization of the product forming transition state due to strain energy. What the important factors are is unclear at present.

Large Surface Metallacycles. The metallacycle studies described above can be extended to other analogs. IDMS spectra of the thermal decomposition products of 1,6-diiodohexane and 6-bromo-1-hexene on Al(100) are shown along with mass spectra for two hexadienes, 1-hexene, and methylenecyclopentane standards in Fig. 5.18. The product evolution temperatures are stated in the figure. As judged by the IDMS data, both of these adsorbates give complex product mixtures. In contrast, we have shown previously that the thermal decomposition of 1-iodehexane yields exclusively 1-hexene (Fig. 5.6 and [5.2, 26]). The 72 eV spectra for both 1,6-diiodohexane and 6-bromo-1-hexene

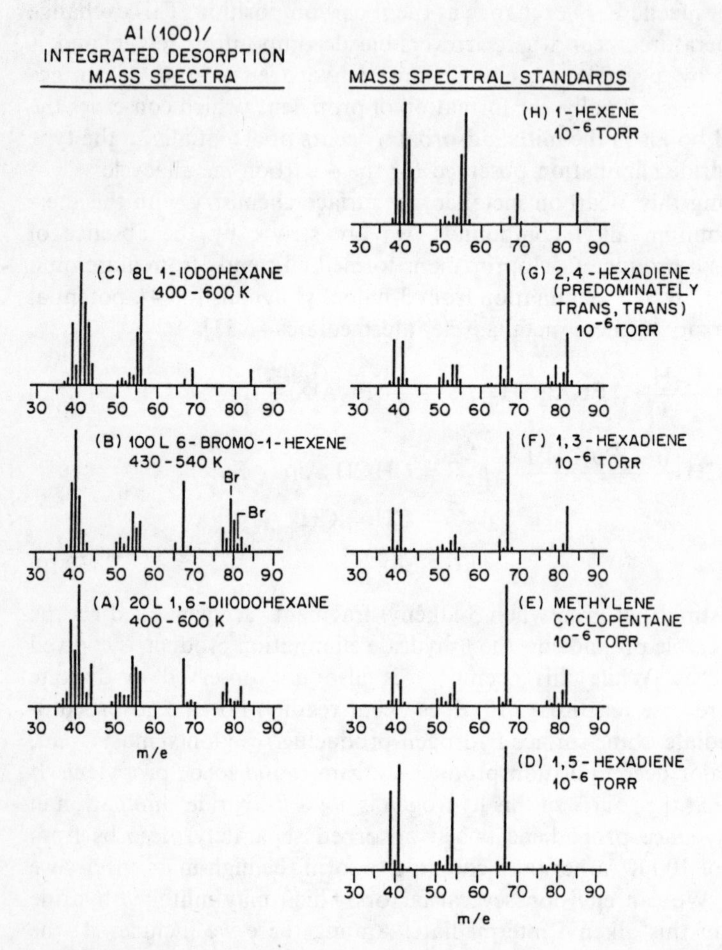

Fig. 5.18. Integrated desorption mass spectra (left) and mass spectral standards (right) at 72 eV ionizer energy for a variety of C_6 hydrocarbons. In the IDMS spectra, the molecule was adsorbed at room temperature and the spectrum of the desorbing species was collected over the temperature range of 400–600 K. As discussed in the text, these spectra and those at 15 eV (not shown) clearly demonstrate that both 1,6-diiodohexane and 6-bromo-1-hexene decompose on Al(100) to form mixtures of products including methylene cyclopentane

adsorbates have ions up to m/e = 84 (note that a portion of the peaks in the 79–82 region for the 6-bromo-1-hexene compound may be due to a small Br and HBr background in the mass spectrometer. Comparing these IDMS spectra with the mass spectral standards shown and others tabulated in data bases [5.29] suggests that both molecules produce a similar mixture of products including both linear and *cyclic* hexenes and dienes. Interestingly, 1,5-hexadiene, the product of a simple β-hydride elimination at either end of the carbon chain in a manner analogous to that of the 4-carbon metallacycle, is *not* one of the major products from the decomposition of 1,6-diiodohexane [5.41]. This conclusion is based on the weak relative intensity of the ion m/e = 81 in the 72 eV IDMS spectrum. The large 82/67 ratio in the 15 eV IDMS spectrum (not shown) of 6-bromo-1-hexene also suggests that 1,5-hexadiene is at best a minor product. Methylenecyclopentane, on the other hand, can account in part for the intensity ratios seen in both the 72 and 15 eV spectra noted above.

The formation of methylenecyclopentane can be explained reasonably by the following steps.

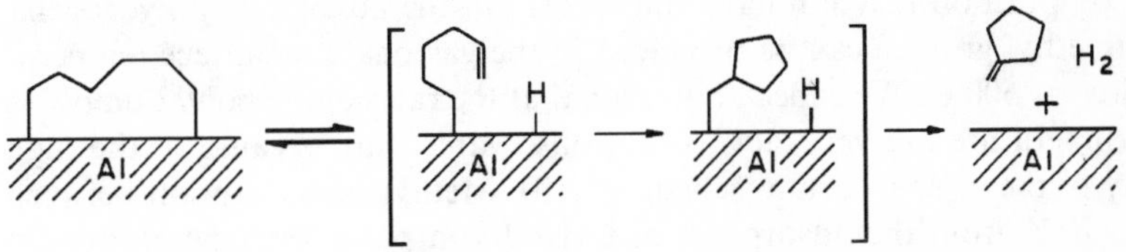

(5.11)

As shown, β-hydride elimination at one end of the chain produces a surface alkenyl moiety whose carbon-carbon double bond can insert into the Al–C bond followed by a second β-hydride elimination to give methylenecyclopentane. Precedence exists for both intermolecular [5.32] and intramolecular [5.42] olefin insertions into Al–C bonds, the latter chemistry in solution being a direct analogue of the surface chemistry we propose above.

In principle, other cyclizations are also possible. For example, addition in a sense opposite to that shown for the conversion of **1** to **2**, would yield a cyclohexyl species bound to the surface (the decomposition of this entity by a β-hydride elimination would yield cyclohexane, a product not present in quantities sufficient to be identified by IDMS). In general, aliphatic cyclizations slightly favor the formation of 5 over 6 membered rings. The dominant effect favoring the formation of **2** here is, however, likely to be steric. The intermediate **2** bonds to the surface via a primary carbon atom; a cyclohexyl moiety would be bound via a secondary carbon center. In organometallic complexes, increased steric bulk at the metal-carbon bond is known to be destabilizing [5.32]. Furthermore, there are many examples known where metal alkyls "isomerize" via β-hydride elimination to yield the least hindered metal alkyl [5.32]. These precedents thus suggest that the absence of cyclohexane as a major product of the decomposition of **1** reflects a strong steric preclusion of the necessary cyclization reaction.

Another potential route from an adsorbed hexenyl species to methylenecyclopentane is radical cyclization. This process, however, requires homolytic Al–C bond cleavage to form the corresponding hexenyl radical. Such an occurrence seems unlikely, since we find experimentally that the products of this reaction are evolved at 520 K, a temperature approximately 100 K below that for the suspected radical formation reactions (see below).

5.4.3 Radical Participation in Aluminum Alkyl Chemistry

A point of some debate in solution organometallic chemistry is the lability of the M–C bond towards homolytic bond scission to form alkyl radical intermediates. Such intermediates, because of their high reactivity, are generally inferred indirectly [5.32]. Further, when aluminum alkyl radical chemistry has been observed in solution, a radical initiator is generally believed to be required. In aluminum alkyl chemistry, it also appears that molecular rather than radical processes dominate. We now turn to a consideration of radical chemistry on aluminum surfaces.

Previous laser-induced fluorescence and resonance-enhanced multiphoton-ionization mass-spectrometry studies of the surface-catalyzed trimethylaluminum decomposition reaction have shown that in the absence of β-hydrogens, radicals (methyl groups) can be produced in the gas phase at surface temperatures as low as 500 K. We expect, however, that the rates of the bond homolysis at this temperature are very slow, being observable only because of the high sensitivity of the laser-detection scheme [5.10]. Methylene desorption between 100 and 170 K from the adsorption of CH_2CI_2 on polycrystalline aluminum surfaces has also been reported [5.28] (although we have been unable to reproduce these results on single-crystal substrates). In the case of methyl radicals (from either the adsorption of trimethylaluminum or methyliodide [5.18]), however, the dominant reaction pathway is dehydrogenation leaving a carbon covered surface and gas-phase hydrogen. These methyl groups can be hydrogenated, at least in part, in the presence of coadsorbed hydrogen atoms and trace amounts of methane have been detected by mass spectrometry [5.10].This hydrogenation reaction is facile at elevated pressures since there are numerous reports of the growth of *clean* aluminum thin films from the pyrolysis of either trimethylaluminum or dimethylaluminum hydride in a hydrogen carrier gas [5.10].

Model studies confirm that radical bond homolysis is indeed a high-barrier process. Thermal desorption and integrated desorption mass spectra for an A1(1 0 0) surface dosed with varying amounts of 1-iodo-2, 2-dimethyl propane ($ICH_2C(CH_3)_3$, neopentyliodide), an alkyl precursor with *no* β-C–H bonds, are shown in Fig. 5.19. From the IDMS data, it appears that m/e = 57, the highest mass fragment of any significant intensity in both the 72 and 15 eV spectra, can be attributed to an ionization cracking fragment of neopentyl radical. This result is somewhat surprising, since the parent ion expected for this radical is m/e = 71. We do observe a mass 71 peak, but it is over an order of magnitude less intense (the literature ratio of masses 71 to 57 for neopentane is less than 0.05 [5.29], indicative of the facile fragmentation of the highly unstable neopentyl-cation).

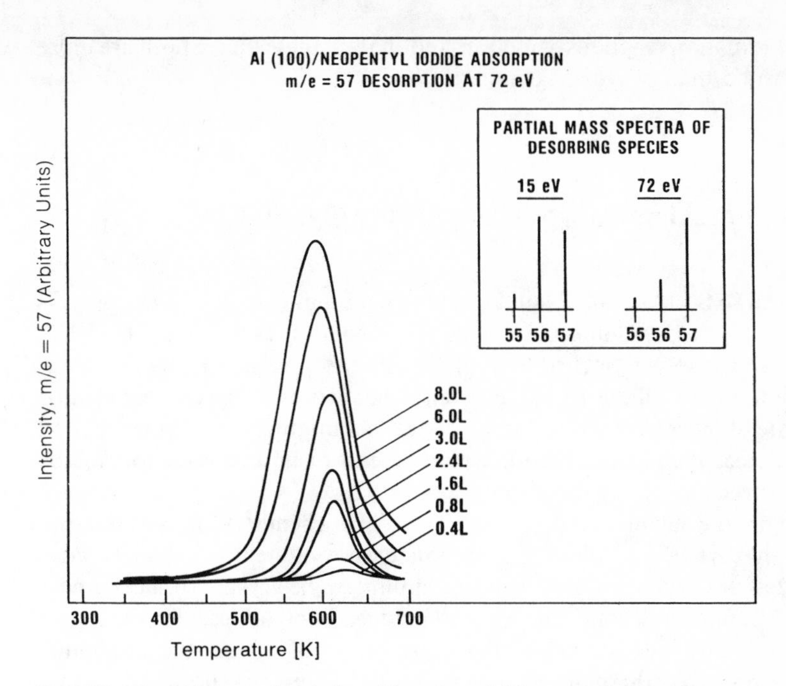

Fig. 5.19. Thermal desorption (*m/e* = 57) and partial integrated desorption mass spectra of the desorbing products for the decomposition of 1-iodo-2,2-dimethylpropane (neopentyliodide) on Al(1 0 0). In all of these experiments, the neopentyliodide was adsorbed at room temperature and the surface was heated at 7 K/s. The IDMS data in the inset were acquired over the temperature range of 530–670 K. These data, as discussed in the text, suggest neopentyl desorption, the product of homolytic aluminum–carbon bond scission

The relative increase of m/e = 56 in the IDMS spectrum recorded at 15 eV ionization energy (Fig. 5.19, insert) we believe to be evidence for the concomitant evolution of isobutylene (m/e = 56) along with neopentyl radical desorption. This latter product may arise from a β-methyl elimination process and has precedence in the thermal decomposition of TIBA on aluminum surfaces at high temperatures [5.4]. We presume isobutene production to be a minor reaction channel, since we do not see large amounts of carbon (which is expected to form on the surface via the dehydrogenation of the adsorbed methyl groups necessarily obtained along with any isobutene production [5.4, 18]) on the surface by Auger electron spectroscopy after the thermal desorption experiment.

The thermal desorption spectra recorded at m/e = 57 as a function of neopentyliodide exposure in Fig. 5.19 are also somewhat unusual. First, for low converages, the desorption temperature is about 100 K higher than was observed for the hydrocarbon products derived from any of the other alkyl halides. Second, the desorption peak continues to grow and shift to lower temperature with increasing gas exposure. The spectra at these larger gas doses are very complex and are, at present, not well understood. These complexities notwithstanding, the lower exposure data reveal that both M–C bond homolysis and

β-methyl elimination reactions can occur and that, in these limits, both are more highly activated than β-hydride elimination.

5.5 Etching of Aluminum Surfaces with Alkyl Halides

As mentioned in Sect. 5.1, alkyl halide etching of aluminum (an efficient process in condensed phases yielding aluminum sesquihalides) is well-known [5.16]. Aluminum can also be "etched" by olefins in the presence of hydrogen to produce aluminum trailkyls [5.32]. For both chemistries, it is well-documented that the state of the aluminum surface is critical in initiating the etching reaction [5.16, 32]. A reasonable presumption is that at least one prerequisite for etching is that the surface oxide be removed.

Based on the facility of this solution-etching chemistry, it is somewhat surprising that we do not observe a preponderance of etching products when alkyl halides are reacted with *clean* aluminum single-crystal surfaces under ultrahigh-vacuum conditions. The carbon–halogen bond (at least for the longer chain alkyl iodides) does readily dissociate (at < 130 K) on the aluminum surface; the preferred thermal reaction pathway for most is, however, surface alkyl decomposition (an *endo*thermic process) and not the efficient formation of organometallic reagents (an *exo*thermic process). In the case of aluminum sesquihalide production, it is known that carbon chains longer than ethyl tend to disproportionate extensively [5.32], but for aluminum alkyls, even isobutyl ligands (formed by the reaction of isobutylene and hydrogen) will etch aluminum with *great facility* at high pressures [5.32].

Several explanations can be advanced to rationalize why the product balances tend to favor decomposition so heavily in the desorption experiments discussed above. Aluminum etching may be inherently slow (even in solution) for the flat, single-crystal (1 1 1) and (1 0 0) surfaces investigated here. Experimental studies on related systems suggest that etching occurs preferentially at low-coordination number surface atoms [5.43], which are certainly a minority in our studies. On the other hand, the activation energies of such condensed phase etching processes must be substantially smaller than the $\lesssim 30$ Kcal/mol barriers we measure for alkyl decomposition given that the rates in many are fast even at 300 K. In order for processes with such disparate activation energies to proceed at comparable rates, the pre-exponential factors would have to differ by many orders of magnitude. Surface defects might account for a factor of ≈ 100, while differences in the molecularity of these processes (unimolecular vs. bimolecular) might produce several more orders of magnitude difference in the relevant pre-exponential terms. Although entropy does not generally favor associative processes at low pressures, the effect of concern here is one rigorously expressed in the entropy of activation and presumably involves coverage dependences of the reaction rate which have not been accounted for. Yet, even with these considerations, it remains difficult to rationalize the sluggish kinetics of etching seen under UHV conditions.

Another explanation may be that a *clean* aluminum surface does not readily etch under *any* conditions. For example, studies of halogen etching of other metals have shown that, at steady state, the near surface region is saturated with halogen [5.44]. If this analogy holds for aluminum, iodine or bromine dissolution may be favored over etching for surfaces bearing only submonolayer qunatities of halogen. Alternatively, other product forming reaction pathways (besides elementary bimolecular reactions between species adsorbed directly on the surface) may be operative in solution-phase processes.

While the extremely facile dissociative adsorption of alkyl iodides on aluminum in UHV would suggest that similar monolayers must also form in solution, in the latter environment there may also occur reactions between this surface phase and other R–X molecules "confined" next to the surface in a hydrodynamic boundary layer [5.17]. These, and perhaps other, reactions may be required to achieve surface coverages of chemisorbed materials which are high enough to effect the removal of metal via a highly coverage-dependent pathway.

We have discussed in an earlier paper concerned with the reactivity of magnesium surfaces [5.45] the notion of whether the product-forming mechanism of the Grignard reaction could involve coupled "inner sphere" reactions (the type we see in UHV) and other "outer sphere" (i.e., Eley-Rideal) processes (the ones occurring through and/or mediated by this first-formed layer). It should also be noted that this type of chemistry might be specific to high pressures and condensed-phase environments where much higher fluxes to the surface can provide significant numbers of collisions with the energetic species which may be necessary to access these boundary-layer pathways [5.17]. In concluding, however, we note that at very high temperatures, etching does occur; Al–X fragments are evolved into the gas phase at temperatures approaching 700 K. This reaction bears little resemblance to the facile ligand assisted pathways discussed above, however.

5.6 Model and Real Systems: A Comparison

The ultimate test of a research paradigm is whether it produces results of value in the understanding of real systems of technological importance. The knowledge we have obtained of aluminum surface chemistry through the use of fundamental UHV studies and model systems has indeed allowed us to develop a good understanding of Chemical Vapor Deposition (CVD) film-growth processes for this important metal. Consider, for example, a kinetic analysis developed around a rate-limiting surface reaction such as β-hydride elimination. We can determine the Arrhenius activation parameters for such a reaction with a fair degree of accuracy using the methods described above (i.e., see Figs. 5.10 and 12). The surface mediated thermolysis of TIBA is a very efficient and potentially important process for CVD metallization of semiconductor devices [5.46, 47]. The question must be addressed if knowledge of the underlying reactions matters for understanding growth in a CVD reactor. The plots shown in Fig. 5.20 argue powerfully that they do. The solid lines are calculated deposition rates [5.4, 5] while the experimentally measured values are indicated by the symbols [5.46, 48]. This rate-temperature profile was obtained using the following expression:

$$\text{Rate of Al deposition} = \frac{An_s \exp(-E_A/RT)}{(An_s/\sigma S_r)\exp(-E_A/RT) + 1} \tag{5.12}$$

where n_s is the number of surface sites, R is the gas constant, T is the growth temperature, and σ is the flux of incident molecules.

The model, which has been described elsewhere [5.4, 5], is based on simple Langmuir-adsorption kinetics, involving no coverage dependent or adjustable parameters with the exception of an assumed reactive sticking probability of near unity. The calculations are adjusted only for the fluxes which pertain at the much higher pressures present in a CVD reactor ($\approx$ 1–10 Torr). The activation parameters determined in TPD studies of TIBA predict these growth rates extremely well, the correspondence being best for the (1 1 1) surface. Most interestingly it is found that CVD aluminum films are heavily (1 1 1) testured [5.47]. The rate–surface structure sensitivity revealed in UHV allows us to

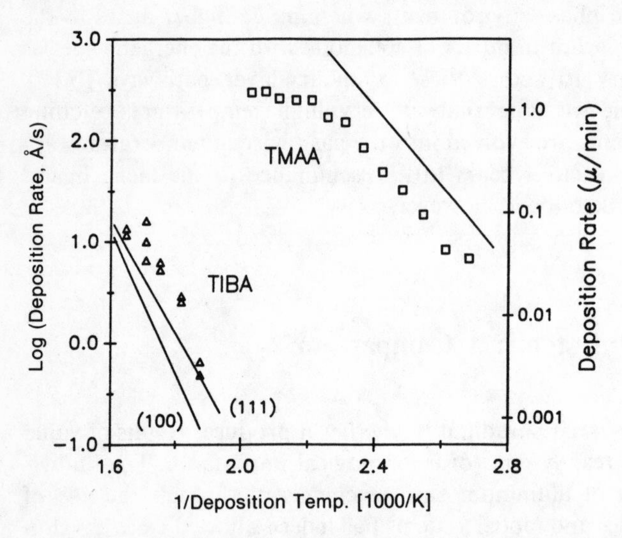

Fig. 5.20. Predicted (solid lines) and measured (symbols) Arrhenius plots [log (deposition rate) vs. 1/deposition temperature] are shown for the growth of aluminum films form the thermal decomposition of triisobutylaluminum (△) trimethylamine alane (□). The experimental data is taken from the work of *Cooke* et al. (TIBA) [5.46] and *Houlding* and *Coons* (TMAA) [5.48]. The development of the model is discussed in the text as well as in [5.4,5 and 50]

understand immediately why this is so. Random nucleation, when propagated, is always overtaken by the faster reaction velocities which obtain on the (1 1 1) surface. How much faster can we deposit a film using this precursor? We know clearly from this work that the limits in a cold-wall CVD reactor are imposed by the intrinsic energetics of a surface β-hydride elimination [5.49]. Does it make sense to raise the pressure beyond the ≈ 10 Torr reactor design used to obtain the data shown? Perhaps, but one should consider that, increasing the pressure from 10^{-4} to 10 Torr only changes the deposition rate by a factor of 2 at 500 K [5.5]! The origin of this effect is only transparent when the intricacies of the surface reaction mechanism are understood and accounted for in a correctly parameterized rate equation.

These kinetic studies allow us to define another striking feature of the TIBA deposition system, namely the autocatalysis which is characteristically seen in high-pressure CVD reactors. It is well known that TIBA can be used as a "selective" source gas: metal can be specifically deposited on "active" regions in the presence of other surfaces which only minimally nucleate film growth (e.g., metal oxides) [5.4, 47]. The thermodynamics of the TIBA deposition system are well known and are summarized in Fig. 5.21. The striking fact shown is that aluminum deposition from the thermal decomposition of TIBA is highly *endo*thermic (> 60 kcal/mol), so much so that it seems, at first glance, to be at odds with the experimentally determined reaction barrier for steady-state film growth on aluminum (≈ 30 kcal/mol) [5.4, 5]. The autocatalysis and high efficiency of this deposition process originate in a subtle feature of the underlying reaction mechanism: exhaustive thermolysis of TIBA results in the release of three equivalents of isobutylene (and a corresponding 3/2 equivalents of H_2). The key point of note here is that the β-hydride elimination steps involved in this net product formation are *kinetically* indistinguishable [5.4]. The "autocatalysis", then, is simply the consequence of surmounting the large reaction endothermicity via three uncoupled steps, each activated by ≈ 30 kcal/mol.

The discussions above provide a good qualitative picture of the mechanistic and kinetic features which distinguish the reactions leading to the facile surface mediated extrusion of aluminum from a TIBA adsorbate. For example, the figure suggests why the surface β-hydride elimination step is irreversible: olefin desorption is only very weakly activated rendering insertion processes kinetically incompetent at low pressures. We emphasize that, while descriptive, they provide only a very broad outline of the relevant processes. Significant reaction features – such as how the metal atoms get placed in the growing lattice, dynamical effects due to varying coverages of adsorbates and/or products, and diffusion rates of intermediates – are still incompletely understood.

The arguments given above, ones justifying fundamental investigations of this specific materials growth process, go well beyond the TIBA system. As shown in the figure, the use of well-defined UHV methods can shed insights on growth processes using other precursors as well. We have studied in our laboratory the utility of various adducts of alane (AlH_3 bound to donor ligands such as trimethylamine, TMAA [5.50, 51] and triethylamine, TEAA [5.51, 52])

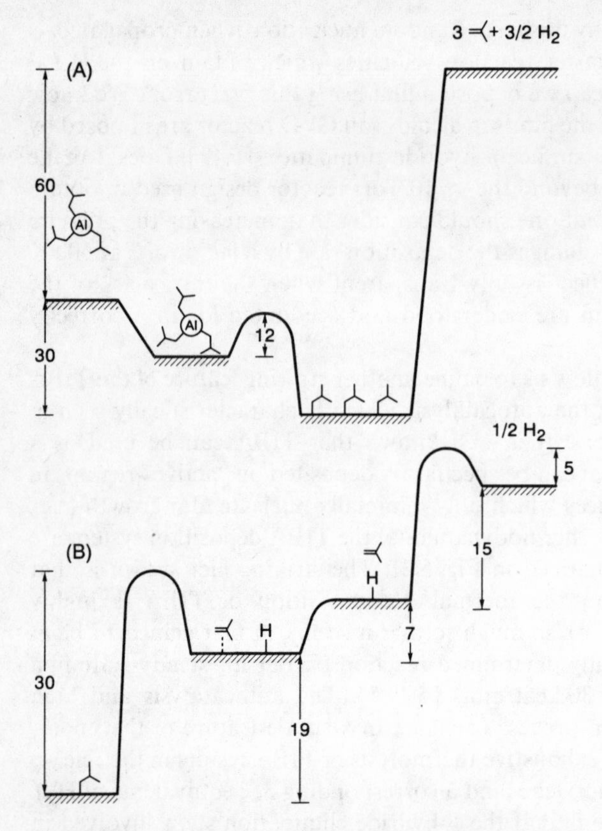

Fig. 5.21. (A) Approximate enthalpies under standard conditions at 298 K for the chemical vapor deposition of aluminum form TIBA. (B) Approximate energetics for β-hydride elimination form adsorbed isobutyl groups and for isobutylene and hydrogen desorption from an aluminum surface. All values are in kcal/mol. A more complete discussion can be found in [5.4]

as CVD aluminum precursors. The rate profiles follow closely what is predicted for the rate limiting bimolecular recombinative desorption of H_2 from a growing A1 surface. The growth kinetics predicted from UHV studies track those measured at higher pressures ($\approx$ 0.6 Torr *total* pressure, including carrier gas) in a cold wall CVD reactor, albeit not as compellingly as that demonstrated for TIBA (Fig. 5.20). We are concerned, however, that the fluxes are not as accurately known for this more complex reagent [5.50]. In addition, care must be used in interpreting the growth rates since an induction period is always present [5.48, 50, 51]. It is interesting to note, however, that unlike aluminum films grown with TIBA, films grown with TMAA are randomly oriented [5.51]. This is in good agreement with our reactive molecular-beam surface-scattering studies which showed that the growth rates on Al(1 1 1) and Al(1 0 0) surfaces were identical [5.50].

5.7 Aluminum Surfaces vs. Aluminum Compounds: A Summary

In this chapter we have considered several specific aspects of alkyl chemistry at various aluminum centers. We now conclude with a general comparison between the thermal reactions of alkyls on aluminum surfaces and in aluminum compounds. To make this comparison, we return to the potential reaction pathways outlined in Fig. 5.1 and consider each in turn, moving clockwise from reductive elimination on the lower right.

Reductive elimination (hydrogenation) reactions do occur at aluminum centers, but are clearly not a favored pathway either on the surface or in aluminum compounds. For aluminum compounds, the inhibiting factor is presumably the lack of accessible oxidation states to permit the required reduction of the coordination number by two ($HAlL_xR$ vs. $AlL_x + RH$). Aluminum trialkyls can be hydrogenated to yield alkanes, but conditions of 420 K and 300 atm are necessary to affect this transformation [5.32]. Under these extreme conditions, the reaction may, in fact, be heterogeneous, and the requirement of "impurities" for these reactions may reflect the inability of aluminum surfaces to readily dissociate hydrogen. Our studies on clean aluminum single-crystal surfaces show that, even when hydrogen atoms are coadsorbed with alkyl intermediates on the surface, chemistries other than hydrogenation are generally favored. Allyl and phenyl moieties are ligand classes that provide notable exceptions. A surface allyl intermediate readily absorbs hydrogen to form a 3-carbon metallacycle (this may be kinetically akin to having very high pressures of olefin present in experiments on hydrogen-covered aluminum surfaces) while phenyls decompose to evolve benzene. This latter chemistry has also been observed in solution and on surfaces at high pressures during the deposition of aluminum from $Al\phi_3$ [5.32]. We also note that trace amounts of methane can be observed from the reaction of adsorbed methyl groups and hydrogen [5.10].

Reductive coupling of alkyl intermediates in aluminum compounds would cause the same unfavorable reduction in coordination number mentioned above for the elimination reactions and is also not generally observed. Many of the reductive couplings reported in the literature have been ascribed to radical chemistry both in solution [5.32] and on surfaces (i.e., the reaction of two adsorbed methylene groups to form ethylene [5.28]). Our studies here suggest that reductive coupling also does not generally occur on the surface, even intramolecularly in the case of metallacycles. Some unidentified coupled products do apparently form, however, during the decomposition of aluminum allyls under hydrogen-deficient conditions and oligomers have been reported during the pyrolytic decomposition of triethylaluminum at higher pressures [5.32]. These latter reactions may represent a polymerization of the olefins rather than an elementary reductive coupling step.

γ-hydride elimination coupled with metathesis and β-alkyl elimination reactions are quite similar in that both evolve the same olefin. Our mass spectrometry studies would not be able to distinguish between these two mechanistic possibilities. Based on the well-established thermal degradation of alkyl ligands in aluminum compounds to evolve olefin (5.13), β-alkyl elimination appears to be the more prevalent path [5.53].

$$Al(CH_2C(CH_3)_3)_3 \rightleftharpoons Al(CH_3)_3 + 3\,(CH_3)_2C{=}CH_2 \uparrow \tag{5.13}$$

This chemistry is particularly well-established for trineopentyl aluminum, both on surfaces at higher pressures and in solution [5.32]. We also see evidence for this chemistry as a secondary pathway in the thermolysis of neopentyl groups on Al(1 0 0) under UHV conditions as well as in the reactive scattering of TIBA from aluminum at surface temperatures above ≈ 600 K [5.4].

Olefin insertion into metal-carbon bonds (the reverse of the β-alkyl elimination reaction in [5.13]) is the basis of the multi-billion dollar polyolefin industry. While this insertion chemistry is generally sluggish at aluminum centers, it will proceed under moderate olefin pressures. Under the low-pressure conditions of our surface experiments here, β-hydride elimination, a pressure-independent chain-termination reaction predominates. In the case of the 6-carbon metallacycle, however, the production of some methylenecyclopentane strongly suggests *intra*molecular olefin insertion into an aluminum–carbon bond (5.11). Such a process may also occur *inter*molecularly as evidenced by the high-mass products formed in some surface allyl decomposition reactions [5.3]. Apparently by tethering the olefin to the aluminum through an alkyl chain, one creates an effective "high pressure" of olefin.

Radical chemistry initiated by homolytic Al–C bond cleavage, as discussed above, is uncommon for aluminum both on surfaces and in discrete compounds. For example, the number of methyl radicals produced from the thermal decomposition of trimethylaluminum may be exceedingly small and may only be detectable because of the very high sensitivity of the laser-based techniques used to study them [5.10]. α-eliminations are also rare on surfaces [5.32] and have not, to our knowledge, been observed for aluminum compounds in solution. By far the most prevalent chemistry of alkyls in aluminum compounds and on aluminum surfaces is β-hydride elimination, and, as noted above, the kinetic similarities are striking.

5.8 Conclusion

The close mechanistic similarities between the thermal decomposition of organoaluminum reagents in solution and for the decomposition of adsorbed alkyls in our surface studies suggests an intriguing point of consideration. It is fairly easy to document that most gas and solution phase β-hydride elimination processes need only one metal atom to effect the transformation. Surfaces clearly offer a wider range of bonding possibilities. One might also expect that either bimolecular or unimolecular processes would predominate on the surface depending on the net effects of diffusion and adsorbate attraction/repulsion [5.54]. In the case of aluminum, however, these factors do not have a dramatic effect on the alkyl reaction pathways observed. Except for secondary chemistries such as reductive elimination and multidentate ligands such as metallacycles, we do not need to invoke either multiple metal sites or surface diffusion to understand or describe the surface chemistry we see. In general, it appears that aluminum alkyl compounds can be assembled with ligands in the proper proximity to mimic aluminum surface chemistry and vice versa. It would thus seem that, in the context of the question posed at the beginning of this chapter, cluster-surface analogies can give very powerful insights into complex surface processes.

References and Notes

5.1 See for example: M.R. Albert, J.T. Yates, Jr: *The Surface Scientist's Guide to Organometallic Chemistry* (Am. Chem. Soc., Washington, D.C., 1987)
E.L. Muetterties: Angew. Chem. **17,** 545 (1978)
E.L. Meutterties: Chem. Soc. Rev. **11,** 283 (1982)
N.D.S. Canning, R.J. Madix: J. Phys. Chem. **88,** 2437 (1984)
B.E. Bent, G.A. Somorjai: Adv. Coll. Intl Sci. **29,** 223 (1989)
5.2 B.E. Bent, R.G. Nuzzo, B.R. Zegarski, L.H. Dubois: J. Am. Chem. Soc **113,** 1137 (1991)
5.3 B.E. Bent, R.G. Nuzzo, B.R. Zegarski, L.H. Dubois: J. Am. Chem. Soc. **113,** 1143 (1991)
5.4 B.E. Bent, R.G. Nuzzo, L.H. Dubois: J. Am. Chem. Soc. **111,** 1634 (1989)
5.5 B.E. Bent, L.H. Dubois, R.G. Nuzzo: MRS Symp. Proc. **131,** 327 (1989)
5.6 G.H. Smuddle, Jr., X.D. Peng, R. Viswanathan, P.C. Stair: J. Vac. Sci. Technol. A **9,** 1885 (1991)
5.7 Thermal dissociation: X.-L. Zhou, J.M. White: Catal. Lett. **2,** 375 (1989) and references cited therein
Photochemical dissociation: X.-L. Zhou, J.M. White: Surf. Sci. **241,** 244 (1991) and references cited therein
Electron beam induced dissociation: X.-L. Zhou, J.M. White: J. Chem. Phys. **92,** 5612 (1990) and references cited therein
5.8 J.B. Benziger, R.J. Madix: J. Catal. **65,** 49 (1980)
F. Zaera: J. Phys. Chem. **94,** 8350 (1990) and references cited therein
5.9 M.B. Lee, Q.Y. Yang, S.T. Ceyer: J. Chem. Phys **87,** 2724 (1985)
S.T. Ceyer: Ann. Rev. Phys. Chem. **39,** 479 (1988)
5.10 D.W. Squire, C.S. Dulcey, M.C. Lin: Chem. Phys. Lett. **116,** 525 (1985), *ibid.*, J. Vac. Sci. Technol. B **3,** 1513 (1985)
D.R. Strongin, P.B. Comita: J. Phys. Chem. **95,** 1329 (1991)
5.11 T.R. Lee, G.M. Whitesides: J. Am. Chem. Soc. **113,** 2568 (1990) and references cited therein
5.12 M.L. Burke, R.J. Madix: J. Am. Chem. Soc. **113,** 3675 (1991); *ibid.* **113,** 4151 (1991)
5.13 R.R. Schrock, G.W. Parshall: Chem. Rev. **76,** 243 (1976)
P.J. Davidson, M.F. Lappert, R. Pearce: Chem. Rev. **76,** 219 (1976)
5.14 G.E. Coates, K. Wade: *Organometallic Compounds*, Vol. 1: *The Main Group Elements* (Metheun, London 1967)

5.15 J.P. Collman, L.S. Heggedus, J.R. Norton, R.G. Finke, *Principles and Applications of Organotransition Metal Chemistry* (University Sci. Books, Mill Valley, CA 1987)
5.16 A.V. Gross, J.M. Mavity: J. Org. Chem. **5,** 106 (1940)
H. Adkins, C. Scanley: J. Am. Chem. Soc. **73,** 2854 (1951)
5.17 See, for example, H.R. Rogers, J. Deutch, G.M. Whitesides: J. Am. Chem. Soc. **102,** 226 (1980)
H.R. Rogers, R.J. Roberts, H.L. Mitchell, G.M. Whitesides: J. Am. Chem. Soc. **102,** 231 (1980)
K.S. Root, J. Deutch, G.M. Whitesides: J. Am. Chem. Soc. **103,** 5475 (1981)
J. Barber, G.M. Whitesides: J. Am. Chem. Soc. **102,** 239 (1980) and references cited therein.
5.18 J.G. Chen, T.P. Beebe, J.E. Crowell, J.T. Yates Jr: J. Am. Chem. Soc. **109,** 1726 (1987)
5.19 R. Zheng, A.J. Gellman: J. Phys. Chem. **95,** 7433 (1991)
5.20 G.A. Crowder, S. Ali: J. Mol. Struc. **25,** 377 (1986)
5.21 See also J. Paul: Phys. Rev. B **37,** 6164 (1988)
A. Winkler, G. Požgainer, K.D. Rendulic: Surf. Sci. **251**/252, 886 (1991)
5.22 B.R. Zegarski, L.H. Dubois: Surf. Sci. Lett. **262,** L129 (1992)
5.23 J.R. Durig, S.E. Godbey, J.F. Sullivan: J. Phys. Chem. **92,** 6908 (1988)
5.24 A.J. Barnes, H.E. Hallam, J.D.R. Howells, G.F. Scrimshaw: JCS Faraday II **69,** 738 (1973)
5.25 A.W.E. Chan, R. Hoffmann: J. Vac. Sci. Technol. A **9,** 1569 (1991)
5.26 L.H. Dubois: Rev. Sci. Instrum. **60,** 410 (1989)
5.27 It is often not possible to ensure uniform operation of a quadrupole mass spectrometer, especially over a protracted series of experiments. The reference spectrum displayed in Fig. 5.7b depicts an example of how valuable IDMS can be, obviating some of these uncertainties. The spectrum shown was recorded with a slightly different transmission function than that used for the collection of the data in Fig. 5.7a: the weighting of the high-mass range is higher in the reference spectrum. Despite these small differences, there is little doubt as to the interpretation of the data.
5.28 A. Mödl, K. Domen, T.J. Chuang: Chem Phys. Lett. **154,** 187 (1989)
5.29 S.R. Heller, G.W.A. Milne: *EPA/NIH Mass Spectral Data Base*, Vol. 1, (EPA/NIH, Washington, DC 1978)
5.30 See, for example, J.A. Beran, L. Kevan: J. Phys. Chem. **73,** 3866 (1969)
5.31 K.W. Egger: J. Am. Chem. Soc. **91,** 2867 (1969); *ibid*, Intl J. Chem. Kinet, **1,** 459 (1969)
5.32 J.J. Eisch: Aluminum in *Comprehensive Organometallic Chemistry*, ed. by G. Wilkinson, F.G.A. Stone, E.W. Abel, (Pergamon, Oxford, 1982) Vol. 1, Chap. 6
T. Mole, E.A. Jeffery: *Organometallic Compounds* (Elsevier, New York 1972)
K. Ziegler: *Organometallic Compounds*, ACS Monograph Series, (Reinhold, New York 1960)
J.K. Kochi: *Organometallic Mechanisms and Catalysis* (Academic, New York 1978)
5.33 L. Melander, W.H. Saunders: *Reaction Rates of Isotopic Molecules* (Wiley, New York 1980)
5.34 See also J.M. Mundenar, R. Murphy, K.D, Tsuei, E.W. Plummer: Chem. Phys. Lett. **143,** 593 (1988)
5.35 See, for example, P.A. Redhead: Vacuum **12,** 203 (1962). Complications arise, however, if some reaction products remain on the surface and block sites during the decomposition process. A. Paul, C. Jenks, B.E. Bent: Surf. Sci. **261,** 233 (1992)
5.36 For a detailed discussion of the application of order plots, see D.H. Parker, M.E. Jones, B.E. Koel: Surf. Sci. **323,** 65 (1990) and references cited therein
5.37 D.H.S. Ying, R.J. Madix: J. Catal **61,** 48 (1980)
R.J. Madix: Colloid and Surfaces Division, Abstract 4, 201st ACS National Meeting, Atlanta, GA (1991)
5.38 G.S. Higashi, K. Ragavachari, M. Steigerwald: J. Vac. Sci. Technol. B **8,** 103 (1990)
5.39 G. Hata: Chem. Com., 7 (1968)
5.40 E. Maslowsky: *Vibrational Spectra of Organometallic Compounds* (Wiley, New York 1977)
5.41 While dienes other than the 1,5-compound are also reasonably (though not completely) consistent with the observed ratios, such products appear less likely based on the negligible amounts of 1,5-hexadiene produced. It is also clear, given that ions at $m/e = 84$ are present, that some portion of the products obtained are in fact a hexene (see above). There thus appears to be a range of complex reaction pathways being followed by these adsorbates. The similarities

noted in the products obtained from either, 1,6-diiodohexane or 6-bromo-1-hexene suggest that, for the C_6 metallacycle, the first step in the decomposition must be a β-hydride elimination. Both adsorbates, however, appear to access (in part) a cyclization reaction channel as well as possible isomerizations (perhaps via a reverse of the β-elimination with an inversion of the stereochemistry-- 1,2 $\rightarrow$ 2,1 etc.)
5.42 R. Reinacker, G.F. Gothel: Angew. Chem. **6,** 872 (1967)
5.43 For leading references on aluminum etching, see W. Kern, C.A. Deckert: In *Thin Film Processes*, ed. by J.L. Vossen, W. Kern (Academic, New York 1988) p. 401
5.44 See, for example, W. Sesselmann, T.J. Chuang: Surf. Sci. **176,** 67 (1986)
5.45 R.G. Nuzzo, L.H. Dubois: J. Am. Chem. Soc. **108,** 2881 (1986)
5.46 M.J. Cooke, R.A. Heinecke, R.C. Stern, J.W.C. Maes: Solid State Technol. **25,** 62 (1982)
5.47 M.L. Green, R.A. Levy, R.G. Nuzzo, E. Coleman: Thin Solid Films **114,** 367 (1984)
R.A. Levy, M.L. Green, P.K. Gallagher: J. Electrochem. Soc. **131,** 2175 (1984)
5.48 V.H. Houlding, D.E. Coons: in *Tungsten and Other Advanced Metals for ULSI Applications in 1990* ed. by G.C. Smith, R. Blumenthal, (Mater. Res. Soc. Pittsburgh, PA 1991) p. 203
5.49 We note that in a *hot*-wall CVD reactor Al growth rates form TIBA as high as 1μ/min have been obtained. T. Kobayashi, A. Sekiguchi, N. Hosokawa, T. Asamaki: Jpn. Appl. Phys. **27,** 11775 (1988)
5.50 L.H. Dubois, B.R. Zegarski, C.-T. Kao, R.G. Nuzzo: Surf. Sci. **236,** 77 (1990)
5.51 M.E. Gross, L.H. Dubois, R.G. Nuzzo, K.P. Cheung: MRS Symp. Proc. **204,** 383 (1991)
5.52 L.H. Dubois, B.R. Zegarski, M.E. Gross, R.G. Nuzzo: Surf. Sci. **244,** 89 (1991)
5.53 W.L. Smith, T. Wartick: J. Inorg. Nucl. Chem. **29,** 629 (1967)
5.54 B.E. Bent, C.-T. Kao, B.R. Zegarski, L.H. Dubois, R.G. Nuzzo: J. Am. Chem. Soc. **113,** 9112 (1991)

6. The Adsorption of Hydrogen at Copper Surfaces: A Model System for the Study of Activated Adsorption

H.A. Michelsen, C.T. Rettner, and *D.J. Auerbach*

One of the most exciting challenges of present-day surface science is the task of developing a detailed microscopic picture of surface chemical reactions. This task involves understanding the intra- and inter-molecular motions of species, as they undergo chemical change at a surface, and understanding the related issues of the energy requirements, energy flow, and energy disposal for these microscopic interactions. Studies directed at describing atomic and molecular motion and the interplay between molecular motion and energy throughout a surface process, such as chemisorption, physisorption, or scattering, define the field of surface chemical dynamics. The descriptions acquired from studies of surface dynamics can range from simple conceptual models, which yield insight into qualitative aspects of molecular interactions, to detailed theories, which provide quantitative information about dynamical processes.

In this chapter we shall discuss one of the most widely studied and best understood reactions in the field of surface chemical dynamics: the dissociative adsorption of hydrogen at copper surfaces and the reverse reaction of recombinative desorption. The hydrogen/Cu system has served as an important model system in the development of our understanding of surface chemical reactions. Here we shall review studies of this system starting with the earliest work performed a century and a half ago. We shall emphasize the concepts and the picture that has emerged from these studies, highlighting the progress made most recently on measuring and understanding state-resolved reaction probabilities.

6.1 Introductory Remarks

Analogous studies of the molecular dynamics of gas-phase processes have enjoyed considerable success recently, particularly in the area of detailed quantitative descriptions of chemical reactions. Not surprisingly, the leading example of this success is the extensively studied model reaction $H + H_2 \rightarrow H_2 + H$. The theoretical description of this reaction extends to full quantum mechanical reactive scattering calculations performed on an accurate *ab initio* potential energy surface to determine state-to-state cross sections and rate constants [6.1]. To date, the $H + H_2$ reaction is the only reaction for which such

Springer Series in Surface Sciences, Vol. 34
Surface Reactions Ed.: R.J. Madix

rigorous quantal scattering calculations have been performed. Performing such calculations for chemical systems involving larger reactive species quickly becomes prohibitively complex and often unnecessary for developing a computational description of the system. For many systems, particularly those involving heavier atoms, more approximate methods are adequate (e.g., quasi-classical calculations performed on semiempirical potential surfaces). These concerns are applicable to the description of surface reactions for which the many degrees of freedom added by the involvement of the surface greatly increase the computational complexity. Furthermore, the many degrees of freedom involved in the surface case make "single-collision" events inherently more difficult to study. For these reasons, and because of the many technical difficulties involved in working with clean surfaces, progress in the field of surface dynamics lags behind that of gas-phase dynamics. Nevertheless, considerable progress has been made in developing an understanding of the dynamics of surface processes; the H_2/Cu system is a good system for displaying this progress.

Although the study of surface dynamics is in its infancy relative to that of gas-phase reaction dynamics, the sophistication of current descriptions of the dynamics of the adsorption of hydrogen on copper verges on that possible for many "well-understood" gas-phase reactions. It is tempting to think of the direct dissociation of H_2 on Cu as the "$H + H_2$" of surface dynamics. That the H_2/Cu system has acquired such distinction is in part because this system is more easily treated theoretically than most other systems. Since the energy spacing between internal states of hydrogen is large, relatively few internal states have populations large enough to contribute significantly to dynamical processes at temperatures that are easily accessed. Furthermore, the electronic structure of hydrogen is simple, easing the task of developing interaction potentials. Copper is a good choice as a target material since it has a filled d-band, which makes its electronic structure easier to model than that of a transition metal possessing d-band holes.

Although theoretically more accessible than other systems, the H_2/Cu system has proven to be challenging experimentally. The adsorption of hydrogen on copper is activated; molecules impinging on the surface must overcome a relatively large activation barrier before dissociating directly (without occupying a molecular chemisorption state) on the surface [6.2]. That this system exhibits a barrier to adsorption is an advantage in studying the adsorption dynamics; that the barrier to adsorption is large, however, makes it difficult to measure. A large number of attempts have been made to measure the activation energy, with widely varying results (Table 6.1).

The present status of the H_2/Cu system as a model system in the study of activated adsorption is *not* a consequence of conclusive early experimental results. To the contrary, there has been a long history of confusing and contradictory results for this system. The first experimental work that we have found dates back almost 150 years. In 1843 *Dumas* [6.3] made measurements of the adsorption of hydrogen on copper as part of a study to determine the stoichiometry of water. A sketch of his apparatus is presented in Fig. 6.1; it used

Table 6.1. Activation energies measured for hydrogen adsorption on copper

Year	Authors and method	E_{act} [kcal/mol]	Surface	Temp[K]
1936	Melville and Rideal [6.26] Pressure change	19.6	Powder	345–443
1948	Rienäcker and Sarry [6.27] p-H_2 conversion	12.44 ± 0.38*	Film	623–893
1948	Kwan and Izu [6.111] Pressure change	20	Film	573–673
1949	Kwan [6.30, 112] Pressure change	20	Powder	573–673
1954	Mikovsky, Boudart, and Taylor [6.113] Isotope exchange	23.1	Foil	583–623
1956	Rienäcker and Vormum [6.114] p-H_2 conversion	11.2 ± 0.6*	Foil	623–893
1956	Eley and Rossington [6.115] p-H_2 conversion	9.6 ± 1.1* 10.3 ± 0.3*	Film Foil	373–573 353–453
1968	Völter, Jungnickel, and Rienäcker [6.116] p-H_2 conversion	9.1 13.5	(1 1 1) face (1 0 0) face	763–853
1971	Cadenhead and Wagner [6.117] Surface oxide reduction and isotope exchange	17 ± 2	Film	521–543
1972	Alexander and Pritchard [6.118] Surface potential change	8.5 ± 1.5*	Film	242–337
1974	Kiyomiya, Momma, and Yasumori [6.119] Isotope exchange	12.3 ± 1.4*	Powder	313–363
1987	Gabis, Kurdyumov, and Mazaev [6.120] Permeation rates	14.1 ± 0.1	Foil	790–1020
1990	Campbell, Domagala, and Campbell [6.110] Surface oxide reduction	14.3 ± 1.4	(1 1 0) face	473–723
1992	Rasmussen et al. [6.22] Temperature programmed desorption	11.4 ± 0.2	(1 0 0) face	218–258

Note: *Where a range of values was given, error bars indicate range.

CuO as a source of oxygen. To ensure that his measurements pertained to the interaction of hydrogen with oxygen and not with copper, he also made measurements of the adsorption of hydrogen on pure copper powder. He concluded that hydrogen adsorbs at low but not high reaction vessel temperature. This conclusion indicated that adsorption was not activated, contrary to our current viewpoint. Although a number of similar studies of this system were made over the next 90 years, there was little evidence that the adsorption of H_2 on Cu was an activated process when, in 1931, *Taylor* [6.4] proposed the idea of activated adsorption using the H_2/Cu system as one example and a year later when *Lennard-Jones* [6.5] proposed a dynamical model for this process also using the H_2/Cu system as an example. These two prominent papers motivated much of the experimental and theoretical work that followed.

Since then considerable progress has been made in understanding activated adsorption, particularly in understanding the details of the adsorption dynamics of the H_2/Cu system. To date more detailed information is available for the

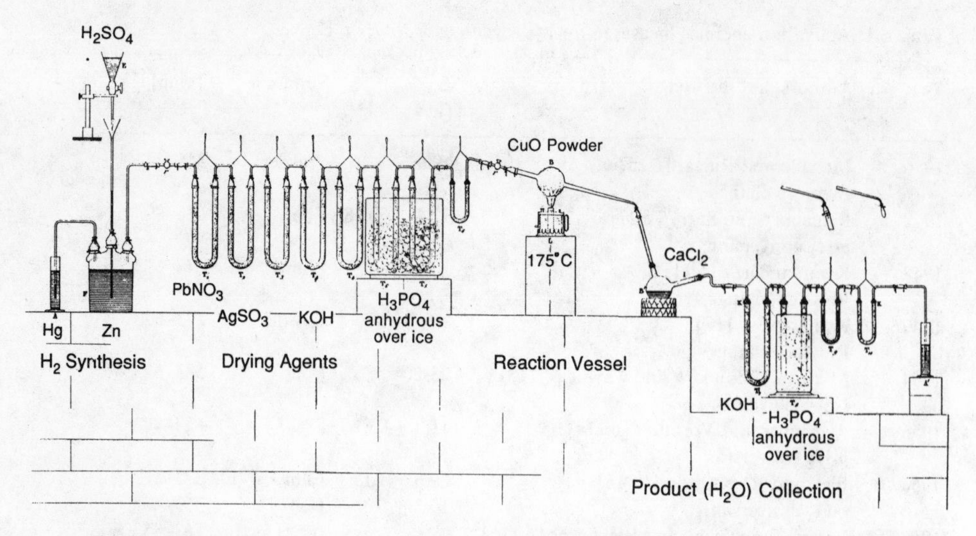

Fig. 6.1. Sketch of the apparatus used by *Dumas* [6.3]. Hydrogen was generated by reacting sulfuric acid with zinc and was dried with a series of dessicants. The stoichiometry of water was then measured by weighing the amount of water produced from the reaction of hydrogen with copper oxide

H_2/Cu system than for any other system. Because this system has served as a model system in the development of gas–surface dynamics, it is instructive to trace the evolution of the field of gas–surface dynamics by tracing the history of experiment and theory concerning the dynamics of hydrogen's interactions with copper.

In this chapter we shall summarize the major developments in the description of adsorption and desorption in the H_2/Cu system, emphasizing the interplay between experiment and theory involved in the process of converging on a detailed picture of the adsorption/desorption dynamics. The chapter is organized as follows. Section 6.2 begins with the earliest dynamical descriptions of the activated adsorption of molecules on metal surfaces, which made use of a one-dimensional interaction potential. This picture is based on separately determining the potential energy of the molecule and the two atoms (resulting from dissociation of the molecule) as a function of their distance from the surface. The barrier to adsorption is defined by the potential at the crossing between the atomic and molecular potential curves. We begin in Sect. 6.2.1 by discussing this one-dimensional picture, the experimental results supporting the idea of a barrier, and the experimental estimates of the barrier height. Because molecular beams can be made to have a narrow range of kinetic energies, adsorption measurements made with molecular beams provide a more direct measure of the barrier height than do measurements of the activation energy made with molecules incident from an equilibrium distribution. In Sect. 6.2.2 we discuss early results of adsorption measurements made with molecular beams. Desorption measurements, when viewed from the perspective afforded by the principle of detailed balance, can also be used to understand the dynamics of adsorption. In Sect. 6.2.3 we present results of desorption measurements, we introduce some of the inconsistencies between these measurements and the early adsorption measurements with regard to detailed balance, and we explain how attempting to resolve some of these inconsistencies naturally leads to a description of the adsorption/desorption dynamics that relies on a second dimension, the intramolecular bond distance of the incident molecule, in addition to the molecule–surface distance.

Section 6.3 extends the one-dimensional description developed in Sect. 6.2 to two dimensions, focusing on the role of the two degrees of freedom of translational and vibrational motion in the adsorption and desorption dynamics. Section 6.3.1 introduces one way of visualizing a chemical system through a two-dimensional Potential Energy Surface (PES). This two-dimensional PES describes the potential energy of the system in terms of the height above the metal surface of the incident molecule or its component atoms and the intramolecular distance between the atoms of the diatomic molecule. The latter part of Sect. 6.3.1 describes how the gross features of such a surface and their location can influence the dynamics and how the topography of the PES can be probed experimentally. Finally, a brief account is given of a 2-D PES currently used to describe the H_2/Cu system. Calculations performed on this PES show that the vibrational energy of the incident molecule should play a significant role in its ability to overcome the barrier to adsorption. In Sect. 6.3.2 we describe several adsorption experiments that confirm the theoretical result presented in Sect. 6.3.1. In Sect. 6.3.3 we present a model for adsorption that allows us to analyze these adsorption data quantitatively, yielding a picture of the dependence of the adsorption probability on the vibrational state, translational energy, and angle. The second part of this subsection discusses the quantitative comparison of adsorption with desorption data via detailed balance and introduces the discovery that surface motion may also play a significant role in the adsorption/desorption process. While the measurements discussed in Sects. 6.3.2 and 6.3.3 give (vibrationally) state-resolved information about adsorption indirectly, state-to-state scattering measurements can provide this information directly. Such measurements are presented in Sect. 6.3.4 along with the surprising result that molecules striking the surface with enough translational energy to overcome the adsorption barrier can be inelastically scattered into higher vibrational states. Section 6.3.5 describes recent state-resolved desorption experiments that surpass molecular beam experiments in being able to provide (via detailed balance) detailed information about the dependence of the adsorption probability on kinetic energy.

In Sect. 6.4 we analyze theoretical and experimental results that address the involvement of degrees of freedom beyond vibration and translation. These degrees of freedom include rotation, molecular orientation, and impact parameter (i.e., where in the unit cell the incident molecule strikes the surface). Recent desorption experiments have shown that the role of rotation in adsorption is complex. For high rotational states J, adsorption is enhanced as the rotational energy is increased; for low J states, however, rotation hinders adsorption. These results are described in Sect. 6.4.3 and are supported by (1) the dependence of vibrational inelastic scattering on rotational state presented in Sect. 6.4.4 and (2) the dependence of adsorption rate on the rotational temperature of thermal reactants presented in Sect. 6.4.2. We conclude in Sect. 6.5 by accepting the challenge of summarizing in a few paragraphs the understanding developed over the past century and a half of the dynamics of the adsorption and desorption of hydrogen on copper.

6.2 A One-Dimensional Description: The Translational Degree of Freedom

6.2.1 The Activation Barrier

Lennard-Jones [6.5], in 1932, sparked theoretical interest in the mechanism for activated, dissociative adsorption when he proposed a model to describe "the nature" of this process. In this classic paper he presented a one-dimensional potential for the interaction of a molecule with a metal that included a crossing between a repulsive molecular state (Curve 1 in Fig. 6.2) and an attractive atomic state (Curve 2). As stressed in this paper, the crossing (Point K) can result in a barrier, the height of which (P) depends on where the crossing takes place. This qualitative picture was based on models proposed to explain predissociation in molecular spectroscopy and, although called tentative at the time [6.5, 6], established the paradigm of activated, dissociative adsorption. This work also suggested the course of study necessary to understand in detail "the nature" or dynamics of such interactions.

Although *Lennard-Jones* used the H_2/Cu system as an example for his model of activated, dissociative adsorption, little was known about this system at the time. Published experimental results were inconclusive and highly controversial in spite of the fact that the first measurement of the adsorption of H_2 on copper had been made nearly 90 years earlier. In 1843 *Dumas* [6.3] concluded that hydrogen adsorbs spontaneously onto copper powder at low temperatures. Although later studies appeared to confirm this observation [6.7,

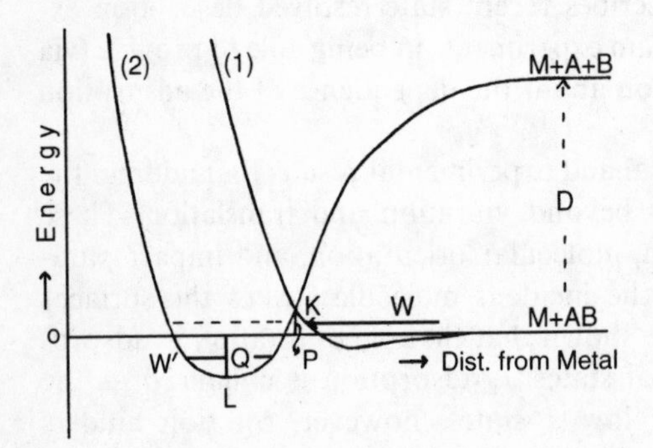

Fig. 6.2. "The interaction of a molecule with a metal". This figure was copied from the classic 1932 paper by *Lennard-Jones* [6.5], which established the conceptual basis for many studies of activated, dissociative adsorption

8], other experiments had shown that hydrogen does not adsorb at temperatures below 491°C [6.9–12]. Since then a number of experiments have confirmed the latter observation that hydrogen does *not* adsorb at low reactant temperatures [6.13–21]. In fact, the most recent studies estimate a reaction probability of $\approx 10^{-14}$ for the adsorption of H_2 on Cu(100) at ≈ 220 K [6.22].

Considering the controversy surrounding this system in 1932 it is likely that *Lennard-Jones* had been inspired by the work of *Taylor* [6.4] published a year earlier. Disregarding the conflicting experimental results, *Taylor* had proposed that for systems for which chemisorption only proceeded at high temperatures, H_2/Cu included, there must be a significant activation barrier to adsorption. Although he acknowledged that there was insufficient evidence in the published data to lead to this conclusion about the H_2/Cu system, *Taylor* supported his claim by quoting unpublished data. Based on these data he asserted that "the activation energy necessary in the case of the hydrogen–copper system is markedly higher than with nickel, platinum and palladium." This assertion has indeed proven to be correct [6.23–25].

The activation energy is a measure of the amount of energy required to surmount a barrier to reaction. If there is a single, classical barrier (independent of impact position in the unit cell, impact angle, tunneling, etc.), then the adsorption will follow the Arrhenius equation. That is, the adsorption rate will be given by $A\exp(-E_a/kT)$, where E_a is the activation energy, k is the Boltzmann constant, T is the reactant temperature, and A is the pre-exponential factor. The Arrhenius activation energy E_a can be determined from measurements of the adsorption rate as a function of T for gas-phase reactants at equilibrium impinging on the surface from all directions. For a system that can be described strictly by a classical, one-dimensional interaction potential, such as that shown in Fig. 6.2, this type of measurement can indeed yield the height of the barrier to adsorption.

It wasn't until 1936, several years after the prominent work of *Lennard-Jones*, that *Melville* and *Rideal* [6.26] made the first measurements yielding a value for the activation energy for adsorption. This measurement was made by filling a glass bulb connected to a vacuum pump and pressure gauge with copper powder. After the copper was cleaned and the bulb evacuated, a calibrated volume of H_2 was let into the chamber, and the pressure was monitored over time at a series of different temperatures. This experiment gave an activation energy of 20 kcal/mol. Twelve years later in 1948 *Rienäcker* and *Sarry* [6.27] confirmed that the adsorption of hydrogen on copper is dissociative as well as activated with a significant activation energy. They used a different approach to measure the activation energy from that used by *Melville* and *Rideal*, which involved measuring the conversion of *para*-hydrogen to *ortho*-hydrogen. *Para*-hydrogen has a different thermal conductivity than does *ortho*-hydrogen and does not convert to *ortho*-hydrogen without first dissociating. By starting with *para*-hydrogen and monitoring the change in thermal conductivity, one can measure the rate of adsorption at temperatures above the desorption temperature. Because these measurements are made at higher temperatures, this approach has two advantages over that used by *Melville* and *Rideal*: 1) the adsorption probability is larger when the reactant gas is at a higher temperature and thus more easily measured, and 2) the higher surface temperature ensures that the rate can be measured without a significant build-up of adsorbed hydrogen on the surface. The value recorded in this way by *Rienäcker* and *Sarry* [6.27], 12–13 kcal/mol, however, did not agree with the value of 20 kcal/mol reported by *Melville* and *Rideal*.

Based on these results, in addition to some of their own measurements, *Couper* and *Eley* [6.28] presented the observation that a large activation energy to hydrogen adsorption is characteristic of metals, such as Cu, Ag, and Au, that possess a filled d-band. This observation confirmed *Taylor's* [6.4] assertion that the activation energy is higher for copper than for Ni, Pt, and Pd, none of which have a filled d-band, and prompted *Trapnell* [6.15] to put forward the theory that was accepted for many years that surface bonds formed with vacant d-orbitals in the metal. Such bonding would be easily achieved in metals that possessed d-holes. In metals such as copper, however, rehybridization would first have to occur, in which d-electrons would be promoted to the s-band. *Trapnell* [6.15] called this rehybridization "d–s electron promotion."

Currently it is believed that the first step to adsorption is overcoming the Pauli repulsion between the ground state of the molecule and the s-electrons at the surface. Pauli repulsion results from enhanced overlap of the $1\sigma_g$ orbital of the molecule with the s-band of the metal surface. The $1\sigma_g$ is a filled valence state, and hence the Pauli principle requires that this level acquire no additional electron density beyond the two electrons already occupying it. If the molecule has enough energy to overcome this repulsion far from the surface and continue along the reaction coordinate toward the surface, interaction of the s-band with the $1\sigma_u$ antibonding orbital allows for a sudden change in electronic configuration as the H_2 molecule dissociates. For metals that have holes in the d-band, Pauli repulsion is alleviated by rehybridization of the s-band with the d-band, such that electron density from the s-band is pushed back into the d-band (in contrast to d–s promotion), resulting in a lower activation energy to adsorption. Since copper has a filled d-band, such rehybridization is impossible; hence the activation energy to hydrogen adsorption is relatively large [6.29].

Although the filled d-band has made the H_2/Cu system theoretically accessible, it is also responsible for the large activation energy that has made the study of this system experimentally challenging. In addition to the measurements of *Melville* and *Rideal* [6.26] and those of *Rienäcker* and *Sarry* [2.27] a large number of measurements of the activation energy (summarized in Table 6.1) have been performed yielding values ranging from 7 to 23 kcal/mol. For the H_2/Cu system these measurements are difficult since the relatively large activation barrier causes the adsorption probability to be very small. Therefore, for most of these measurements, pressures must be high and dosing times long to measure the adsorption probability. *Kwan* [6.30] made measurements with dosing times as long as a week. These conditions can lead to coadsorption of impurities, or, in some cases, to a build-up of adsorbed hydrogen, which can influence the rate of adsorption.

6.2.2 Early Adsorption Measurements with Molecular Beams

A more direct way of probing the barrier to adsorption involves controlling the incidence conditions using molecular-beam techniques. In these experiments a molecular beam is directed at the surface while the kinetic energy and angle of incidence are controlled. This approach allows the adsorption probability to be recorded as a function of incidence energy, particularly at higher energies where the adsorption probability may be larger and therefore measured more accurately. In contrast to measurements of the activation barrier in which energies of incident gas can be described by equilibrium distributions, measurements made with molecular beams generally involve narrow translational energy distributions. Measuring the adsorption probability as a function of incidence kinetic energy, E_i, and angle relative to the surface normal, θ_i, can give the translational threshold to adsorption, the amount of translational energy required to overcome the barrier. Such measurements can also give an indication of the effective corrugation of the molecule–surface potential in the adsorption process.

In 1974, *Balooch* et al. [6.31] reported the results of a pioneering study in which a molecular beam was used to measure activated adsorption probabilities for the first time. They measured the adsorption of hydrogen on Cu(1 1 1), Cu(1 0 0), and Cu(3 1 0) as a function of incidence kinetic energy and angle, and

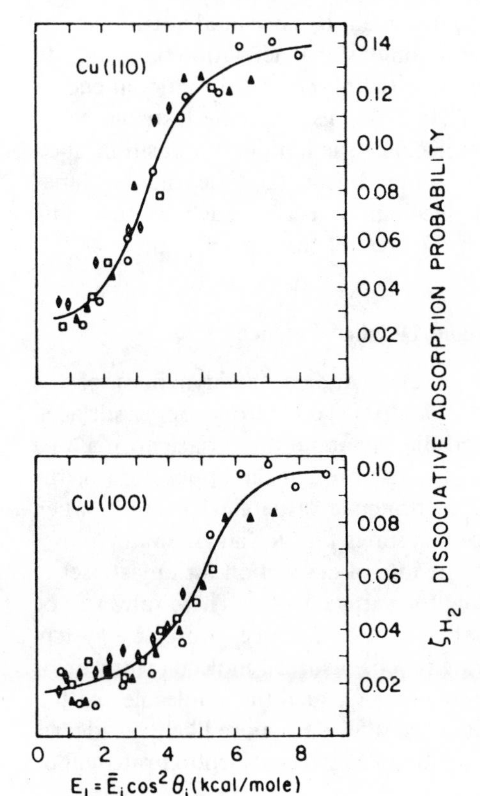

Fig. 6.3. Dissociative adsorption probabilities for H_2 on Cu. Values are given (right ordinate) for two faces of Cu, as indicated. From [6.41] with permission

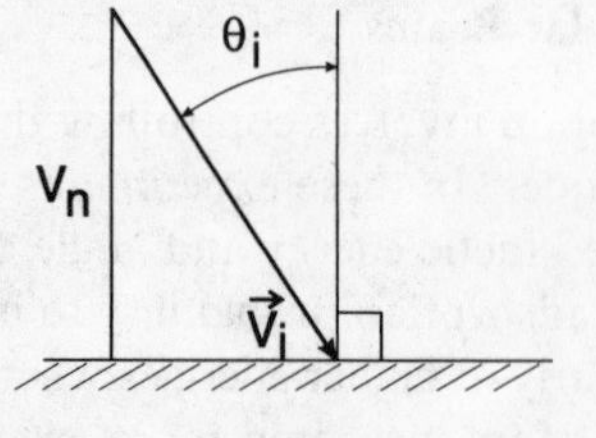

$$E_n = \frac{1}{2} m v_n^2 = \frac{1}{2} m |v_i|^2 \cos^2(\theta_i)$$

Fig. 6.4. Angle of incidence and "normal energy". The vector V_i represents the velocity vector of an incident molecule. V_n is the component of V_i normal to the surface. The angle θ_i is the incidence angle

found a dramatic increase in the adsorption probability as the energy of incidence was increased, followed by a saturation of the adsorption probability at higher energies. This behavior is described by an S-shaped (or sigmoidal) adsorption curve as a function of kinetic energy (Fig. 6.3). Further, results recorded at different angles of incidence were found to depend only on the translational energy associated with momentum normal to the surface. Similar behavior has since been reported for a number of other systems [6.24, 32–37]. In these cases the results are said to scale with the so-called "normal energy", $E_n = E_i \cos^2 \theta_i$ (Fig. 6.4). If the adsorption were controlled by a single, classical, one-dimensional barrier, the curve describing the dependence of the adsorption probability on the incidence kinetic energy would be expected to have the shape of a step function instead of the S-shape observed. *Balooch* et al. attributed the observed behavior to a distribution of one-dimensional activation barriers with a height distribution given by differentiating the observed translational energy dependence of adsorption probability. They claimed that the distribution of barriers could be associated with the molecular orientation and position upon impact with the surface. Although not pointed out at the time, the "S-shape" could also be the result of quantum mechanical effects, such as tunneling through the barrier and reflection from the crest of the barrier [6.38].

6.2.3 Early Desorption Measurements and Detailed Balance

The group of *Stickney* and co-workers also studied the dynamics of the recombinative desorption of H_2 [6.39, 40] and HD [6.31] from copper surfaces. They found angular distributions peaked sharply about the surface normal in a manner consistent with the adsorption results through the application of the principle of detailed balance [6.41]. The principle of detailed balance, when applied to adsorption/desorption processes, states that for a system at equilibrium the rate of adsorption is equal to the rate of desorption for any subset of molecules of the equilibrium flux striking the surface [6.42]. These rates can be related to adsorption/desorption probabilities. For instance, consider a system for which the adsorption is direct (having no chemisorbed molecular precursor states) and activated, and depends directly on E_n, such that molecules with a large velocity component normal to the crystal face are more likely to adsorb. The principle of detailed balance suggests that since the adsorption rate will be higher for molecules at small incidence angles (which have higher normal energies), the desorption rate should *also* be higher at these angles. The angular dependence of *adsorption* is then reflected by an angular distribution of *desorption* flux that is more sharply peaked (has higher flux) at angles close to the surface normal. Previously it had been shown that the principle of detailed balance could be applied to relate adsorption/reflection measurements to measurements of peaked angular distributions in desorption made on systems under quasi-equilibrium conditions [6.43–46]. Through the application of detailed balance *Cardillo* et al. showed that the peaked angular distributions measured were consistent with their adsorption measurements (Fig. 6.5), apparently confirming that the principle of detailed balance could be applied under the quasi-equilibrium conditions established in a molecular–beam experiment. As we shall see in Sect. 6.3.2, however, the adsorption measurements have since been found not to be reproducible. Recent studies indicate that the barrier to dissociation is much higher than suggested by the molecular beam study of *Balooch* et al. The consistency of the adsorption and desorption measurements via detailed balance reported by *Cardillo* et al. appears, in retrospect, to be somewhat fortuitous. We shall see in Sect. 6.3.3 that detailed balance does, nevertheless, apply to the adsorption and desorption of D_2 on Cu(1 1 1).

Given the apparent agreement between the angular distribution and adsorption measurements, it was expected that the velocity distribution would similarly be consistent with the adsorption measurements when compared via

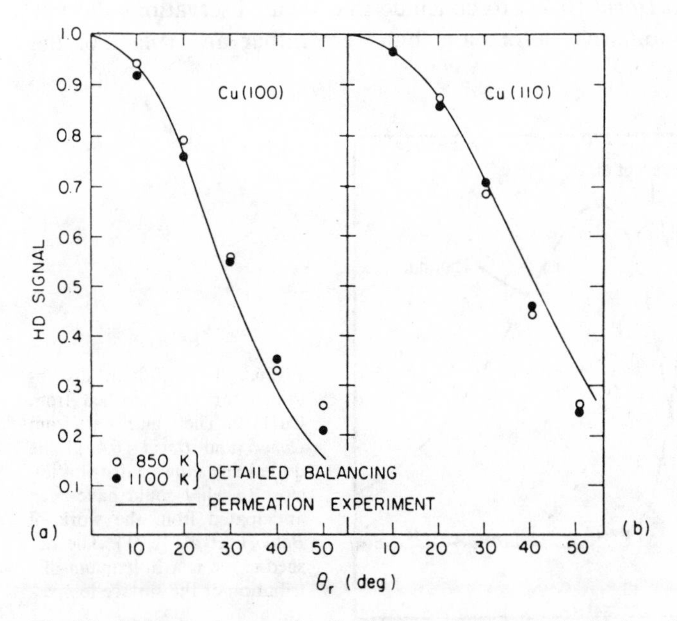

Fig. 6.5a, b. Desorption angular distributions. Calculated distributions (lines) are compared with the measured distribution from post-permeation, recombinative desorption of H_2 from two faces of copper, as indicated. From [6.41] with permission

detailed balance. Measurements made by *Comsa* and *David* [6.47], however, revealed that mean energies of D_2 desorbed from Cu(1 0 0) and Cu(1 1 1) were 0.68 and 0.64 eV, approximately four times the mean energy of a thermal distribution at the surface temperature and far in excess of the value of ≈ 0.3 eV anticipated from the work of *Balooch* et al. [6.31, 47]. The x-marks in Fig. 6.6 represent the time-of-flight measurements recorded by *Comsa* and *David* [6.47]. The dotted curve demonstrates a distribution that could have been expected from the work of *Balooch* et al. This distribution appears at a later time corresponding to a lower energy for the desorbed molecules. The dashed curve depicts a Maxwell–Boltzmann distribution at the surface temperature of 1000 K, which represents the anticipated distribution for a system with no adsorption barrier.

An additional surprise resulting from the work of *Comsa* and *David* involved the angular dependence of the measured velocity distributions. The mean energy was found to be independent of desorption angle (Fig. 6.7). This result is an apparent contradiction of the predictions by detailed balance. Since the adsorption probability was found to scale with E_n (i.e., the beam energy required to overcome the barrier increased as the angle of incidence increased) [6.31], one might naively expect, on the basis of detailed balance, that the kinetic energy of desorbed molecules should increase sharply as the desorption angle increases. The anticipated increase in kinetic energy with angle is shown by the dashed curve in Fig. 6.7. The apparent disagreement between the magnitude and angular dependence of the desorption energy with the detailed balance predictions led *Comsa* and *David* [6.47] to conclude that their observations did not apply to true recombinative desorption, but were rather an artifact of the

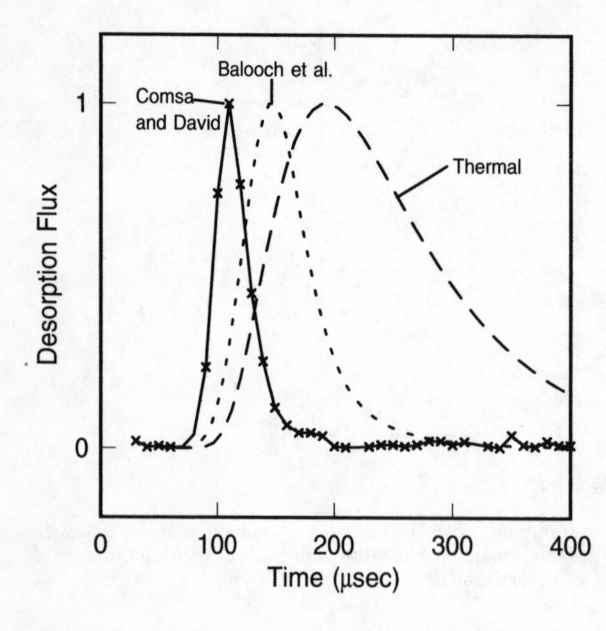

Fig. 6.6. Time-of-flight distributions for D_2 desorbed from Cu(1 0 0). The x-marks are from *Comsa* and *David* [6.47]. The dotted line shows a distribution close to what could have been anticipated from the work of *Balooch* et al. [6.31]. The dashed curve is a Boltzmann distribution at the surface temperature

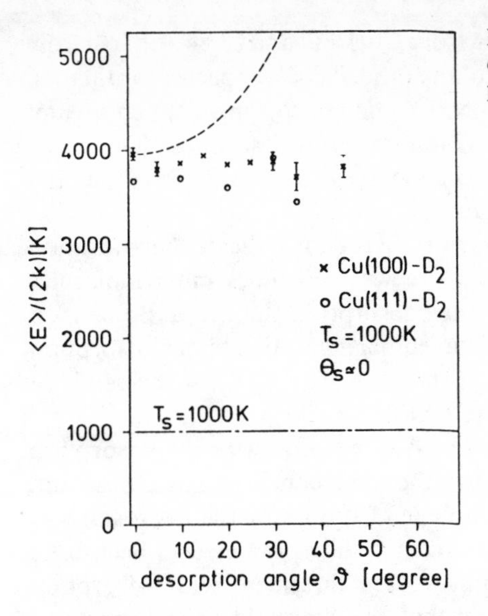

Fig. 6.7. Desorption mean energy. Mean energy of D_2 molecules desorbed from two faces of copper, as indicated. From [6.47] with permission

method by which they supplied the hydrogen to the surface, i.e., via atomic permeation. In this method molecules are supplied to the back side of a thin crystal at high temperature, where they dissociate and diffuse through the crystal as atoms before recombining on the front face and desorbing. They proposed that molecules adsorbed directly populate a *surface* chemisorbed state, whereas molecules desorbed after permeation originate from recombination of atoms in a *subsurface* state (Fig. 6.8).

Further difficulties of the one-dimensional barrier picture in adequately describing adsorption and desorption for the hydrogen–copper system arose in connection with the first state-resolved measurements on this system. *Kubiak* et al. [6.48] measured the internal state distributions of H_2 and D_2 desorbed after permeation through Cu(1 1 0) and Cu(1 1 1). For H_2 they observed a population

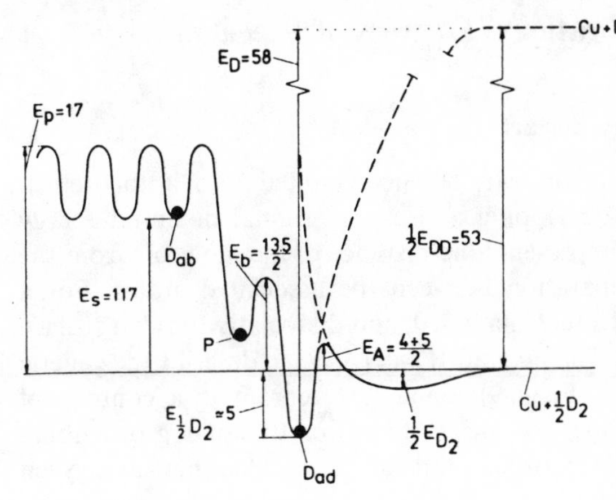

Fig. 6.8. Hypothetical potential energy diagram for subsurface species. One-dimensional potential energy diagram for D_2/Cu and D/Cu illustrating the possibility of post-permeation desorption from a region of elevated potential in the sub-surface region. (Energies are in kcal/mole/D atom.) From [6.47] with permission

ratio, $P(v=1)/P(v=0)$, of 0.052 ± 0.014 from Cu(1 1 0) and 0.084 ± 0.030 from Cu(1 1 1). These ratios are well in excess of the ratio, 0.0009, expected for thermal equilibrium at the surface temperature (850 K) demonstrating a much greater than statistical disposal of energy into vibration upon desorption. This effect was clearly outside of the one-dimensional barrier picture, which does not include the vibrational degree of freedom.

Through detailed balance, the results of *Kubiak* et al. indicate that vibration enhances adsorption. Simply put, since there were 50 times more molecules populating the $v=1$ vibrational state after desorption than in a Boltzmann population at the surface temperature, detailed balance indicates an adsorption probability 50 times higher for $H_2(v=1)$ than for $H_2(v=0)$, for the range of incidence conditions relevant to the system at 850 K.

Kubiak et al. were unable to reconcile this prediction with the adsorption data of *Balooch* et al. They were thus led to the same conclusion as *Comsa* and *David* that the mechanism for desorption depended on how the hydrogen was originally supplied to the surface. More recent results presented in Sect. 6.3.2 bring these desorption measurements into closer agreement with adsorption measurements. These recent results show that the threshold to adsorption is much higher than deduced from the results of *Balooch* et al. and are consistent with both velocity distributions measured by *Comsa* and *David* and the internal state distributions of *Kubiak* et al. Although the reason for the low thresholds given by the experiments of *Balooch* et al. still remains unresolved, this agreement removes the need for separate mechanisms for post-permeation desorption and desorption after direct adsorption. Nonetheless, the striking results of *Kubiak* et al. showing a large excess population of vibrationally excited molecules desorbed from the surface clearly showed that the interaction of H_2 with copper is not strictly one-dimensional. That is, the vibrational degree of freedom must be taken into consideration in the dynamical description of this system.

6.3 A Two-Dimensional Description: The Translational and Vibrational Degrees of Freedom

6.3.1 The 2-D Potential Energy Surface

The vibrational degree of freedom can be incorporated into a theoretical description of this system by developing a two-dimensional potential energy surface, where one coordinate represents the distance of the molecule from the surface, and the other, the separation between the associated atoms. For a system for which adsorption is direct, activated, and dissociative, such a surface resembles those constructed by *Polanyi* and *Wong* [6.49] to describe the generic gas-phase reaction of A + BC (Fig. 6.9). Each line represents a contour of constant potential energy for the A + BC system. For the analog describing adsorption, A represents the metal surface, and BC, the incident molecule. Such a surface is characterized by:

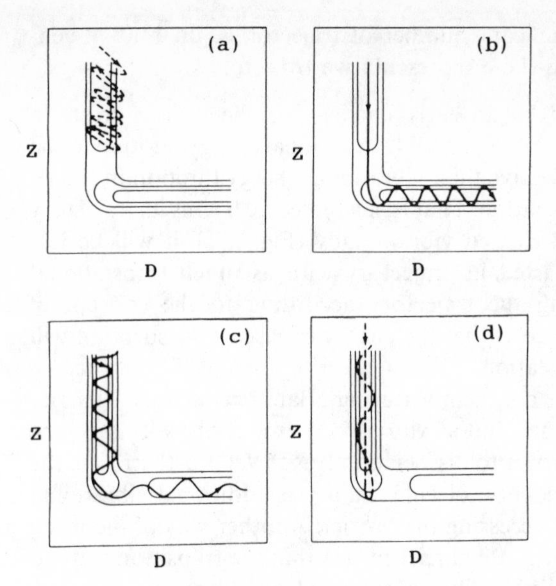

Fig. 6.9a–d. Two-dimensional potential energy surfaces for a diatomic molecule–surface interaction. The figure illustrates the effect of the location of the barrier on the translational and vibrational energy requirements for dissociation. "Z" is the distance between the incident molecule and the solid surface. "D" is the internuclear distance between the geminate atoms of the incident molecule. Adapted from [6.121]

1) A potential valley positioned along an axis marking the unperturbed molecular internuclear separation. This feature is the entry or reagents' valley or entrance channel to adsorption. At distances far from the surface (large Z) the influence of the surface on the molecule is small. At large Z the shape of the PES is thus defined by the gas-phase molecular potential, which is usually very well determined from gas-phase spectroscopic studies.
2) A potential valley similarly positioned along an axis marking the equilibrium adsorbate–surface separation. This feature is the exit or products' valley or exit channel to adsorption (the entrance channel to desorption). At large separation between atoms (large D) the shape of the valley corresponds to the equilibrium condition of atoms bonded to the surface and, although difficult with hydrogen, can in principle be determined by the application of suitable surface spectroscopies such as Electron Energy Loss Spectroscopy (EELS).
3) A barrier positioned in between the exit and entrance channels. Since this intermediate region is not defined by equilibrium characteristics of the system, the topography of this region is the least easily determined. The adsorption/desorption dynamics, however, are the most sensitive to the topography of this region, particularly to the location of the crest of the activation barrier. Demonstrating how the barrier region controls the dynamics of adsorption and desorption can be accomplished by examining several suggestive trajectories, such as the classical trajectories added to the potential surfaces in Fig. 6.9.

These trajectories were obtained by solving the equations of motion. The results of these solutions can be visualized easily, however, by picturing the physical trajectories of a ball allowed to roll on the surface without friction. Although a complete study requires that a wide range of initial conditions be sampled for a statistically significant number of trajectories, qualitative generalizations can be drawn from these representative trajectories:

1) For a system characterized by an early barrier, i.e., the barrier crests in the entrance channel to adsorption, initial translational energy (motion in the *Z*-direction) will be more effective than vibrational energy (motion in the *D*-direction) in leading to dissociation. That is, if a molecule begins its trajectory along the entrance channel excited vibrationally (Fig. 6.9a), it will be less likely to react than if it started its trajectory with as much translational energy (Fig. 6.9b). Reversing the trajectory according to the concept of microscopic reversibility suggests that the energy released in desorption will go predominantly into translation.
2) The situation is different for a system featuring a late barrier, i.e., a barrier that crests in the exit channel. Initial vibrational excitation will assist the molecule in accessing the barrier to dissociation (Fig. 6.9c). Furthermore, the zero-point energy of the incident molecule can act as additional vibrational energy available to assist in accessing the barrier. Another way of picturing this process is by noting that a late barrier means that the transition state to dissociation will be distinguished by an extended equilibrium molecular-bond length. Vibrational motion may help the molecule access the transition state and is thus translated into motion along the reaction coordinate. Initial translational energy, on the other hand, may not be as effective in helping the molecule surmount the barrier if the barrier is located too far in the exit channel (Fig. 6.9d). By arguments of microscopic reversibility, a large fraction of the energy of the desorption product will go into vibration.

Halstead and *Holloway* [6.50] have addressed the issue of the influence of PES topographies on H_2 dissociation dynamics in some detail. It is clear that a qualitative idea of the barrier location and an estimate of its height can be obtained from experimental measurements of (1) state-specific adsorption probabilities that demonstrate the relative effectiveness of vibrational and translational energy in surmounting the barrier to adsorption and (2) the partitioning of energy into vibration and translation during desorption. A more quantitative estimate of the topography of the PES can be gained by starting with a trial semi-empirical PES, performing appropriate dynamical calculations with the surface to compare with state-resolved adsorption and desorption measurements, modifying the surface, and iteratively completing this process until the calculations converge as closely as possible to the experimental results.

In 1985 *Harris* and *Andersson* [6.29] constructed a two-dimensional potential energy surface for the hydrogen/Cu system from total energy calculations of an H_2Cu_2 cluster (Fig. 6.10). This surface has since been parameterized and used in a number of calculations concerning the dynamics of H_2 scattering, adsorbing, and desorbing from Cu surfaces [6.51]. This PES includes two regions, an entrance channel controlling the trajectories of incoming H_2 molecules and an exit channel defining the states of the adsorption products. These two regions are separated by a barrier representing the repulsive interaction between the

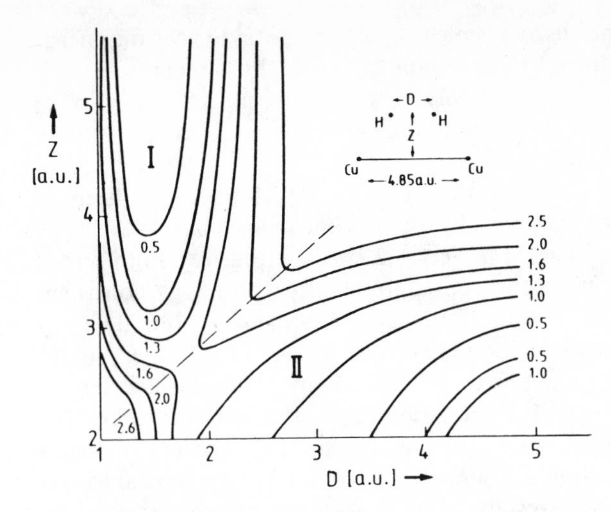

Fig. 6.10. Energy contours for Cu H. Results are shown as a function of D, the H–H separation, and Z, the distance between H_2 and Cu_2. Energies are in eV relative to the value at $Z = \infty$, $D = 1.4$. From [6.29] with permission

molecule and the surface that results from the Pauli repulsion between the ground state of the molecule, the lσ_g state, and the s electrons at the Cu surface. The barrier derived from this relatively simple cluster calculation is on the order of 1 eV. This calculated barrier height is inconsistent with the much smaller values of 0.3–0.4 eV estimated from the adsorption measurements of *Balooch* et al. [6.31] described in Sect. 6.2.2. The calculated barrier height is, however, consistent with the desorption measurements of *Comsa* and *David* described in Sect. 6.2.3 [6.47], which give a mean energy of desorption of 0.68 eV for D_2 from Cu(1 0 0). Taking the zero-point energy into consideration (0.19 eV), this value gives an estimate for the barrier height of 0.87 eV. Another test of the surface, of course, is how well calculations performed using it can predict the results of *Kubiak* et al., which are sensitive to the location of the barrier. Values of the vibrational population ratios calculated [6.52] with this surface are larger than those measured. For H_2 the calculated population ratio, $P(v = 1)/P(v = 0)$, is large, 0.33, compared to the measured values of 0.08 ± 0.03 for Cu(1 1 1) and 0.05 ± 0.01 for Cu(1 1 0). The agreement is better for D_2: 0.48 compared with the measured values of 0.35 for Cu(1 1 1) and 0.24 for Cu(1 1 0). A very simple qualitative interpretation would suggest that the calculated barrier position is slightly too far toward the exit channel (to adsorption). Nevertheless, the agreement is remarkably good considering the simplicity of a theoretical model based on a 4-atom cluster. Not surprisingly, these results have inspired a number of new experimental and theoretical studies, which will be discussed in the next sections.

6.3.2 Recent Adsorption Measurements with Molecular Beams

In 1989 two groups reported the results of new studies of the adsorption of hydrogen on copper. *Anger* et al. [6.53] measured the adsorption probability as a function of energy and angle of incidence for H_2 on three faces: Cu(1 1 1),

Cu(1 1 0), and Cu(1 0 0). They used a heated supersonic beam to vary the energy of incident H_2 and monitored the amount adsorbed by temperature programmed desorption measurements. *Hayden* and *Lamont* [6.54] performed similar measurements for the energy and angular variation of the adsorption probability of H_2 on Cu(1 1 0).

Figure 6.11 displays the results of *Anger* et al. [6.53]. Though quantitatively very different, in some ways these results are qualitatively similar to the early adsorption measurements discussed in Sect. 6.2.2 [6.31] (Fig. 6.3). Both sets of data show a sharp increase in the adsorption probability with energy, consistent with the idea of activated adsorption. Both show approximate normal-energy scaling, consistent with the predictions of a one-dimensional barrier model. (There are some small departures from normal-energy scaling in the data of Fig. 6.11, especially for the Cu(1 1 1) data–a point we shall return to later.) The differences between the two sets of data are also striking. The newer results reveal no measurable adsorption at low energy, in contrast to the constant value of about 2% shown in the early results. More importantly, the new data have a higher onset energy and a lower overall adsorption probability throughout the range of energies studied. Since the adsorption probability increases in roughly an exponential manner up to the highest energies studied, it is not possible to assign a mean barrier height from the inflection point of the adsorption vs. energy curve, as was done for the early data [6.31, 41]. Nonetheless, the new data clearly indicate that the barrier must be higher than previously supposed.

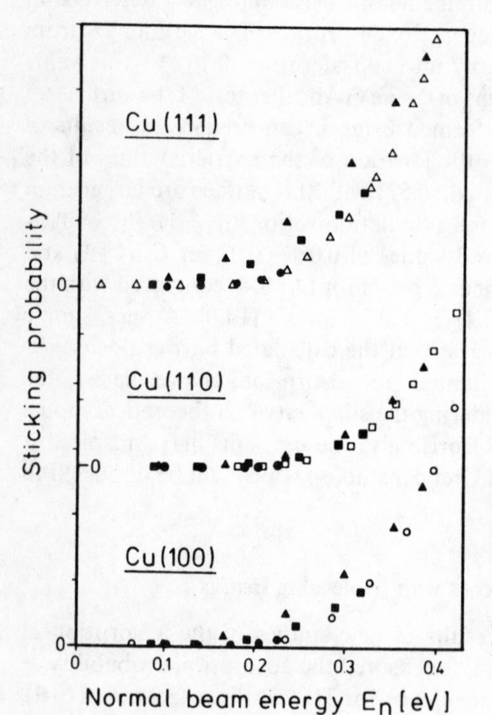

Fig. 6.11. Adsorption probability for H_2 on three faces of Cu. Results are shown for various energies and angles of incidence plotted as a function of the normal energy E_n. Beam energies are: solid triangles, 0.4 eV; solid squares, 0.33 eV; solid circles 0.23 eV. The open symbols are for normal incidence and variable beam energy. From [6.53] with permission

It is also interesting to compare the data of Fig. 6.11 with predictions based on electronic-structure calculations. The data show a rapid increase in adsorption probability at ≈ 0.2 eV. Taking into account the zero-point energy of H_2, this behavior suggests that the PES has a barrier of ≈ 0.47 eV. In contrast, electronic-structure calculations give values ranging from 0.8 eV for a jellium model [6.55, 56] to 1.3–2.0 eV for cluster models [6.29, 57, 58]. Thus there is still a considerable discrepancy between the barrier inferred from these experiments and theoretical estimates of the barrier. A possible explanation for this discrepancy was suggested by *Harris* [6.38] in terms of the role of vibrationally excited H_2 in the dissociation process. This explanation is discussed in the following subsection.

Both theory and experiment suggest that the role of vibrationally excited molecules must be carefully considered in the construction of a picture of the dissociative adsorption of hydrogen on Cu. We discussed the role of vibration briefly in Sect. 6.2.3 in connection with the measurements of *Kubiak* et al. and in Sect. 6.3.1 with the description of the 2-D PES and how molecular vibration may assist a molecule in overcoming a barrier to dissociative adsorption. Figure 6.9c shows a schematic classical trajectory of a vibrationally excited molecule on a PES featuring a late barrier, and Fig 6.9a depicts a similar trajectory on a PES with an early barrier. In these calculations, the vibrationally excited molecule surmounts the barrier when the barrier is late but not when the barrier is early. The efficacy of vibrational energy in promoting adsorption strongly depends on the location of the barrier. Thus the question of whether we expect vibrationally excited molecules to play an important role in adsorption is very much system dependent.

Hand and *Holloway* [6.59] have performed calculations for dissociation on a model of the H_2/Cu(1 0 0) system, using the 2-D PES described in Sect. 6.3.1. Figure 6.12 displays the results of their calculations for the dissociative adsorption probability as a function of translational energy for the $v = 0$ and $v = 1$ vibrational state of H_2. The dissociative adsorption probability follows an S-shaped curve for both states, but the results reveal that the $v = 1$ curve is shifted to lower energy by ≈ 0.25 eV. Although we cannot expect these curves

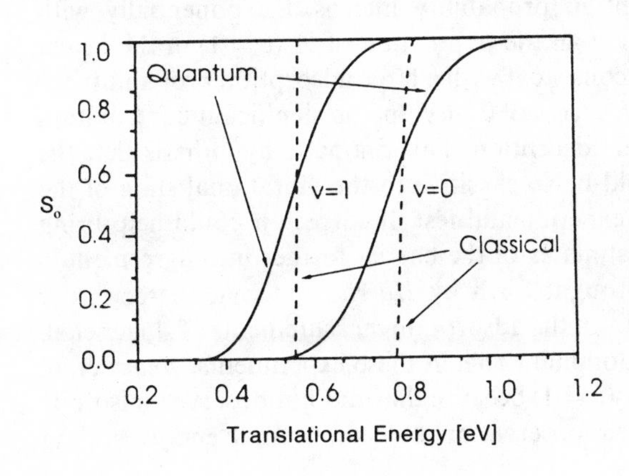

Fig. 6.12. Quantum mechanical and classical calculations of dissociative adsorption. The adsorption probability, S_0, for H_2 molecules is shown for initial vibrational states $v = 0$ and $v = 1$. From [6.59] with permission

to be quantitatively correct, they indicate that vibrationally excited molecules can play a dominant role in adsorption under certain conditions. For example, consider molecules with a kinetic energy of 0.5 eV. The results of Fig. 6.12 indicate that the adsorption probability for $H_2(v = 1)$ is about 0.4, but the adsorption probability of $H_2(v = 0)$ is essentially zero.

For other systems [6.60–63] it has been demonstrated experimentally that vibrational and translational energy are approximately equally effective in activating dissociative adsorption. For the hydrogen/copper system, the most convincing experimental evidence discussed thus far for enhanced adsorption of vibrationally excited molecules comes from the state-resolved desorption measurements of *Kubiak* et al. [6.48] (Sect. 6.2.3), which show that the population of vibrationally excited molecules is far in excess of the thermal equilibrium value. By detailed balance, these results support the hypothesis that vibrational excitation plays an important role in promoting dissociative adsorption for this system. Thus both experimental and theoretical arguments suggest that vibrationally excited H_2 may exhibit enhanced probability of dissociation on Cu.

The kinetic energy threshold for adsorption of $H_2(v = 0)$ molecules calculated by *Hand* and *Holloway* [6.59, 64] (Fig. 6.12) is clearly much higher than the onset of sticking observed by *Anger* et al. (Fig. 6.11). From the work of *Hand* and *Holloway*, then, we might suspect that the adsorption probabilities measured by *Anger* et al. were due, in some large part to the adsorption of molecules in vibrationally excited states. This point was first specifically noted by *Harris* [6.38] in regard to his own very similar calculations. *Harris* suggested that the adsorption measurements of *Anger* et al. were dominated by the adsorption of hydrogen molecules in the first excited vibrational state, $H_2(v = 1)$. For now we shall consider the *total* energy of an $H_2(v = 1)$ molecule to be the sum of its kinetic energy and vibrational energy, i.e., vibrational spacing (0.52 eV) plus zero-point energy (0.27 eV). Thus, $H_2(v = 1)$ molecules with ≈ 0.2 eV kinetic energy have sufficient total energy to overcome a barrier of ≈ 1 eV. *Harris* pointed out that since *Anger* et al. varied the kinetic energy of the incident beam of H_2 by changing the nozzle temperature, the population of $H_2(v = 1)$ must have increased exponentially with the beam energy (nozzle temperature) in their experiments. Since the adsorption probability increased exponentially with energy and was always smaller than the population of $H_2(v = 1)$ in the beam, *Harris* argued, the results were consistent with a large adsorption probability for $H_2(v = 1)$ for kinetic energies in excess of 0.2 eV and *no* significant contribution from $H_2(v = 0)$ to the observed adsorption. This dramatic hypothesis that the probability of adsorption would be so sensitive to the vibrational state of the incident molecule required an experimental test. If correct, it could help bring theoretical and experimental estimates of the barrier further into agreement.

Prior to the hypothesis brought forth by *Harris*, however, there were a number of indirect indications from the adsorption measurements of *Anger* et al. that $H_2(v = 1)$ did not play a dominant role in these experiments. *Anger* et al. [6.53] discounted the role of $H_2(v = 1)$ because they did not observe an isotope effect in their experiment. Their observation of near normal energy scaling

(Fig. 6.11) also suggested that $H_2(v = 1)$ could not dominate the results. These data exhibit deviations from normal energy scaling, however, that are in the right direction to indicate a vibrational effect [6.38], but the deviations seemed too small for $H_2(v = 1)$ to have a dominant role. We now understand how both the deviations from normal energy scaling and the lack of a large isotope effect can be understood in terms of the role of vibrational states in quantitative sticking probability models [6.65, 66] (Sect. 6.3.3a), but at the time, these observations seemed to be inconsistent with *Harris's* hypothesis.

A more direct experimental test of the role of vibrational excitation was clearly required. The problem with the experiments discussed so far is that both the translational energy and vibrational temperature of the beam are proportional to the nozzle temperature. This coupling makes it very difficult to probe the role of vibrational excitation. The "seeded" supersonic beam technique provides a simple method for overcoming this difficulty. This technique allows for the separate variation of translational and vibrational energy of the molecules in a beam by the use of gas mixtures. It has been employed by a number of groups to examine the relative efficacy of vibrational and translational energy in the promotion of dissociative chemisorption [6.34, 36, 60, 61, 63, 67–72].

For example, in an expansion of D_2 from a nozzle at 300 K there are a sufficient number of collisions to cool the translational and partially ($\approx 20\%$) the rotational degrees of freedom but an insufficient number of collisions to relax vibration [6.73, 74]. As a result, the mean translational energy is about $5.4/2\ kT$ or 0.07 eV, and the vibrational temperature is ≈ 300 K. If we heat the nozzle to 600 K, the mean translational energy increases to ≈ 0.14 eV and the vibrational temperature to 600 K. If we make a dilute mixture of a D_2 in H_2, the velocity of the beam will be determined by the velocity of the H_2 molecules. Effectively, the D_2 seeded in a lighter gas such as H_2 becomes aerodynamically accelerated in the expansion. If we expand this mixture from a 300 K nozzle, we attain a mean kinetic energy of 0.14 eV (i.e., the same as that of the pure D_2 beam from a 600 K nozzle) but a vibrational temperature of only ≈ 300 K. If we want a high vibrational temperature and a low kinetic energy, we can seed the D_2 in a heavier gas while the nozzle is hot. For example, D_2 seeded dilutely in Ne and expanded from a 1600 K nozzle will have a mean translational energy of only 0.07 eV (the same as that of a pure beam from a 300 K nozzle) but a vibrational temperature of ≈ 1600 K.

Hayden and *Lamont* [6.75] applied the seeded-beam technique to the study of the hydrogen/Cu system. Their results for the dissociative adsorption of H_2 on Cu(1 1 0) are shown in Fig. 6.13. Four sets of measurements are shown, one for a pure H_2 beam with variable nozzle temperature and three for seeded beams at three different nozzle temperatures. The open circles represent measurements for the pure H_2 beam for which the beam energy was varied by varying the nozzle temperature. The three sets of solid points represent measurements for fixed nozzle temperatures of 1085, 1120, and 1150 K. For these curves, the incidence kinetic energy was controlled by seeding in helium, i.e., starting with pure H_2 and moving to increasingly dilute mixtures of H_2 in He, lowering the

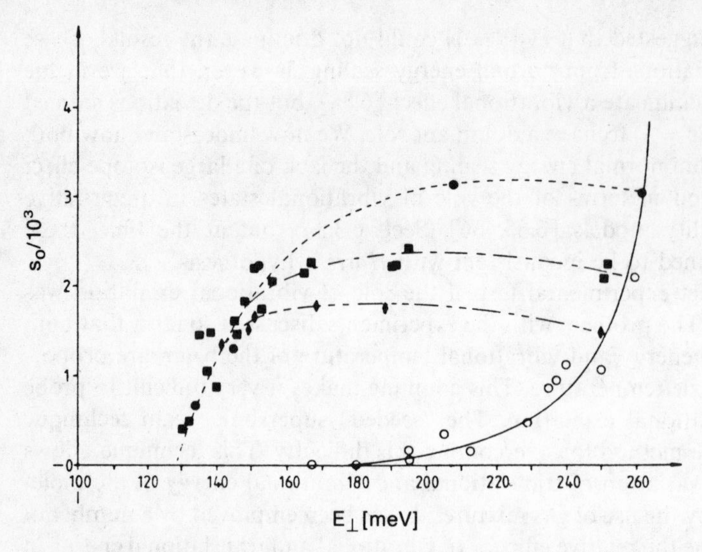

Fig. 6.13. The dissociative adsorption probability, S_0, for H_2 on Cu(1 1 0). Open circles and solid line are pure H_2 beam results obtained at various nozzle temperatures. Solid symbols are for H_2–He seeded beams with nozzle temperatures T_n as follows: circle, 1150 K; square, 1100 K; diamond 1085 K. From [6.75] with permission

energy by up to a factor of two. (Increasing the energy by seeding is, of course, not possible in this case, simply because there is no gas that is lighter than H_2.) *Hayden* and *Lamont* interpreted the variation in adsorption probability S_0 with nozzle temperature for a fixed translational energy E_i, as being due to changes in the population of $H_2(v = 1)$ in the beam. In fact, they suggested that the constant value of S_0 for $E_i > 150$ meV indicates that *only* $H_2(v = 1)$ is contributing to adsorption in their experiment.

The conclusion that $H_2(v = 1)$ dominates the results of hot nozzle beam experiments was immediately challenged and tested with other seeded beam experiments. Measurements were reported by *Berger* and *Rendulic* [6.70] and *Hayden* and *Lamont* [6.71] for H_2 and D_2 on Cu(1 1 0), and H_2 and D_2 on Cu(1 1 1) [6.69], all of which showed some degree of enhancement of adsorption by vibrational energy. There was, however, disagreement over the extent of this enhancement and in particular over the conclusion that the role of $H_2(v = 1)$ "dominates" adsorption measurements with hot nozzle beams. In part this dispute was the result of a lack of agreement between the experimental results obtained by different groups, and in part it was the result of a poorly framed question. The relative role of ground state and vibrationally excited molecules in a given experiment depends on the conditions of that experiment. If we perform an experiment at sufficiently low translational energy, for sufficiently high vibrational temperature, vibrationally excited molecules will dominate the results. On the other hand, if we perform the experiment at high enough translational energy, ground-state molecules will dominate. We shall not explore this dispute in detail here since a more quantitative treatment of the role of vibration is required and is presented in Sect. 3.3.

The preceding discussion highlights the importance of employing seeded molecular beams to achieve a very wide range of incidence conditions in order to unravel the effects of translational and vibrational energy. As we have already noted, seeding H_2 in a lighter gas to increase its kinetic energy is not possible, limiting the highest kinetic energy available to about 0.45 eV. For D_2 it is possible to seed in H_2 to increase the kinetic energy. *Rettner* et al. [6.72] used this technique to extend the energy range up to 0.85 eV and thus to obtain the most complete data currently available for direct dissociative adsorption in the hydrogen/Cu system. Figure 6.14 depicts the adsorption probability, S_0, of D_2 on Cu(1 1 1) for beams with kinetic energies, E_i, up to 0.85 eV, vibrational temperatures up to 2100 K, and incidence angles from 0° to 60°. For a surface temperature of 120 K, S_0 increases from below 10^{-6} to 0.34 over the range studied.

Data covering such a wide range allow a number of important conclusions to be drawn directly from the experimental results without the use of detailed modeling. The dramatic dependence of S_0 on both E_n and T_n is an indication that both translational and vibrational energy play an important role in overcoming the barrier to adsorption. Whether the results are dominated by molecules populating a ground or excited vibrational state depends entirely on the incidence conditions. For example, the value of S_0 for D_2 adsorption on Cu(1 1 1) at $E_n \approx 0.25$ eV (Fig. 6.14) increases with T_n roughly in proportion to the $D_2(v = 2)$ population. Thus, molecules in the $v = 2$ and higher vibrational

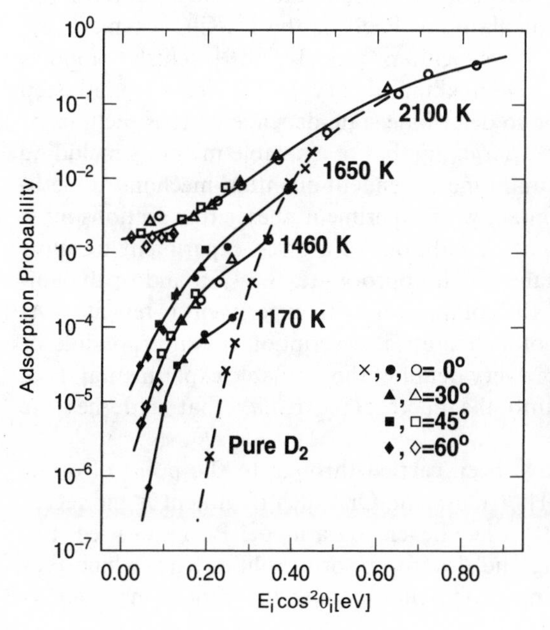

Fig. 6.14. Initial adsorption probability for D_2 on Cu(1 1 1). Results are plotted as a function of the energy associated with motion normal to the surface for the incidence angles indicated. Data are shown for seeded beams using four nozzle temperatures, as indicated, and a solid line is drawn (to guide the eye) through each set of data points recorded at the same nozzle temperature. The x-marks (followed by the dashed curve) represent data for pure D_2 expanded from a nozzle at temperatures ranging from 875 to 2100 K. From [6.72] with permission

states dominate the results at low energies. On the other hand, at the highest energies the populations of all vibrationally excited species are too small to account for the observed S_0 values. The percentage of molecules populating excited vibrational states is $\approx$ 14% at 2100 K, while the adsorption probability is more than twice that value at 0.85 eV, indicating that most of the adsorption must be due to $D_2(v = 0)$. Indeed, the increase in S_0 in the range of 0.6–0.8 eV must reflect the increase in the dissociation of $D_2(v = 0)$ over this energy range.

The picture that emerges from this discussion is that $D_2(v = 3)$ and more highly vibrationally excited D_2 determine S_0 at the lowest incidence kinetic energies, and contributions from $D_2(v = 2)$, $D_2(v = 1)$, and $D_2(v = 0)$ successively become significant as the translational energy is increased. A detailed analysis is required to extract information on the energy dependence of S_0 separately for each vibrational state. Such an analysis is presented in the following section.

6.3.3 Quantitative Treatment of Adsorption and Desorption Data

We have emphasized that the dissociation of hydrogen on Cu has become a model system for both experimental and theoretical studies of gas–surface chemistry. Comparison of experimental results with theoretical predictions requires a quantitative treatment of the various measurements or a simulation of experimentally observed quantities based on calculated differential cross sections. The basic methodology typically used in the comparison of theoretical calculations with data is straightforward. The first step is to develop a trial potential energy surface, which may be based in an *ab initio* electronic structure calculation, a semi-empirical method, or a qualitative model. Several approaches have been used to calculate the PES for the H_2/Cu system. These approaches include the use of a jellium model [6.56], cluster models [6.29, 57, 58, 76], and the effective-medium theory [6.77]. The second step involves dynamical calculations to determine state-dependent cross sections or rate constants. Again there are a large number of available methods including the use of classical trajectories and time-dependent quantum-mechanical methods [6.78]. Finally, to make contact with experiment, these cross sections must be summed or convolved over the conditions of a given experiment to allow comparison with measured quantities. If appropriate, the PES and/or dynamical methods are adjusted, and the comparison with experiment is repeated. At this stage in the development of a theoretical description it is not possible or appropriate to try to reproduce every detail of the available experimental data. The goal is to gain insight into the microscopic reality that underlies the experimental results.

Few theoretical studies have been carried through to the point of direct comparison with data for the H_2/Cu system. One such treatment is the recent work of *Küchenhoff* et al. [6.79]. They developed a model PES and used it to calculate state-resolved sticking and desorption probabilities from which they calculated the observed sticking probabilities under the conditions used by

Berger et al. [6.69], in their measurements of the adsorption of H_2 and D_2 on Cu(1 1 1). The agreement is good and gives some insight into the reasons that the isotope effect observed by *Berger* et al. is so small. *Küchenhoff* et al. [6.79] stressed that there are competing factors contributing to this phenomenon. Since D_2 has a higher (Boltzmann) probability of occupying vibrationally excited states because of its smaller vibrational spacing, the enhancement of reactivity for vibrationally excited molecules may allow D_2 to adsorb more readily than H_2. On the other hand, the zero-point energy and vibrational spacing for D_2 is smaller than for H_2, which makes its vibrational energy smaller than that of H_2, and the enhancement of reactivity for molecules with higher vibrational energy may favor H_2 over D_2. In addition, the heavier mass of D_2 over H_2 inhibits its ability to tunnel through the barrier. Because of the large mass difference, we should expect a large difference in the results for H_2 and D_2. *Küchenhoff* et al. found that the sticking probabilities calculated for molecules in *individual vibrational states* do show a large isotope effect. The sticking probability functions for $H_2(v = 1)$ and $D_2(v = 1)$, for example, are sigmoidal. The inflection point for the D_2 curve, however, comes at a considerably higher energy than does the inflection point for H_2. Thus at a given energy, an $H_2(v = 1)$ molecule will have a higher adsorption probability than a corresponding $D_2(v = 1)$ molecule. This strong isotope effect in the state-specific sticking probabilities, however, is offset by another factor when applied to the experiments of *Berger* et al. In these (and other) experiments, the comparison between the adsorption probability for D_2 and H_2 is made for beams with the same nozzle (and hence vibrational) temperature. Thus the contribution to adsorption from molecules in the $v = 1$ state must be multiplied by the Boltzmann factor for the population of those molecules in the beam. The Boltzmann factor governing the population of $D_2(v = 1)$ is larger than the Boltzmann factor for $H_2(v = 1)$, and the competition between this factor and the state-specific sticking probabilities causes the adsorption of H_2 and D_2 to be nearly the same under the conditions maintained in the experiments of *Berger* et al.

The comparison between theory and experiment presented by *Küchenhoff* et al. is impressive for the experiments they consider. Their calculated sticking probability curves do not, however, reproduce some of the more recent experimental data – especially seeded beam data covering a large dynamic range such as that of Fig. 6.14.

a) Sticking Probability Models

An alternate approach to the theoretical methods described above is to start with models describing how the sticking probability varies with energy, angle, and quantum state and to use these models to analyze the adsorption data [6.65, 66]. Desorption data can be treated with the same models by use of the principle of detailed balance. This approach involves replacing state-specific sticking probability curves derived from a PES and appropriate dynamical calculation with a parameterized form for the sticking probability. Using a parameterized form to describe the behavior of the sticking probability with kinetic energy has the clear advantage of allowing the sticking probability curves to be adjusted to agree with experimental data more directly and with far less computation than the order method. In this way, one may hope to establish reliable state-resolved sticking probability curves from the experimental data. Considerable qualitative insight can be gained directly from these state-resolved sticking curves. Such curves allow the results of different types of experiments to be compared and checked for compatibility. Furthermore, the state-resolved sticking curves make the comparison with the results of dynamical calculations for a given PES easier since they eliminate the need to sum over the range of internal states and convolve over the range of conditions appropriate for comparison with a given experiment.

In this approach, we take the sticking probability to be a sum of terms for the vibrational states involved, each weighted by a Boltzmann factor representing the population of the corresponding vibrational state. We ignore rotational-state effects at this point, assuming them to be considerably smaller than those associated with changes in vibrational energy. We shall discuss the role of rotational energy in Sect. 6.3.5. We thus assume the adsorption probability to have the form

$$S(v, E_e, T_n) = \sum_v F_B(v, T_n) S_0(v, E_e), \tag{6.1}$$

where $F_B(v, T_n)$ is the Boltzmann factor for state v at nozzle temperature T_n, and E_e is an effective energy given by

$$E_e = E_i \cos^n(\theta_i). \tag{6.2}$$

For $n = 2$, E_e corresponds to E_n. The form of $S_0(v)$ is chosen to be a flexible form that can reproduce S-shaped curves such as those computed from quantum-mechanical calculations of adsorption (Fig. 6.12). Curves qualitatively similar to those of Fig. 6.12 frequently arise in the study of activated adsorption for light molecules. The inflection point in the curve is related to the kinetic energy required to activate the adsorption process. The width, or range of energies over which the curve rises, is related to a number of factors including (1) quantum-mechanical effects such as tunneling through the barrier and reflection from the crest of the barrier and (2) a distribution of barrier heights resulting from a dependence of the barrier height upon the position of impact of the incident molecule with respect to the surface unit cell and the orientation angle of the molecule at impact. A convenient form for this adsorption function has been suggested by Harris [6.38]:

$$S_0(v) = \frac{A}{2}\left[1 + \tanh\left(\frac{E_e - E_0(v)}{W(v)}\right)\right], \tag{6.3}$$

where $E_0(v)$ and $W(v)$ are parameters used to adjust, for each state, what we refer to as the threshold energy (i.e., the inflection point of the rising curve) and the width. The adjustable parameter A controls the level at which the adsorption probability saturates at higher energies.

This model has been used successfully by *Michelsen* and *Auerbach* [6.66] to perform a critical examination of many of the data on dissociative adsorption and associative desorption for the hydrogen/Cu system. The treatment includes data on adsorption vs. energy measured with pure beams at normal and variable angle of incidence, seeded-beam adsorption data, and angular, velocity, and internal-state distributions for desorption. As discussed above, the data of *Anger* et al. [6.53] on the energy and angular variation of H_2 adsorption on Cu(1 1 1) (Fig. 6.11) exhibits small but consistent deviations from normal energy scaling. Figure 6.15 shows a comparison of the model calculation with the data. The calculation involves adjusting the values of $E_0(v)$ and $W(v)$ for $v = 0$ and $v = 1$ to agree with the data. The parameters obtained are listed in Table 6.2. We thus learn that the deviations from normal energy scaling can be explained in terms of the role of these two vibrational states in the adsorption. The filled symbols in Fig. 6.15 display points measured with a constant value of E_i with the angle of incidence varied in order to vary the normal energy E_n. For a given E_n, higher values of sticking are found for higher values of E_i as a result of the higher population of $H_2\,(v = 1)$ in those beams.

This model has also been used by *Rettner* et al. [6.72] in the treatment of the D_2/Cu(1 1 1) adsorption data of Fig. 6.14. Figure 6.16 exhibits the form of the $S_0(v)$ terms of this sticking probability model with four vibrational states, where the parameters were adjusted to reproduce the adsorption data. A comparison of the model calculation with the data, shown in Fig. 6.17, indicates that the model does an excellent job of reproducing the data. Thus for the D_2/Cu(1 1 1) system, it is possible to extract *state-resolved* information giving the variation of

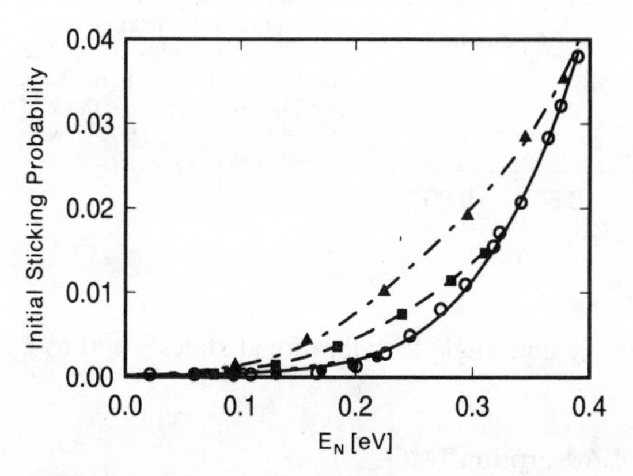

Fig. 6.15. Normal incidence and angular adsorption data for Cu(1 1 1). S_0 is plotted against $E_n = E_i \cos^2 \theta_i$. The points are from *Anger* et al. [6.53], and the lines correspond to the model described in Sect. 6.3.3a. The open circles and solid line refer to normal incidence data at variable beam energy. The filled symbols and broken lines are for one incidence energy and different incidence angles for different "normal energy": triangles and dot-dashed curve, 0.40 eV; squares and dashed curve, 0.33 eV; circles and dotted curve, 0.23 eV. From [6.66] with permission

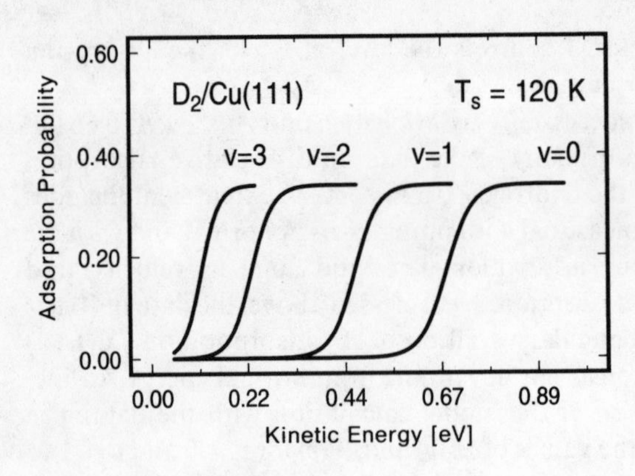

Fig. 6.16. Form of the sticking probability vs. effective energy. Results are shown for the vibrational states 0–3 of D_2 as indicated. The curves for each vibrational state were determined using the data of Fig. 6.14

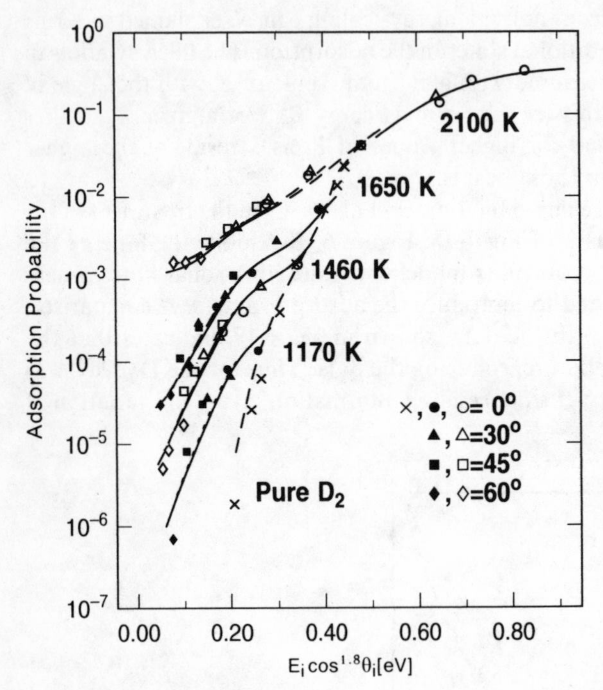

Fig. 6.17. Comparison of the results of the sticking probability model with data. The model is described in Sect. 6.3.3a, and the data are those of Fig. 6.14. From [6.72] with permission

adsorption probability with energy and angle for vibrational states $v = 0$ to $v = 3$.

b) Quantitative Comparison of Adsorption Data

We can also use the sticking probability model to facilitate a comparison among the results of different kinds of experiments on the same system. For example, *Anger* et al. [6.53] have reported measurements of the angular and energy variation of the adsorption probability of H_2 on Cu(1 1 0), and *Hayden* and

Lamont [6.71, 75] have reported the results of seeded-beam experiments on the same system. The sticking-probability model gives a good description of each of these experiments separately. The comparison of the fit of the model with the results of *Anger* et al. has already been displayed in Fig. 6.15. The corresponding comparison of the fit with the data of *Hayden* and *Lamont* is given in Fig. 6.18. Although each of these two fits (one for the data of *Anger* et al. [6.53] and one for the data of *Hayden* and *Lamont* [6.71, 75]) *individually* describes the data well, unfortunately the parameters derived from the two experiments are significantly different. The comparison is given in Table 6.2. Both the determined threshold energies, $E_0(1)$, and widths, $W(1)$, are markedly different. The differences, especially for $W(1)$, are so large that we regard these measurements as incompatible.

Table 6.2 also presents a comparison of parameters for the H_2/Cu(1 1 1) system determined from the measurements of *Anger* et al. [6.53] with those of *Auerbach* et al. [6.80]. In this case, the agreement is better than it was in the previous example, but there are significant differences, particularly in values of the widths, $W(v)$, and the saturation value, A.

c) Desorption and the Role of Surface Motion

As discussed in Sect. 6.2.3, dynamical characteristics associated with adsorption can be used to predict or understand related characteristics associated with desorption. For instance, the angular dependence of the adsorption probability has been shown to correspond, via the principle of detailed balance, to angular distributions of desorbed molecules for a number of systems [6.43–46]. In this subsection we give examples of (1) how we use the information gained from

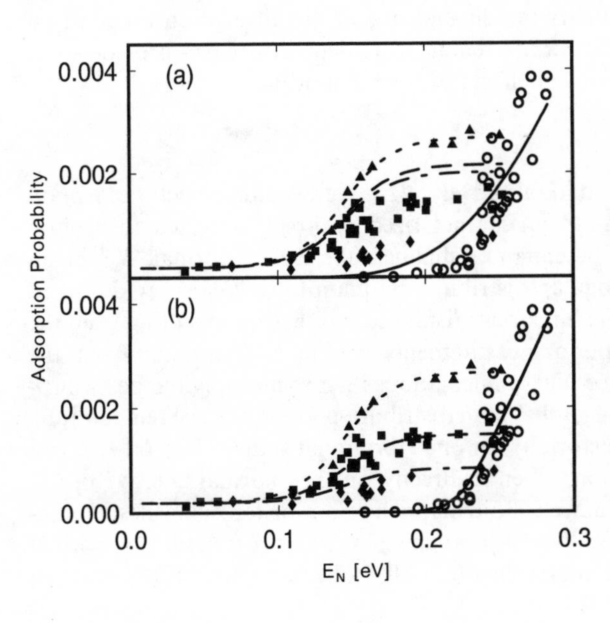

Fig. 6.18. Adsorption probability for H_2 on Cu(1 1 0). The points are data from *Hayden* and *Lamont* [6.71], and the lines are a fit of the sticking probability model to these data. (b) In fitting the data, the nozzle temperatures were adjusted to 1150, 1000, and 870 K for the circles, diamonds, and triangles, respectively. (a) The temperatures originally reported for these curves were 1150, 1100, and 1085 K, respectively. From [6.66] with permission

Table 6.2. Adsorption functions for H_2 on Cu(1 1 0) and Cu(1 1 1). The table displays a comparison of the parameters of the adsorption functions determined from various experiments on the same system

System	Authors and method	$E(0)$ [eV]	$W(0)$ [eV]	$E_0(1)$ [eV]	$W(1)$ [eV]	$E_0(2)$ [eV]	$W(2)$ [eV]	A
H_2	Anger, Winkler, and Rendulic [6.53] Angular variation of adsorption	0.57	0.12	0.26	0.10	*	*	0.22
Cu(1 1 0)	Hayden and Lamont [6.71] Seeded beam adsorption	*	*	0.15	0.03	*	*	0.45
H_2	Anger, Winkler, and Rendulic [6.53] Angular variation of adsorption	0.62	0.15	0.23	0.09	*	*	0.55
Cu(1 1 1)	Auerbach, Rettner, and Michelsen [6.80] Seeded beam and angular variation	0.63	0.07	0.29	0.05	0.10	0.04	0.20

Note: *indicates that the experiment provides no information on this parameter

adsorption measurements of D_2 on Cu(1 1 1) (described above) to predict quantitatively desorption angular and velocity distributions and (2) how comparison of these predictions with experimental *desorption* measurements can provide additional information about *adsorption* dynamics, particularly about the influence of surface motion on the adsorption process. Since the direct adsorption measurements (Sect. 6.3.2) were made at low surface temperature (120–190 K), and the desorption data (Sect. 6.3.5) were recorded at much higher surface temperatures (850–1000 K), making a quantitative comparison between these measurements highlights the dependence of the adsorption function on surface temperature. Such a surface temperature dependence can provide some understanding of the dynamical effects of surface motion.

Angular Distributions

In Sect. 6.2.3 we described qualitatively why we would expect desorption angular distributions for a system characterized by direct, activated adsorption that depends on E_n to exhibit enhanced flux near the surface normal. We can, in addition, predict these angular distributions quantitatively for D_2 desorbed from Cu(1 1 1), knowing the functions (displayed in Fig. 6.16) resulting from the fit of (6.1) to the adsorption measurements of Fig. 6.14 and applying the principle of detailed balance. These calculations involve multiplying the adsorption function by a Maxwell–Boltzmann distribution at the surface temperature. The procedure is shown pictorially for one vibrational state in Fig. 6.19. Recall that the adsorption function depends directly on the "normal energy" or E_n. Figure 6.19 displays the Maxwell–Boltzmann distribution for molecules impinging on the surface at three incidence angles plotted as a function of normal energy. At larger incidence angles the Maxwell–Boltzmann distribution has less

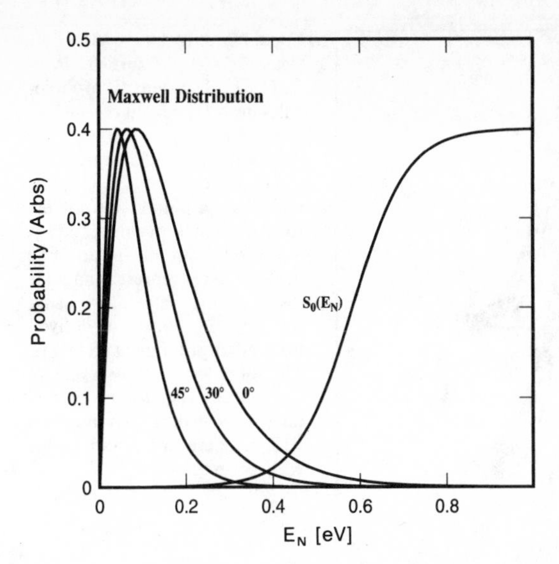

Fig. 6.19. Overlap of a thermal distribution and adsorption functions. A Maxwell distribution representing a temperature of 1000 K is plotted as a function of "normal energy", E_n, for several angles as indicated. A representative adsorption function is also shown to demonstrate that the overlap of the Maxwell distribution with the adsorption function decreases with increasing angle

"normal energy", and therefore its overlap with the adsorption function also pictured in Fig. 6.19 is reduced. This reduction in the overlap at larger angles of the Maxwell–Boltzmann distribution with the adsorption function leads to the peaked angular distributions observed.

Performing this calculation using the function $S_0(E_i, \theta_i)$ parameterized from adsorption measurements of D_2/Cu(1 1 1) (Fig. 6.16) produces the angular distributions shown by the dashed curves in Fig. 6.20. The experimental distributions recorded by *Rettner* et al. [6.81] are presented as circles. Both experimental values and predicted distributions are shown for three surface temperatures, 370, 600, and 800 K. The experimental values for 370 K were obtained using a flash desorption method for a coverage of 0.07 ML. The data for 600 and 800 K were obtained by exposing the surface to an atomic beam of D atoms at grazing incidence (70–80° relative to the surface normal) and recording the desorption density with a rotatable mass spectrometer at the angles indicated. In this later case the surface coverage remained below 0.1% ML throughout.

Comparison of the predicted distributions (dashed curves) with the measured distributions (open circles) reveals that for all three surface temperatures the predicted distributions are too narrow to reproduce the measured distributions. In addition, the discrepancy between the predicted and experimental distributions increases with increasing temperature. One way of accounting for this discrepancy is by assuming that the model function S_0, derived from adsorption measurements made at 120 K, is temperature dependent. An aspect of this function that is manifested in the breadth of the angular distribution is the parameter $W(v)$ representing the width of the function. If $W(v)$ is allowed to increase with increasing temperature, the calculated distributions perfectly reproduce the measured distributions. The solid curves in Fig. 6.20 show the

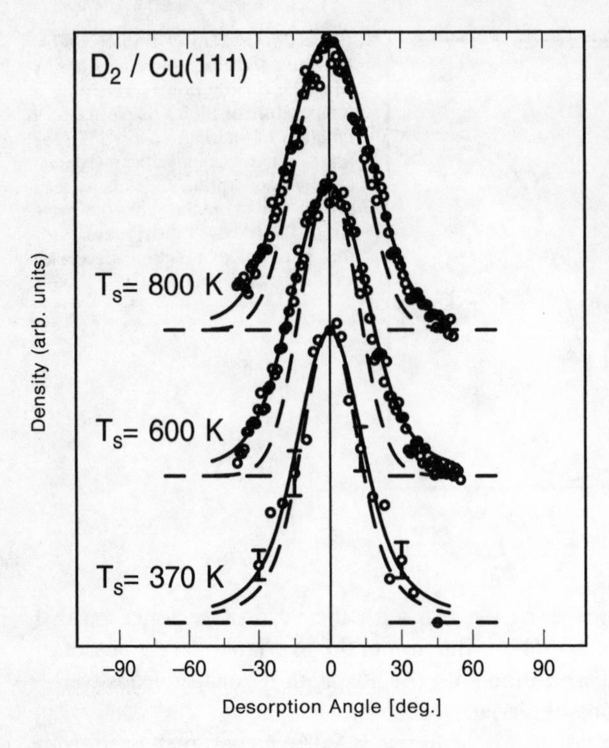

Fig. 6.20. Angular distributions of D_2 desorbed from Cu(1 1 1). The density of molecules desorbing from the surface is plotted against angle at the three surface temperatures indicated. The circles are the experimental data points, and the error bars represent one standard deviation from the mean. The dashed curves represent angular distributions calculated using the sticking function derived from adsorption measurements. The solid curves show results of the same calculations performed with widths that increase with temperature. From [6.89] with permission

agreement with the experimental distributions obtained when the width is allowed to increase with surface temperature.

Figure 6.21 depicts how the shape of the function deduced for $v = 0$ changes with surface temperature. This figure demonstrates that the increase in the width

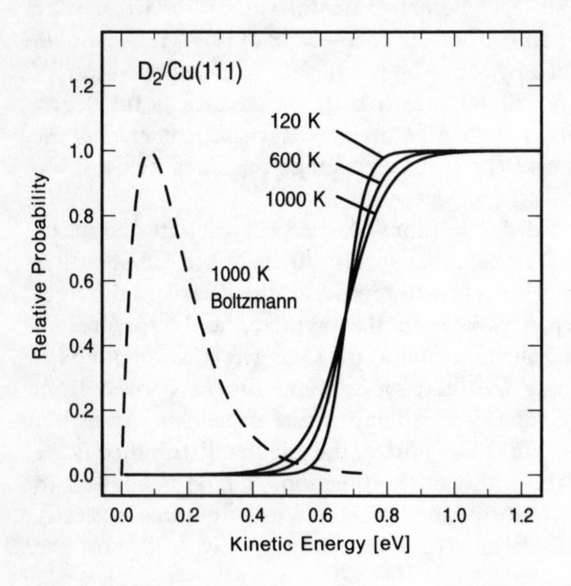

Fig. 6.21. Dependence of adsorption functions on surface temperature: adsorption functions used to fit desorption data for several surface temperatures (as indicated) for the $D_2(v = 0)$/Cu(1 1 1) system. The dashed curve shows a Maxwell–Boltzmann distribution of energies corresponding to a temperature of 1000 K. From [6.89] with permission

at higher surface temperatures causes the tail of $S_0(E_e)$ to extend to lower energy. A Maxwell–Boltzmann distribution for $T_s = 1000$ K is displayed in Fig. 6.21 for comparison and shows that much of the overlap occurs with the low energy tail of $S_0(E_e)$, making the angular distribution sensitive to this region of the model function. That is, an increase in the low-energy tail of $S_0(E_e)$ represents increased adsorption at lower "normal energy", which leads to enhanced adsorption at larger angles where the mean "normal energy" of the Maxwell–Boltzmann distribution is lower and, hence, to broader angular distributions [6.66].

The observation that increasing the width of the model adsorption function fully accounts for the increase in the breadth of the angular distribution provides some new insight into the dynamics of adsorption. An example of this increase in the width is shown graphically in Fig. 6.21 for $S_0(E_e, v = 0)$ at 1000 K relative to the curve appropriate for a surface temperature of 120 K. The increase in the width with temperature is significant, on the order of the increase in the mean Debye energy of the transverse mode with temperature. This suggests that surface motion is an important factor in the adsorption dynamics. The most obvious way in which such motion can influence a direct adsorption (and thus desorption) process is through its effect on the energy available to overcome the dissociation barrier. *Hand* and *Harris* [6.82] have performed detailed theoretical studies of the effect of energy transfer on the effective barrier to dissociation. They modified the model potential energy surface described in Sect. 6.3.1 to include a harmonic oscillator representing motion of the surface lattice. With masses appropriate for the dissociation of N_2 on Fe they demonstrated that the molecule can lose energy to or gain energy from the surface during collision with the surface. If the molecule loses energy to the surface, the dissociation probability will be reduced. Likewise, interaction with an excited surface atom can enhance the dissociation probability.

Recently, *Harris* et al. [6.68] and *Luntz* and *Harris* [6.83] performed an extensive theoretical study that accounted for the surface temperature dependence of the dissociative chemisorption observed for methane on W [6.33, 84, 85], Rh [6.86], Ni [6.87, 88], and Pt [6.36, 68] assuming that the dissociation is direct and activated. In a simplified form of the model that they referred to as the "surface mass model" they included motion of the surface lattice by representing the surface atom as a free mass (instead of a harmonic oscillator) moving with a velocity normal to the surface. They assumed that the incident molecules interact with a distribution of surface atoms with velocities given by a Boltzmann distribution. If an incoming molecule interacts with a surface atom moving *towards* the incoming molecule, the effective collision energy will be larger than the initial incidence energy, and *the adsorption probability will be enhanced.* On the other hand, if the surface atom is moving *away* from the incoming molecule upon collision, the effective collision energy will be reduced, and *the adsorption probability will decrease. Michelsen* et al. [6.89] have followed the approach of *Harris* and *Luntz* to assess the degree to which the effects of such surface motion can account for the proposed temperature dependence of the adsorption function for D_2 on Cu(1 1 1). The results of

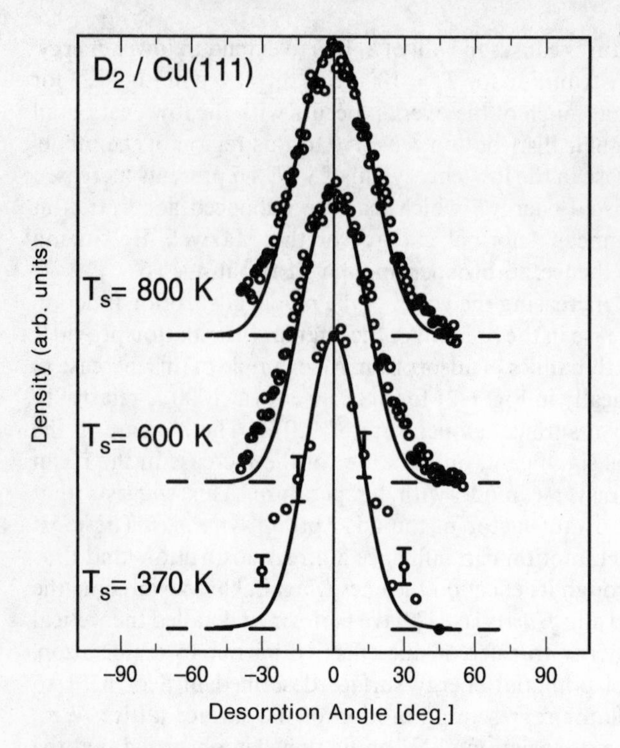

Fig. 6.22. Angular distributions of D_2 desorbed from Cu (1 1 1). The density of molecules desorbed from the surface is plotted against angle at the three surface temperatures indicated. The circles are the same data shown in Fig. 6.20. The solid curves are the result of a calculation based on the adsorption function derived from adsorption measurements and convolved with a Maxwell–Boltzmann distribution of surface atom energies. From [6.89] with permission

these calculations are presented as solid curves in Fig. 6.22. These results demonstrate that, although the calculated distributions are slightly too narrow to reproduce the measured distributions, including surface motion in this way qualitatively accounts for the observed increase in the breadth of the angular distributions with T_s. The conclusion made by *Michelsen* et al. is that there is an additional effect of surface temperature that is not accounted for in the surface mass model. Such an effect could involve a modification of the potential energy surface, particularly of the barrier region, by perturbation of the local electronic structure of the surface with movement of the surface atom.

Mean Energy vs. Desorption Angle

The concept of detailed balance applied to derive the mean energies of desorbed molecules at various desorption angles shows surface temperature effects similar to those discovered for angular distributions of desorbed molecules [6.89]. Figure 6.23 depicts the results of such calculations for D_2 desorbed from Cu(1 1 1) [6.89] along with the experimental results of *Comsa* and *David* [6.47]

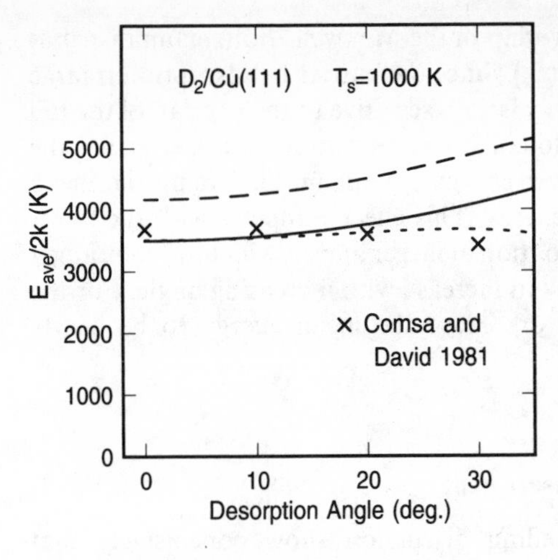

Fig. 6.23. Mean energy vs. desorption angle for D_2 desorbed from Cu(1 1 1). The x-marks are from *Comsa* and *David* [6.47]. The dashed line shows the prediction based on the adsorption function derived from adsorption measurements. The dotted curve is the result of the same calculation performed with larger (surface temperature dependent) widths. The solid curve is the result of a calculation based on the adsorption function derived from adsorption measurements and convolved with a Maxwell–Boltzmann distribution of surface atom energies. From [6.89] with permission

(x marks). Using $S_0(E_e)$ derived directly from the adsorption measurements in Fig. 6.14 gives values (dashed curve) that are much too large at all angles. As with the angular distributions, employing the surface-mass model to account for surface temperature effects brings the calculated values (solid line) into much better agreement with the experimental values. The agreement is not perfect, however, particularly at higher desorption angles. Including surface-temperature effects by simply increasing the width of $S_0(E_e)$ (dotted curve) gives nearly perfect agreement. The conclusion is the same as that derived from the angular distributions. That is, the adsorption process is dependent on surface temperature, the effects of which can be reproduced with an increase in the width of the adsorption function $S_0(E_e)$. Considering the motion of the surface atoms via the surface-mass model nearly accounts for this temperature dependence. There is, however, not surprisingly, some additional surface-temperature dependence not accounted for with this simple model.

As discussed in Sect. 6.2.3, the fact that the mean energy is independent of the desorption angle is an apparent contradiction of the principle of detailed balance, considering that the adsorption probability depends strongly on the incidence angle. Such a feature might be expected if the adsorption depended on the total incidence energy, E_i, instead of the "normal energy", $E_n = E_i \cos^2\theta$. That is, the dependence on E_n requires that molecules adsorbed at larger angles relative to the surface normal have higher total energies than those adsorbed at small angles. Likewise, molecules desorbed at these larger angles would be expected to desorb with higher energies. It would be anticipated that the mean energy would increase dramatically with increasing angle (see the dashed curve in Fig. 6.7). Such would be the case for a classical system with a single, one-dimensional barrier to adsorption for which $S_0(E_e)$ has the shape of a step function with respect to E_n. The key to understanding the observed lack of dependence of the mean energy on desorption angle rests in understanding the behavior of the low energy tail of the adsorption function. The mean desorption energy can be computed from the overlap of the Maxwell–Boltzmann distribution with the adsorption function $S_0(E_e)$. Since the barrier to adsorption is large for the H_2/Cu system, the mean energy is very sensitive to the overlap of the tail of S_0 with the tail of the Maxwell–Boltzmann distribution. At larger angles the relative overlap of these tails at lower energies is enhanced, forcing the mean energy to decrease with increasing energy. This effect competes with the lower E_n of the Maxwell–Boltzmann distribution at larger angles, which, as mentioned above, causes the mean (total) energy to increase with increasing angle. For the H_2/Cu system these competing factors cause the mean energy to be nearly independent of angle [6.66].

6.3.4 State-Resolved Scattering Measurements

The results presented in the preceding discussion show conclusively that vibrational states play an important role in the adsorption and desorption of hydrogen at copper surfaces. With the exception of the state-resolved desorption experiments of *Kubiak* et al. [6.48], this information comes from indirect methods and careful modeling. In this section we discuss measurements in which the internal state distributions of hydrogen molecules are measured before and after the molecules are scattered from a copper surface. From such measurements one can determine the loss or gain of molecules in individual quantum states resulting from both inelastic and adsorption processes. These measurements generally confirm the picture developed in previous sections on the strong influence of the vibrational state on the adsorption dynamics. They also show an interesting new phenomenon – direct vibrational excitation arising from the approach of the hydrogen to the transition state to dissociation.

a) Reflection Probability Measurements

Hodgson et al. [6.90] have reported measurements of vibrationally excited H_2 scattered from Cu(1 1 1). Specifically they were interested in determining the probability of detecting vibrationally excited molecules after collision with the surface. Measuring the intensity of a given state in the incident and scattered beam is not sufficient to determine the reflection probability because the scattered beam is broadened into a lobe by incoherent scattering. Accurate integration over the lobe is difficult. To solve this problem, *Hodgson* et al. took the ratio of the intensity of H_2 ($v = 1, J = 1$) in the scattered and incident beam, and compared this ratio to the corresponding ratio for $H_2(v = 0, J = 1)$. They checked that the angular lobes were identical for the two states. Below the threshold for adsorption for ground-state molecules this ratio of ratios gives an estimate of the reflection probability for vibrationally excited H_2, provided inelastic scattering can be ignored. Diffraction studies have shown that although rotational energy exchange with the surface is significant for HD scattered from a metal surface [6.91], it is minimal for H_2 [6.92, 93].

Table 6.3. Adsorption functions for D_2 on Cu(1 1 1). The table displays a comparison of the parameters of the adsorption functions determined from various experiments on this system

System	Authors and method	$E_0(0)$ [eV]	$W(0)$ [eV]	$E_0(1)$ [eV]	$W(1)$ [eV]	$E_0(2)$ [eV]	$W(2)$ [eV]	A
D_2 Cu(1 1 1)	Rettner, Auerbach, and Michelsen [6.72] Molecular beam direct adsorption	0.677	0.118	0.466	0.114	0.231	0.109	0.339
D_2 ($J = 2$) Cu(1 1 1)	Michelsen, Rettner, and Auerbach [6.97] State-resolved desorption	0.60	0.11	0.40	0.11	0.24	0.12	*
	Michelsen, Rettner, and Auerbach [6.97] State-resolved desorption	0.63	0.11	0.40	0.12	0.24	12	*
D_2 ($J = 4$) Cu(1 1 1)	Hodgson et al. [6.94] Survival probabilities	*	*	0.35	0.15	*	*	0.5

Note: *indicates that the experiment provides no information on this parameter

For a beam of energy $E_i = 0.34$ eV at an angle of incidence of 45° *Hodgson* et al. reported a survival probability of 0.67 ± 0.1. A sticking probability of ≈ 0.3 for $E_n = 0.17$ eV seems high compared with estimates by *Michelsen* and *Auerbach* [6.66] using the sticking-probability model described in Sect. 6.3.3a. Similar measurements of the survival probability of $D_2(v = 1,\ J = 4)$ as a function of kinetic energy [6.94] have yielded a translational threshold of 0.35 eV and a width of 0.15 eV, which, as shown in Table 6.3, are not in good agreement with the values determined by molecular beam adsorption measurements.

b) Inelastic Scattering Measurements: Vibrational Excitation

Rettner et al. [6.95] have introduced a different approach to study the loss or gain of molecules in a given quantum state as a function of velocity of the incident molecules. The principle of the method is illustrated schematically in Fig. 6.24. A beam of H_2 or D_2 is chopped to produce a narrow pulse. The source conditions are deliberately chosen to give a beam with a broad velocity distribution. During the flight to the surface, the beam disperses in time so that fast molecules arrive at the surface earlier than slow ones. The molecules are detected using a pulsed laser (≈ 200 nm) to ionize molecules populating a specific v, J state. By varying the chopper-laser delay, the Time-Of-Flight (TOF) distribution can be determined, either for the incident beam, i.e., *before* it has scattered from the surface, or for molecules scattered from the surface. This TOF distribution of the scattered beam is recorded close to (≈ 3.5 mm) from the surface to minimize the dispersion resulting from translational-energy exchange with the surface. The basic idea of the experiment is simple. A comparison of the TOF distributions of the incident and scattered molecules will show losses due

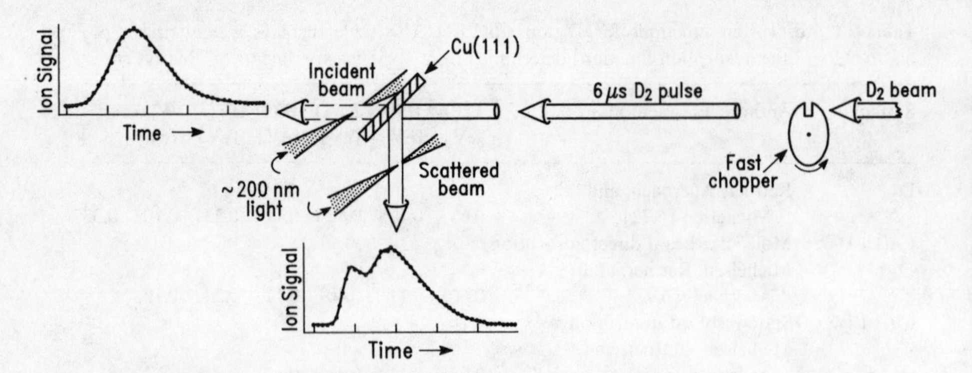

Fig. 6.24. Schematic drawing of the state-resolved scattering experiment. The molecular beam is chopped and allowed to disperse before scattering from the surface. A single rovibrational state of the scattered distribution is selected by the ionization wavelength of the laser. The delay between the chopper trigger and the laser pulse is varied to yield an ion signal as a function of time

to adsorption and inelastic scattering out of that state and gains due to transitions into that state.

Consider first incident molecules that are in the ground vibrational state. At high kinetic energy some of these will adsorb, and some will be excited to the $v = 1$ state. In addition, a fraction will be elastically scattered. There may also be rotational energy transfer. Figure 6.25 depicts the results for $D_2(v = 0, J = 0)$. The dashed curve displays the time spread for this state in the incident beam. The results are the same for D_2 scattered from a surface saturated with D atoms. The points are results for D_2 scattered from a clean surface. Above an incidence

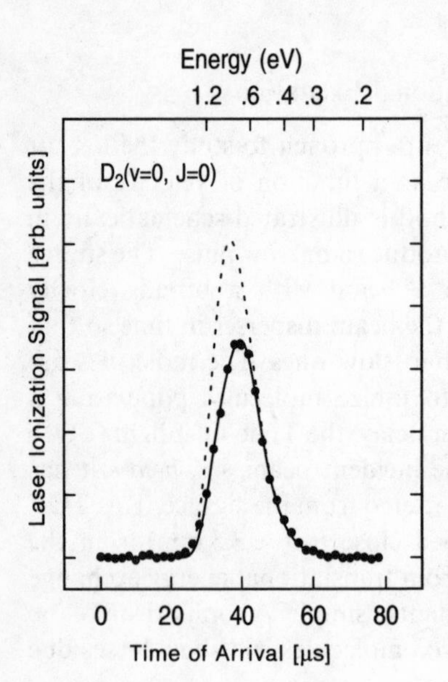

Fig. 6.25. State-specific scattering measurements, $v = 0$: time-of-flight distributions of $D_2(v = 0, J = 0)$ molecules detected after scattering from a Cu(1 1 1) surface. The solid line corresponds to a fit of the product of the incident distribution (indicated as a dashed curve) with an excitation function that has a threshold to excitation at high kinetic energies, in addition to adsorption function as discussed in Sect. 6.3.3a. From [6.95] with permission

kinetic energy of ~ 0.6 eV there is a loss of $D_2(v = 0, J = 0)$. A detailed comparison with the previous results for this system (Figs. 6.14, 16, and 17) indicates a larger loss than expected for dissociation but only at high energy. This excess loss at high energy is consistent with the availability of inelastic channels at high energy but not at low energy.

Those molecules that are inelastically scattered into $v = 1$ will appear as an enhancement in the $D_2(v = 1)$ TOF distribution. This enhancement will appear at an early time in the TOF distribution because this process can only occur for high-energy molecules, which are detected in the early part of the TOF distribution. Figure 6.26 displays the result for $D_2(v = 1, J = 4)$. The top panel shows data for the incident beam (D_2 scattered from a D-saturated surface). The middle panel exhibits data for the clean surface, and the lower panel shows a comparison of the two. The molecules scattered from the clean surface reveal a loss of intensity above ≈ 0.4 eV, which is consistent with the results of the non-state-selected adsorption experiment described in Sect. 6.3.2. Above ≈ 0.6 eV, a second peak in the TOF distribution emerges. *Rettner* et al. attributed this peak to *vibrational excitation* of high energy $D_2(v = 0)$ molecules. *Hodgson* et al. [6.94] have reported similar results for $D_2(v = 1)$ scattering from Cu(1 1 1), which they also interpreted in terms of vibrational excitation. That this vibrational excitation does not occur on scattering from a D-saturated surface suggests that it is not a direct "mechanical" effect. Rather, the excitation results from the chemical interaction of the molecules with the surface. For a

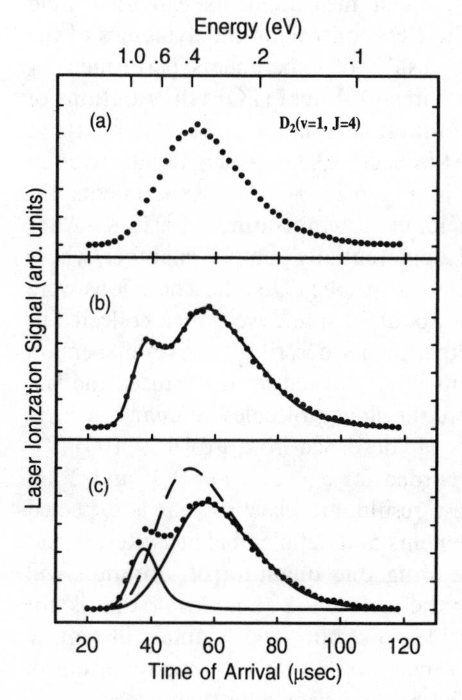

Fig. 6.26. State-specific scattering measurements, $v = 1$. Time-of-flight (TOF) distributions of $D_2(v = 1,\ J = 4)$ detected after scattering from a Cu(1 1 1) surface: (a) scattering from a D-saturated surface, which is indistinguishable from the TOF distribution of the incident beam: (b) scattering from the bare surface; (c) a comparison of (a) and (b) scaled to agree at long times. The solid line on (b) corresponds to a fit of the product of the incident distribution (indicated as a dashed curve on (c)) with an excitation function that has a threshold to excitation at high kinetic energies, in addition to an adsorption function as discussed in Sect. 6.3.3a. From [6.95] with permission

PES like that shown in Fig. 6.10 the D–D bond will stretch as the molecule approaches the barrier. Some molecules will tunnel through or go over the barrier and go on to dissociate. Others will be scattered. For the scattered molecules, the stretched bond can lead to vibrational excitation.

Calculations performed by *Darling* and *Holloway* [6.96] confirm the conclusion that the molecule must enter a region of the potential where the D–D bond is stretched. These calculations demonstrate that the location of the barrier determines the probability of vibrational excitation. Thus, the observation of vibrational excitation provides a new means of probing the PES and the microscopic dynamics of the H_2/Cu system. Recall that the transition state for a late (or even an intermediate) barrier is distinguished by an extended internuclear distance or bond length. Hence, the observation of strong vibrational excitation tells us directly that the barrier is in a region of the potential where the D–D bond is already stretched, i.e., that the barrier is an "intermediate" or "late" barrier.

6.3.5 State-Resolved Desorption Measurements: $S_0(v, E_i)$ via Detailed Balance

As noted previously in the context of measurements of internal-state distributions of desorbed molecules, state-resolved desorption measurements can provide considerable insight into the efficacy of specific internal states in adsorption. As discussed in Sect. 6.3.1, such measurements can also yield information about the topography of the PES controlling the dynamics of the system. In this section we discuss the results of experiments performed by *Michelsen* et al. [6.97, 98], in which the Time-Of-Flight (TOF) distributions of molecules desorbed into individual quantum states were recorded and analyzed using a method similar to that described in Sect. 6.3.3. A schematic drawing of the experimental procedure is shown in Fig. 6.27. In these experiments D_2 molecules desorbed from a crystal held at a temperature of 925 K. After desorbing from the crystal, they entered a differentially pumped chamber, where they were ionized by a laser tuned to select a specific v, J state. These ions were allowed to drift in a field-free region for about 2 cm and were then collected at the detector. As with the scattering experiments described above, dispersion over the flight distance in which the ions were allowed to drift caused the fast molecules to arrive at the detector before the slow molecules. Similar measurements have recently been performed for H_2 desorbed from Pd [6.99, 100].

Examples of TOF distributions recorded for $J = 2$ of $v = 0$, 1, and 2 are depicted in Fig. 6.28. Qualitatively these results are exactly what is expected based on the direct adsorption measurements and detailed balance. Recall that on average an incident molecule possessing one quantum of vibration will require less kinetic energy to adsorb on the surface than a molecule that is not vibrationally excited. Likewise, a molecule possessing two quanta will require less kinetic energy than an incident molecule possessing only one quantum of vibrational energy. According to detailed balance we would then expect mole-

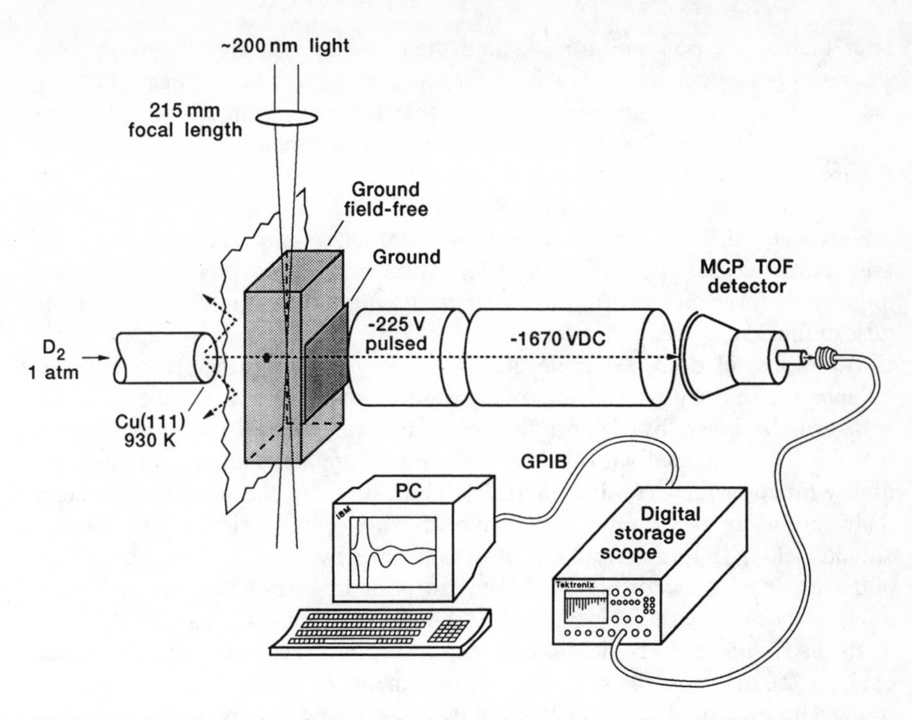

Fig. 6.27. Schematic drawing of state-resolved desorption experiment. From [6.98] with permission

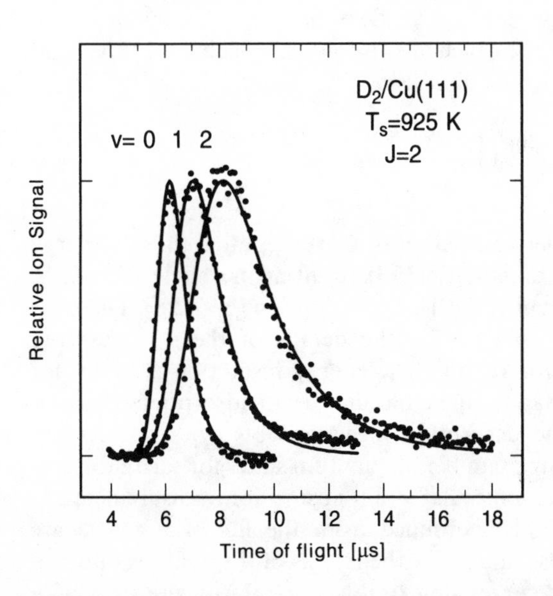

Fig. 6.28. Desorption time-of-flight distributions vs. vibrational state. Time-of-flight distributions obtained for D_2 desorbed from Cu(111) in $J = 2$ of $v = 0$, 1, and 2. The lines represent fits to the distributions using (6.4) to represent the adsorption functions for the three vibrational states. From [6.97] with permission

cules desorbed with one quantum in vibration to have less kinetic energy than molecules desorbed into $v = 0$. Similarly we would expect molecules desorbed into $v = 2$ to possess less kinetic energy than those desorbed into $v = 1$. This trend is shown in Fig. 6.28. Molecules desorbed into $v = 2$ arrive at the detector later than those populating $v = 1$ and thus possess less kinetic energy. Not surprisingly $v = 1$ molecules arrive later and possess less kinetic energy than $v = 0$ molecules. *Michelsen* et al. reported a mean energy (summed over all states) of 0.58 ± 0.05 eV for $T_s = 925$ K [6.97], which is in agreement with the results of *Comsa* and *David* who reported a value of about 0.63 eV for D_2 desorbed from Cu(1 1 1) at a surface temperature of 1000 K [6.47].

In Sect. 6.3.3c we demonstrated how quantitative information gained from adsorption measurements, i.e., $S_0(v, E_i)$, could be used to predict desorption angular and velocity distributions. Similarly, quantitative information about adsorption, $S_0(E_i)$, can be derived from TOF measurements (i.e., velocity distributions) of desorbed molecules. According to the principle of detailed balance, the velocity distributions of molecules desorbed into a specific quantum state will be given by the product of a flux-weighted Maxwellian velocity distribution associated with the surface temperature and the adsorption probability function for that state, $S_0(v, E_i)$. Thus, dividing the measured velocity distribution by the known flux-weighted Maxwellian velocity distribution should yield $S_0(E_i)$ numerically. A more convenient way of deriving, compiling, and comparing this information about adsorption involves fitting the data using an appropriate functional form for $S_0(E_i)$. For the purposes of comparing these TOF distributions with the direct beam adsorption measurements of *Rettner* et al. [6.72], and with the scattering measurements of *Hodgson* et al. [6.94], we fit the data presented in Fig. 6.28 using (6.3) to give the parameters presented in Table 6.3. These parameters are in good agreement with those determined from direct adsorption measurements (for E_0, see Sect. 6.3.3b) and angular distributions (for $W(T_s)$, see Sect. 6.3.3c), which are also given in Table 6.3. The solid lines in Fig. 6.28 represent fits to the data using

$$S_0(v) = \frac{A}{2}\left[1 + \mathrm{erf}\left(\frac{E_e - E_0(v)}{W(v)}\right)\right]. \tag{6.4}$$

This functional form is nearly identical to that of (6.3) except that the low-energy tail falls off much more quickly than that of (6.3), resulting in a slightly better fit in the low-energy (longer-time) tail of the TOF distribution [6.97, 98]. These fits give the most reliable results to date for the details of the state-resolved adsorption function, $S_0(v, E_i)$, for H_2/Cu. The energy resolution is better for these desorption experiments than for the molecular beam adsorption measurements, and the analysis of the desorption measurements does not require deconvolving contributions from a number of quantum states for each measurement as does the analysis of the molecular beam adsorption measurements.

The functions $S_0(E_n, v, J = 2)$ determined from the fits of Fig. 6.28 are depicted in Fig. 6.29. As was also shown by the results of the analysis of direct adsorption data (Fig. 6.16), the curves shift to lower translational energy with increasing v. It was assumed previously, however, that the parameter A was independent of v, whereas Fig. 6.29 displays functions with different values of A.

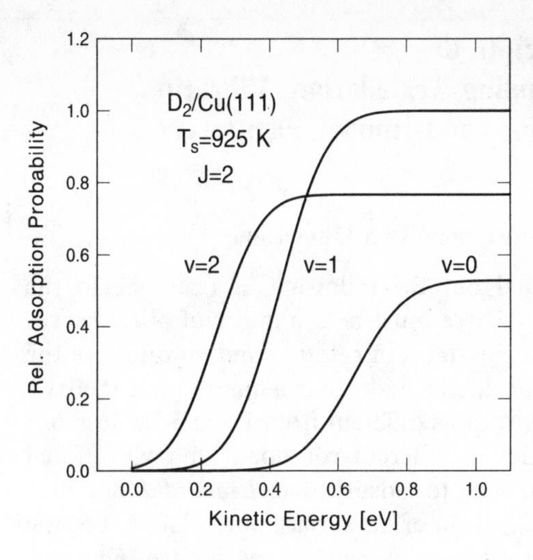

Fig. 6.29. Kinetic energy dependence of the adsorption probability. The internal-state-dependent adsorption probability function for D_2 on Cu(1 1 1), $S_0(E_n, v, J)$, is plotted against E_n for a single rotational state ($J = 2$) and several vibrational states ($v = 0$, 1, and 2). From [6.98] with permission

The relative values of this parameter are determined from the relative integrated intensities rather than the shapes of the TOF distributions. The curves in Fig. 6.28 have been rescaled so that their maximum values are the same. Integrating the unnormalized curves gives the relative populations in each quantum state. Correcting the intensities of each state for flux and summing over rotational states, $J = 0$–8, *Michelsen* et al. [6.98] obtained the value of 0.20 $\pm$ 0.03 for the $D_2(v = 1)/D_2(v = 0)$ population ratio and 0.0096 $\pm$ 0.0018 for the $D_2(v = 2)/D_2(v = 0)$ ratio [6.98]. The Boltzmann ratios for these states are 0.0101 for $D_2(v = 1)/D_2(v = 0)$ and 0.000122 for $D_2(v = 2)/D_2(v = 0)$, making enhancement a factor of 20 for $v = 1$ and a factor of 79 for $v = 2$ over that of a thermal distribution at 925 K. Predicting these ratios using detailed balance and the functions $S_0(E_n, v, J)$ derived from the shape of the TOF distributions gives values of 0.108 for $D_2(v = 1)/D_2(v = 0)$ and 0.0068 for $D_2(v = 2)/D_2(v = 0)$. The discrepancies between the measured and calculated values suggest that the parameter A varies with v. *Michelsen* et al. [6.98] have determined the relative values of $A(v)$ to be $A(0)$:$A(1)$:$A(2)$ = 0.54 $\pm$ 0.16:1.00:0.77 $\pm$ 0.18, as is shown in Fig. 6.29.

Note that for these state-resolved desorption measurements we can determine $S_0(E_i)$ and relative populations as a function of rotational state in addition to vibrational state. Rotation is also resolved for the scattering measurements described in the previous section. Since it is difficult to gain independent control of rotational energy in molecular beam experiments (e.g., by stimulated Raman pumping of the incident molecules into a single rovibrational state [6.101–104]), we have ignored thus far the role of rotation in the discussion of experimental data. These state-resolved experiments, however, reveal a surprising richness in the role of rotational motion in the dynamics of adsorption, which we shall highlight in Sect. 6.4.3 and 6.4.4.

6.4 A Multidimensional Description: The Degrees of Freedom Including Translation, Vibration, Rotation, Molecular Orientation, and Impact Parameter

6.4.1 Theoretical Descriptions for More than Two Dimensions

In 1932 *Lennard-Jones* [6.5] pointed out the following in reference to the interactions represented in Fig. 6.2: "There must be a number of other curves between (1) and (2) corresponding to the states of rotation and vibration of the molecule at infinity. These must be considered in detail in a quantitative theory". Since these "other curves" cross in locations different from Point K in Fig. 6.2, this statement implies that molecules in different rotational and vibrational states will experience different barriers to adsorption. *Lennard-Jones* also commented on the effects of the corrugation of the surface potential, that is, the variation in the potential with the position of impact on the surface. *Lennard-Jones* was ahead of his time not only in presenting the basic aspects of the interaction potential in the one-dimensional picture, but also in appreciating the finer points, such as the influence on the barrier height of the internal states of the reactant and its position of impact on the surface.

The 2-D theoretical description we have so far discussed applies only to a single orientation, the collinear approach of H_2 to a Cu_2 dimer. Specifically, this PES is for the approach of H_2 with its molecular axis parallel to that of the Cu_2. The spacing between the two Cu atoms is approximately that of the average Cu–Cu distance at the Cu(1 0 0) face, and the dissociation takes place from bridge to atop sites. Although many of the qualitative conclusions we have presented are consistent with calculations on this potential, such 2-D calculations do not contain information about the anisotropy of the potential, and thus do not treat the influence of molecular orientation or the effect of rotational motion on adsorption. Calculations in higher dimensions will generally give considerably smaller dissociation probabilities, because of the inclusion of polar orientations unfavorable to dissociation [6.105].

A number of attempts have been made recently to extend the theoretical description to higher dimensions. Calculations made by *Nielsen* et al. [6.106] showed that important dynamical effects are lost in calculations that fix the molecular orientation. This study also demonstrated that the rotational state distributions of scattered molecules may reflect the topography of the PES. This latter point has also been made by *Holloway* and *Jackson* [6.107] and by *Chang* and *Holloway* [6.108], who have shown that valuable information on the dynamics of dissociation and on the geometry of the transition state is contained in the alignment moments and rotational distributions of the scattered molecules.

Engdahl et al. [6.109] have compared dynamical calculations for the H_2/Cu system in two and six dimensions. They concluded that all six molecular degrees of freedom must be included in order to understand the trends in reactivity from one Cu face to another. This group employed a PES computed using the effective medium theory in which the mean electron density of the surface is derived, and the energy of embedding the adsorbate in an electron gas of this density is calculated. Their 6-D calculations indicate that the barrier to dissociation is slightly lower on Cu(1 1 0) than on Cu(1 1 1), consistent with the analysis of the data of *Anger* et al. [6.53] presented by *Michelsen* and *Auerbach* [6.66]. This result is in contrast to their 2-D calculation, which yields a much lower barrier for the (1 1 0) face compared to dissociation on the (1 1 1) face. It is clear that understanding the multidimensional aspects of the H_2/Cu system is an important step in developing a complete understanding of the dynamics. In particular, calculations at fixed molecular orientation will obviously not be able to evaluate the role of rotational motion in overcoming the barrier to dissociation. As we shall see in the following subsections, this motion can have a strong influence both on dissociation and on the probability of vibrational excitation.

6.4.2 Activation Energy Measurements: The Effect of Rotation on Adsorption Rate

One way of gaining some information about the influence of rotation on adsorption involves measuring the rate of adsorption as a function of the rotational temperature of reactant molecules at equilibrium. Such measurements have been made by *Campbell* et al. [6.110] who studied the adsorption of hydrogen on Cu(1 1 0). These workers recorded the adsorption rate as a function of surface temperature and buffer-gas pressure. The buffer gas served to control the degree of vibrational and rotational energy in the reacting H_2. This approach is based on the fact that vibrational energy transfer requires many more collisions than does rotational energy transfer, and rotational, many more than translational. *Campbell* et al. controlled the equilibration rate of each degree of freedom by varying the pressure of the buffer gas. Their results suggest that the adsorption probability increases as energy in the rotational degree of freedom increases.

6.4.3 State-Resolved Desorption Measurements: The Role of Rotation in Adsorption and Desorption

The most detailed information to date concerning the role of rotation in adsorption/desorption dynamics has come from state-resolved desorption measurements. Some of these measurements were described in Sect. 6.3.5 in the discussion of the role of vibration in adsorption/desorption dynamics. Measurements similar to those shown in Fig. 6.28 for different vibrational states were recorded by *Michelsen* et al. [6.97, 98] for a number of different rotational states for several vibrational states. As with vibration, if rotational motion enhances or hinders adsorption by changing the effective translational barrier to adsorption, the kinetic energy of molecules desorbed from the surface will be different for each rotational state.

The results of *Michelsen* et al. distinctly show a dependence of the TOF distribution on rotational state. This dependence can be seen in Fig. 6.30 in which is displayed the TOF distributions for even rotational state J (up to $J = 14$) of $v = 0$ for D_2 desorbed from Cu(1 1 1). As J is increased from 6 to 14, the mean kinetic energy of the desorbed molecules decreases, implying that the translational barrier to adsorption *decreases with increasing J*. On the other hand, as J decreases from 6 to 0, the mean kinetic energy also decreases, suggesting that the translational barrier *increases with increasing J*.

As may be expected, these trends can be seen clearly in the quantitative analysis of these data. Using (6.4) in fits described in Sect. 6.3.5 to evaluate the data for each rovibrational state, *Michelsen* et al. [6.97, 98] have determined $S_0(E_n, v, J)$ for a series of rovibrational states of D_2 on Cu(1 1 1). The results for several rotational states for $v = 0$ are depicted in Fig. 6.31. Similarly to the behavior observed for different vibrational states, these curves representing different rotational states shift with respect to translational energy. Unlike the behavior with vibrational state, however, the shift with J is nonmonotonic. The compiled results of this analysis are presented in Fig. 6.32, which shows the translational threshold as a function of rotational state J for vibrational states $v = 0, 1$, and 2. The results demonstrate behavior with increasing J at low J that is strikingly different from that at high J. At low J ($J = 0$–5) as J is increased, the translational threshold (the effective barrier to adsorption) increases while at high J ($J > 5$), as J is increased, the translational threshold decreases. These trends mean that at low J rotational motion *hinders* adsorption, but at high J rotational motion *enhances* adsorption.

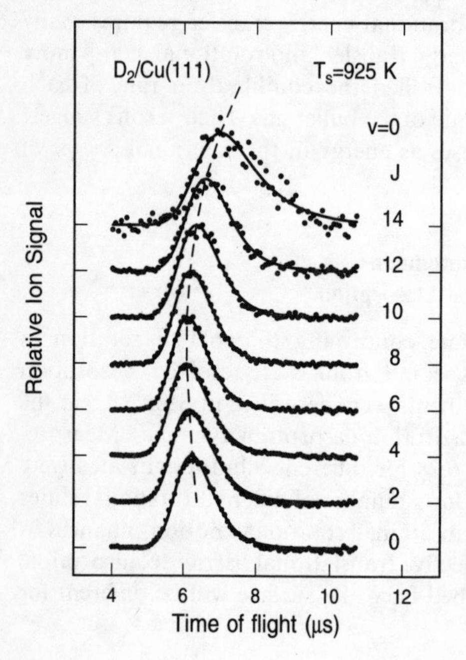

Fig. 6.30. Desorption time-of-flight distributions vs. rotational state. Time-of-flight distributions obtained for D_2 desorbed from Cu(1 1 1) in different rotational states of the ground vibrational state. The lines show fitted distributions based on the adsorption function given in (6.4). From [6.97] with permission

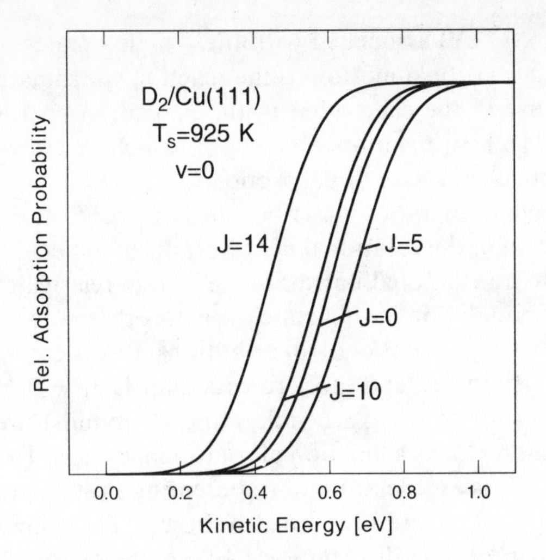

Fig. 6.31. Kinetic energy dependence of the adsorption probability. The internal-state-dependent adsorption probability function for D_2 on Cu(1 1 1), $S_0(E_n, v, J)$, is plotted against $\dot{E}_n$ for selected rotational states of the ground vibrational state. From [6.98] with permission

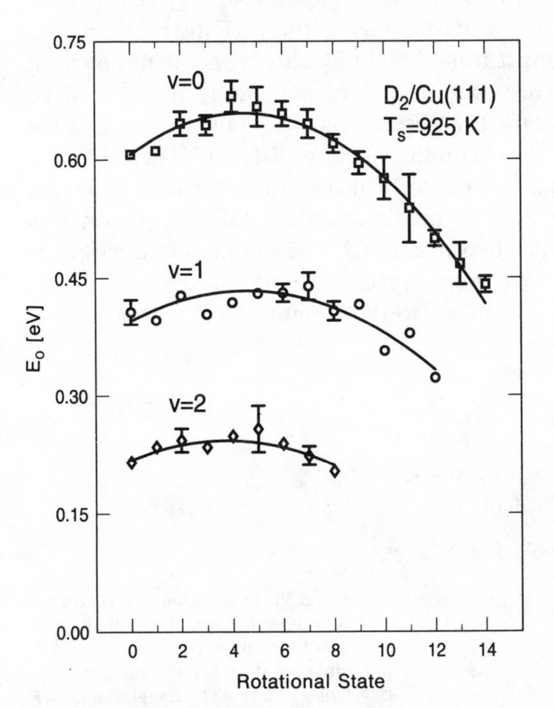

Fig. 6.32. State-specific translational energy requirements for adsorption. Values of the translational threshold, E_0, are obtained from fits to D_2 time-of-flight distributions for different rovibrational states. From [6.98] with permission

Michelsen et al. [6.97, 98] suggest that the predominant effect at low J is a steric one in which increased rotation prevents the incident molecules from following the lowest energy path to adsorption. This effect is apparent in the calculations of *Cruz* and *Jackson* [6.105]. A different mechanism may explain the trend observed at high J. The decrease in the effective translational barrier with increasing J at high J suggests that the molecules can use some fraction of the large rotational energy (up to 0.7 eV) associated with these high J states to overcome the barrier by coupling rotational motion to the reaction coordinate. This mechanism may be the cause of the effect seen in the activation energy measurements of *Campbell* et al. [6.110] mentioned above, in which increased rotational energy appears to increase the rate of adsorption.

As with the vibrational population ratios described in Sect. 6.3.5, these rotational effects should be reflected in the rotational distributions of molecules desorbed from the surface. A low translational barrier for one state relative to another should result in a higher population in that state over the other relative to the Boltzmann populations at T_s. Rotational distributions recorded by *Kubiak* et al. [6.48] for $D_2(v = 1)$ (triangles) and those recorded for $D_2(v = 0, 1,$ and 2) by *Michelsen* et al. [6.98] (open squares, circles, and diamonds) are exhibited in Fig. 6.33 as a Boltzmann plot as a function of rotational energy. The agreement between the two data sets is excellent. Lines representing Boltzmann rotational distributions at the surface temperatures are also shown. (These lines do not represent the Boltzmann vibrational distribution.) As may be expected, the relative populations reflect the trends observed in the TOF distributions and translational barriers. Decreasing J from $J = 5$ to 0 shows an enhancement in population over the Boltzmann distribution, as does increasing J from 5 to 12 or 14. Also shown in Fig. 6.33 (x marks) is a prediction via detailed balance of the rotational distribution for $v = 0$ based on the functions $S_0(E_n, v, J)$ derived from the TOF distributions. The shape of the TOF distribution yields $E_0(v, J)$ and $W(v, J)$ but does not give $A(v, J)$. The predicted rotational distribution was calculated assuming the $A(v, J)$ is independent of J. The validity of this assumption is confirmed by the nearly perfect agreement with the experimentally determined distribution. Thus, in contrast to the variation of A with v, A does not change with J.

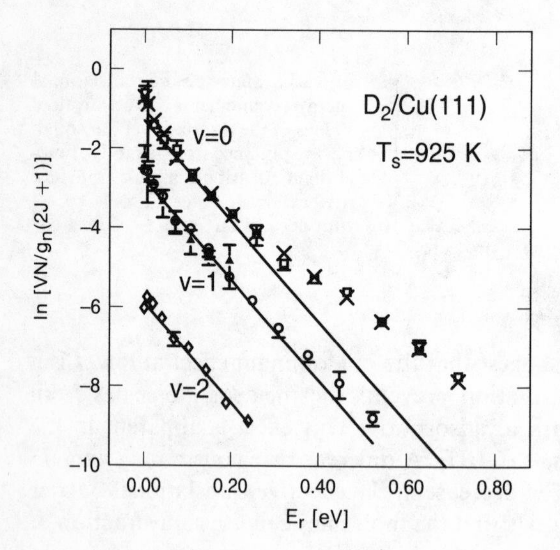

Fig. 6.33. Desorption rotational state distributions. Boltzmann plots of rotational distributions for several vibrational states recorded by *Kubiak* et al. [6.48] (triangles) and *Michelsen* et al. [6.98] (open squares, circles, and diamonds). The lines represent the rotational Boltzmann distributions at the surface temperature. The x marks represent the rotational distribution for $v = 0$ predicted via detailed balance

The influence of J on the dynamics of adsorption and desorption leaves considerable room for speculation. Understanding these interactions, at least as well as we understand the role of the vibration on adsorption, will require further coordinated efforts between experiment and theory.

6.4.4 Inelastic Scattering Measurements: The Influence of Rotation on Vibrational Excitation

The scattering results of *Rettner* et al. for Cu(1 1 1) also reveal that there is a significant effect of rotational state on the probability of vibrational excitation [6.95]. This effect can be seen in Fig. 6.34, where it is shown that the bimodal character of the TOF distribution becomes pronounced as the rotational state is increased from $J = 0$ to 5. More specifically, the height of the early (high energy) component decreases relative to the height of the later peak with increasing J. This effect is predominantly due to a decrease in the inelastic component scattered into $v = 1$ from $v = 0$. That is, the probability of vibrational excitation upon scattering decreases with increasing J from $J = 0$ to 5. The decrease in the bimodality with increasing J could in part be due to an increase in the elastically scattered component, i.e., a decrease in the adsorption probability with increasing rotational state, however the quality of the data was insufficient to establish this conclusively.

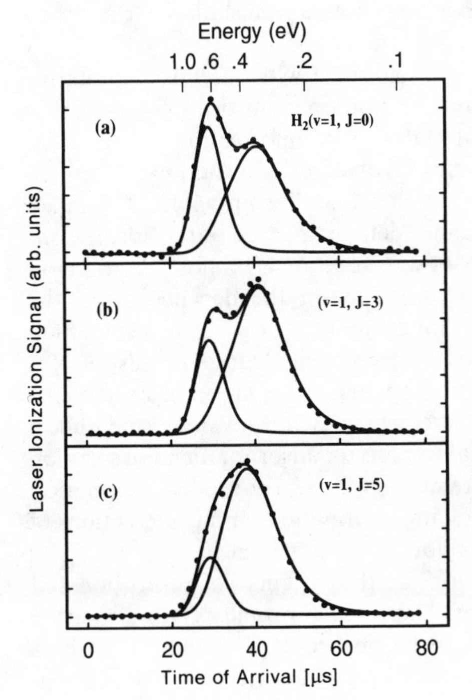

Fig. 6.34. Vibrational excitation vs. rotational state. Time-of-flight distributions of $H_2(v = 1)$ molecules detected after scattering from a Cu(1 1 1) surface. Results are displayed for $J = 0$, $J = 3$, and $J = 5$, as labelled. In each case the solid lines correspond to the fit and to the two separate components of the fit. From [6.95] with permission

These results show that increasing rotational state reduces the probability of vibrational excitation. The probability of vibrational excitation is related to the adsorption probability in that both processes require the incident molecule to access the reactive region of the PES, the transition state. Hence, a decrease in vibration excitation with an increase in rotation implies that such an increase in rotation *inhibits* the incident molecule in accessing the transition state to adsorption. This picture is consistent with the influence of rotation on the adsorption probability described above.

6.5 Summary

The dissociative adsorption of hydrogen at copper surfaces and the reverse reaction of associative desorption are currently the most completely understood reactions in surface chemistry. The main features of this reaction can be understood in terms of the existence of an activation barrier between the physisorbed state of molecular hydrogen and the chemisorbed state of the dissociated atoms. This barrier is ≈ 1 eV high and located in an intermediate to late region along the reaction path to dissociation. The location of the barrier leads to a significant coupling of translational and vibrational and, to some extent, rotational motion to the reaction coordinate. As a consequence these forms of energy are effective in promoting dissociative adsorption. Furthermore, vibrational excitation of molecules that scatter from the surface is highly probable if the incident molecules possess enough translational energy to approach the barrier.

The use of molecular-beam techniques coupled with careful, quantitative modeling has allowed us to learn about the variation of adsorption probability with energy for individual vibrational states. This information is crucial in establishing a description of the dynamics involved in adsorption and desorption. State-resolved desorption measurements and the principle of detailed balance have allowed us to develop a more detailed picture of the adsorption/desorption dynamics that includes the role of rotational motion. In addition, these measurements provide better information on the dependence of the adsorption probability on kinetic energy for ground state ($v = 0$) molecules than currently attainable with molecular-beam techniques. More details of this picture have been added by (1) desorption angular-distribution measurements made as a function of surface temperature, which have provided some understanding of the effects of surface thermal motion on adsorption and desorption and (2) state-to-state scattering measurements, which have yielded information about the barrier to adsorption by supplying information on the interactions of scattered molecules with the reactive region of the potential.

Fertile ground remains for future work on the dynamics of adsorption and desorption for the hydrogen/Cu system. There remain some details about the dynamics of this system that have yet to be probed experimentally. Little is known experimentally about the influence of molecular orientation on adsorption or about orientational effects in desorption. Likewise, there are many other aspects of the dynamics to be explored, such as the influence of surface structure, defects, and impact parameter. The development of an accurate, multidimensional potential energy surface and corresponding dynamical calculations is also a rich area for future work.

Acknowledgements. In preparing this chapter, we have benefited greatly from generous input, stimulating discussions, and manuscripts provided prior to publication from many people. We would particularly like to think C.T. Campbell, J. Harris, B.E. Hayden, S. Holloway, A.C. Luntz, R.J. Madix, J.K. Nørskov, K.D. Rendulic, and R.N. Zare. We would also like to thank the ONR for partial support of H.A.M. under grant #N00014-91-J-1023.

References

6.1 W.H. Miller: Recent advances in quantum mechanical reactive scattering theory. Annu. Rev. Phys. Chem., ed. by H.L. Strauss (Annual Reviews, Palo Alto, CA, 1990) Vol 41
6.2 Although there is no evidence for a molecular chemisorbed state of hydrogen on a copper surface, H_2 has been found to physisorb to Cu(100), with a binding energy of $\sim$ 20 meV
6.3 J.B.A. Dumas: Ann. de Chim. et de Phys. III **8,** 189 (1843)
6.4 H.S. Taylor: J. Am. Chem. Soc. **53,** 578 (1931)
6.5 J.E. Lennard-Jones: Trans. Faraday Soc. **28,** 333 (1932)
6.6 W.G. Pollard: Phys Rev. **56,** 324 (1939)
6.7 A.F. H. Ward: Proc. Roy. Soc. A **133,** 506 (1931)
6.8 R.A. Beebe, G.W. Low, E.L. Lincoln, S. Goldwasser: J. Am Chem. Soc **57,** 2527 (1935)
6.9 W. Hampe: Z. Anal. Chem. **13,** 352 (1874)
6.10 A. Sieverts: Z. Physik. Chem. **77,** 591 (1911)
6.11 H.S. Taylor, R.M. Burns: J. Am. Chem. Soc **43,** 1273 (1921)
6.12 W.E. Garner, F.E.T. Kingman: Nature **126,** 352 (1930)
6.13 J.A. Allen, J.W. Mitchell: Disc. Faraday Soc. **8,** 360 (1950)
6.14 O. Beeck: Disc. Faraday Soc. **8,** 118 (1950)
6.15 B.M.W. Trapnell: Proc. Roy. Soc. A **218,** 566 (1953)
6.16 G.C.A. Schuit, L.L. van Reijen: Adv. in Catalysis (Academic, New York (1958) p. 242
6.17 J. Pritchard, F.C. Tompkins: Trans. Faraday Soc. **56,** 540 (1960)
6.18 D.O. Hayward, B.M.W. Trapnell: Chemisorption (Butterworths, London 1964)
6.19 R.N. Lee, H.E. Farnsworth: Surf. Sci. **3,** 461 (1965)
6.20 D.A. Cadenhead, N.J. Wagner: J. Phys. Chem. **72,** 2775 (1968)
6.21 O. Leypunsky: Acta Physicochim URSS **5,** 271 (1936)
6.22 P.B. Rasmussen, P.M Holmblad, H. Christoffersen, P.A. Taylor, I. Chorkendorff: Surf. Sci., **287,** 79 (1993)
6.23 A.V. Hamza, R.J. Madix: J. Phys. Chem. **89,** 5381 (1985)
6.24 K.D. Rendulic, G. Anger, A. Winkler: Surf. Sci. **208,** 404 (1989)
6.25 A.C. Luntz, J.K. Brown, M.D. Williams: J. Chem. Phys. **93,** 5240 (1990)
6.26 H.W. Melville, E.K. Rideal: Proc. Roy. Soc. A **153,** 77 (1936)
6.27 G. Rienäcker, B. Sarry: Z. Anorg. Chem. **257,** 41 (1948)
6.28 A. Couper, D.D. Eley: Disc. Faraday Soc. **8,** 172 (1950)
6.29 J. Harris, S. Andersson: Phys. Rev. Lett. **55,** 1583 (1985)
6.30 T. Kwan: Bull. Chem. Soc. Jpn. **23,** 73 (1950)
6.31 M. Balooch, M.J. Cardillo, D.R. Miller, R.E. Stickney: Surf. Sci. **46,** 358 (1974)
6.32 C.T. Rettner, L.A. DeLouise, D.J. Auerbach: J. Chem. Phys. **85,** 1131 (1986)
6.33 C.T. Rettner, H.E. Pfnür, D.J. Auerbach: Phys. Rev. Lett. **54,** 2716 (1985)

6.34 M.B. Lee, Q.Y. Yang, S.T. Ceyer: J. Chem. Phys. **87,** 2724 (1987)
6.35 A.V. Hamza, H.-P. Steinrück, R.J. Madix: J. Chem. Phys. **86,** 6506 (1987)
6.36 A.C. Luntz, D.S. Bethune: J. Chem. Phys. **90,** 1274 (1989)
6.37 G.R. Schoofs, C.R. Arumainayagam, M.C. McMaster, R.J. Madix: Surf. Sci. **215,** 1 (1989)
6.38 J. Harris: Surf. Sci. **221,** 335 (1989)
6.39 T.L. Bradley, R.E. Stickney: Surf. Sci. **38,** 313 (1973)
6.40 M. Balooch, R.E. Stickney: Surf. Sci. **44,** 310 (1974)
6.41 M.J. Cardillo, M. Balooch, R.E. Stickney: Surf. Sci. **50,** 263 (1975)
6.42 More Information about the principle of detailed balance can be found in a review written by G. Comsa, R. David: Surf. Sci. Rept. **5,** 145 (1985)
6.43 G. Comsa: J. Chem. Phys. **48,** 3235 (1968)
6.44 W. Van Willigen: Phys. Lett. **28A,** 80 (1968)
6.45 R.L. Palmer, J.N. Smith, Jr., H. Saltsburg, D.R. O'Keefe: J. Chem. Phys. **53**, 1666, (1970)
6.46 R.L. Palmer, D.R. O'Keefe: Appl. Phys. Lett. **16,** 529 (1970)
6.47 G. Comsa, R. David: Surf. Sci **117**, 77 (1982)
6.48 G.D. Kubiak, G.O. Sitz, R.N. Zare: J. Chem. Phys. **83,** 2538 (1985)
6.49 J.C. Polanyi, W.H. Wong: J. Chem. Phys. **51,** 1439 (1969)
6.50 D. Halstead, S. Holloway: J. Chem. Phys. **93,** 2859 (1990)
6.51 S. Holloway: J. Phys, Condens, Matter **3,** S43–S54 (1991)
6.52 J. Harris, S. Holloway, T.S. Rahman, K. Yang: J. Chem. Phys. **89,** 4427 (1988)
6.53 G. Anger, A. Winkler, K.D. Rendulic: Surf. Sci. **220,** 1 (1989)
6.54 B.E. Hayden, C.L.A. Lamont: Chem. Phys. Lett. **160,** 331 (1989)
6.55 In the jellium model the semi-infinite positive ion lattice is represented by a uniform positive background charge and an associated electron density profile extending into the vacuum region, which creates an electrostatic surface dipole
6.56 P.K. Johansson: Surf. Sci. **104,** 510 (1981)
6.57 P. Madhaven, J.L. Whitten: J. Chem. Phys. **77**, 2673 (1982)
6.58 P.E.M. Siegbahn, M.R.A. Blomberg, C.W. Bauschlicher: J. Chem. Phys. **81**, 1373 (1984)
6.59 M.R. Hand, S. Holloway: J. Chem. Phys. **91,** 7209 (1989)
6.60 C.T. Rettner, H.E. Pfnür, H. Stein, D.J. Auerbach: J. Vac. Sci. Tech. **6,** 899 (1988)
6.61 C.T. Rettner, H.E. Pfuñr, D.J. Auerbach: J. Chem. Phys. **84**, 4163 (1986)
6.62 S.T. Ceyer, J.D. Beckerle, M.B. Lee, S.L. Tang, Q.Y. Yang, M.A. Hines: J. Vac. Sci. Technol. A **5,** 501, (1987)
6.63 C.T. Rettner, H. Stein: J. Chem. Phys. **87,** 770 (1987)
6.64 M.R. Hand, S. Holloway: Surf. Sci. **211/212,** 940 (1989)
6.65 H.A. Michelsen, D.J. Auerbach: Phys. Rev. Lett. **65,** 2833 (1990)
6.66 H.A. Michelsen, D.J. Auerbach: J. Chem. Phys. **94,** 7502 (1991)
6.67 M.P. D'Evelyn, A.V. Hamza, G.E. Gdowski, R.J. Madix: Surf. Sci. **167,** 451 (1986)
6.68 J. Harris, J. Simon, A.C. Luntz, C.B. Mullins, C.T. Rettner: Phys. Rev. Lett. **67,** 652 (1991)
6.69 H.F. Berger, M. Leisch, A. Winkler, K.D. Rendulic: Chem. Phys. Lett. **175,** 425 (1990)
6.70 H.F. Berger, K.D. Rendulic: Surf. Sci. **253,** 325 (1991)
6.71 B.E. Hayden, C.L.A. Lamont: Surf. Sci. **243,** 31 (1991)
6.72 C.T. Rettner, D.J. Auerbach, H.A. Michelsen: Phys. Rev. Lett. **68**, 1164 (1992)
6.73 R.J. Gallagher, J.B. Fenn: J. Chem. Phys. **60,** 3492 (1974)
6.74 J.B. Anderson, R.P. Andres, J.B. Fenn: High intensity and high energy molecular beams, in *Advances in Atomic and Molecular Physics*, ed. by D. R. Bates, I. Estermann (Academic, New York 1965) Vol. 1
6.75 B.E. Hayden, C.L.A. Lamont: Phys. Rev. Lett. **63,** 1823 (1989)
6.76 J.E. Müller: Surf. Sci. **272,** 45 (1992)
6.77 J.K. Nørskov: J. Chem. Phys. **90,** 7461 (1989)
6.78 S. Holloway: Quantum effects in gas-solid interactons, in *Dynamics of Gas-surface Collisions*, ed. by C.T. Rettner, M.N.R. Ashfold (Royal Soc. Chemistry, Cambridge, 1991)
6.79 S. Küchenhoff, W. Brenig, Y. Chiba: Surf. Sci. **245,** 389 (1991)
6.80 D.J. Auerbach, C.T. Rettner, H.A. Michelsen: Surf. Sci. **283,** 1 (1993)

6.81 C.T. Rettner, H.A. Michelsen, D.J. Auerbach, C.B. Mullins: J. Chem. Phys. **94,** 7499 (1991)
6.82 M.R. Hand, J. Harris: J. Chem. Phys. **92,** 7610 (1990)
6.83 A.C. Luntz, J. Harris: Surf. Sci. **258,** 397 (1991)
6.84 H.F. Winters: J. Chem. Phys. **62,** 2454 (1975)
6.85 T.-C. Lo, G. Ehrlich: Surf. Sci. **179,** L19 (1987)
6.86 S.G. Brass, G. Ehrlich: Surf. Sci. **187,** 21 (1987)
6.87 T.P. Beebe, Jr., D.W. Goodman, B.D. Kay: J. Chem. Phys. **87,** 2305 (1987)
6.88 I. Chorkendorff, I. Alstrup, S. Ullmann: Surf. Sci. **227,** 291 (1990)
6.89 H.A. Michelsen, C.T. Rettner, D.J. Auerbach: Surf. Sci. **272,** 65 (1992)
6.90 A. Hodgson, J. Moryl, H. Zhao: Chem. Phys. Lett. **182,** 152 (1991)
6.91 J.P. Cowin, C.-J. Yu, S.J. Sibener, J.E. Hurst: J. Chem. Phys. **75,** 1033 (1981)
6.92 J. Lapujoulade, Y. Le Cruer, M. Lefort, Y. Lejay, E. Maurel: Surf. Sci. **103,** L85 (1981)
6.93 G. Boato, P. Cantini, R. Tatarek: J. Phys. F. **6,** L237 (1976)
6.94 A. Hodgson, J. Moryl, P. Traversaro, H. Zhao: Nature **356,** 502 (1992)
6.95 C.T. Rettner, D.J. Auerbach, H.A. Michelsen: Phys. Rev. Lett. **68,** 2547 (1992)
6.96 G.R. Darling, S. Holloway: J. Chem. Phys. **97,** 734 (1992)
6.97 H.A. Michelsen, C.T. Rettner, D.J. Auerbach: Phys. Rev. Lett. **69,** 2678 (1992)
6.98 H.A. Michelsen, C.T. Rettner, D.J. Auerbach, R.N. Zare: J. Chem. Phys. **98,** 8294 (1993)
6.99 L. Schröter, G. Ahlers, H. Zacharias, R. David: J. Electr. Spectr. Rel. Phen. **45,** 403 (1987)
6.100 L. Schröter, Ch. Trame, R. David, H. Zacharias: Surf. Sci. **272,** 229 (1992)
6.101 R.W. Minck, R.W. Terhune, W.G. Rado: Appl. Phys. Lett. **3,** 181 (1963)
6.102 P. Lallemand, P. Simova: J. Molec. Spect. **26,** 262 (1968)
6.103 R.L. Farrow, D.W. Chandler: J. Chem. Phys. **89,** 1994 (1988)
6.104 J. Arnold, T. Dreier, D.W. Chandler: Chem. Phys. **133,** 123 (1989)
6.105 A. Cruz, B. Jackson: J. Chem. Phys. **94,** 5715 (1991)
6.106 U. Nielsen, D. Halsted, S. Holloway, J. K Nørskov; J. Chem, Phys. **93,** 2879 (1990)
6.107 S. Holloway, B. Jackson: Chem. Phys. Lett. **172,** 40 (1990)
6.108 X.Y. Chang, S. Holloway: Surf. Sci. **251,** 935 (1991)
6.109 G. Engdahl, B.I. Lundqvist, U. Nielsen, J.K. Nørskov: Phys. Rev. B, **45,** 11362 (1992)
6.110 J.M. Campbell, M.E. Domagala, C.T. Campbell: J. Vac. Sci. Technol A **9,** 1693 (1991)
6.111 T. Kwan, T. Izu: Catalyst **4,** 28 (1948)
6.112 T. Kwan: J. Res. Inst. Catalysis **1,** 95 (1949)
6.113 R. J. Mikovsky, M. Boudart, H.S. Taylor: J. Am. Chem. Soc. **76,** 3814 (1954)
6.114 G. Rienäcker, G. Vormum: Z. Anorg. Chem. **283,** 287 (1956)
6.115 D.D. Eley, D.R. Rossington: The parahydrogen conversion on copper, silver, and gold, in *Chemisorption*, ed. by W.E. Garner (Butterworths., London 1956)
6.116 J. Völter, H. Jungnickel, G. Rienäcker: Z. Anorg. Chem. **360,** 300 (1968)
6.117 D.A. Cadenhead, N.J. Wagner: J. Catalysis **21,** 312 (1971)
6.118 C.S. Alexander, J. Pritchard: J. Chem. Soc. Faraday Trans. I, **68,** 202 (1972)
6.119 M. Kiyomiya, N. Momma, I. Yasumori: Bull. Chem. Soc. Jpn. **47,** 1852 (1974)
6.120 I.E. Gabis, A.A. Kurdyumov, S.N. Mazaev: Poverkhnost **12,** 26 (1987)
6.121 R.B. Bernstein: *Chemical Dynamics via Molecular Beam and Laser Techniques* (Oxford Univ. Press. New York (1982)

7. Kinetics and Dynamics of Alkane Activation on Transition Metal Surfaces

C.B. Mullins and *W.H. Weinberg*

The initial reaction with and subsequent chemical rearrangements of saturated hydrocarbons on transition metal surfaces are vitally important to industrial catalysis [7.1] and other processes. Indeed, much applied and fundamental research [7.2–9] has been conducted on these chemical systems which has provided a significant increase in our understanding of these complex phenomena. However, an accurate predictive capability, the ultimate level of understanding, is far out of reach and will remain so, for the near term.

Part of what has been learned is the following: two distinct reaction mechanisms can contribute to the initial dissociative chemisorption of alkanes on well-characterized single-crystalline metal surfaces, namely, a trapping-mediated channel and a direct channel. The trapping-mediated mechanism is characterized by the following phenomena for alkanes: (1) a gas-phase molecule collides with the surface and transfers sufficient kinetic energy to the solid and/or the internal molecular degrees of freedom to be trapped in the physical adsorption well of the surface, where it subsequently accommodates or equilibrates with the surface; (2) there is a kinetic competition between the dissociative reaction of this accommodated molecule with the surface and the desorption of the molecule from the surface. Hence, the trapping-mediated mechanism is typically a strong (exponential) function of surface temperature. In contrast, a direct dissociative chemisorption mechanism occurs during the initial collision of the gas-phase molecule with the surface. Thus, the probability of dissociation is primarily dependent on the kinetic energy and internal energy of the impinging molecule (and to a lesser extent upon the temperature of the surface). In this chapter these mechanisms and the supporting experimental data are discussed, in detail, as they apply to the reaction of saturated hydrocarbons on metal surfaces.

7.1 Background

We will limit our focus in this chapter to fundamental studies of the initial (in the limit of zero-surface coverage) dissociative chemisorption of alkanes on well-characterized single crystalline transition-metal surfaces. These studies have employed both molecular beam scattering [7.9–14] and "bulb" surface chemical

Springer Series in Surface Sciences, Vol. 34
Surface Reactions Ed.: R.J. Madix

reaction techniques [7.15–24]. Excellent treatises have been written regarding molecular-beam reactive surface scattering techniques [7.5, 9, 25, 26], bulb methods [7.9, 27], and standard surface analysis tools [7.28]; we refer the interested reader to these documents for more detail. Briefly, in the case of beam studies, molecular beams generated by supersonic jets are directed at the metal substrate to expose the surface to a flux of nearly mono-energetic molecules. The outcome of the resulting collisions with the surface is determined either by (1) counting the molecules that reflect from the surface per unit time and comparing that to the impingement rate (Fig. 7.1), or (2) measuring the surface coverage of molecules that have adsorbed and comparing that to the beam exposure. The first method, commonly referred to as the reflectivity method of *King* and *Wells* [7.26] is very accurate but only useful for adsorption probabilities greater than approximately 0.03. The second method typically employs Auger electron spectroscopy [7.28], thermal desorption mass spectrometry, or a surface reaction/titration technique to determine the surface coverage. This value of the surface coverage is then compared to an estimated value for the beam exposure to determine the adsorption probability. The kinetic energy, and to some extent the internal energy, of the supersonic beams is varied by dilutely "seeding" the species of interest in a carrier gas and/or varying the nozzle temperature [7.29]. Both the average value and the distribution of kinetic energies are determined by time-of-flight techniques [7.29]. In bulb experiments the flux of molecules to the surface is provided by the ambient gas leaked into the chamber containing the crystal. For these experiments the second method, from above, is used to measure the surface coverage of adsorbed species, and the molecular exposure is determined from the partial pressure-time history in the chamber.

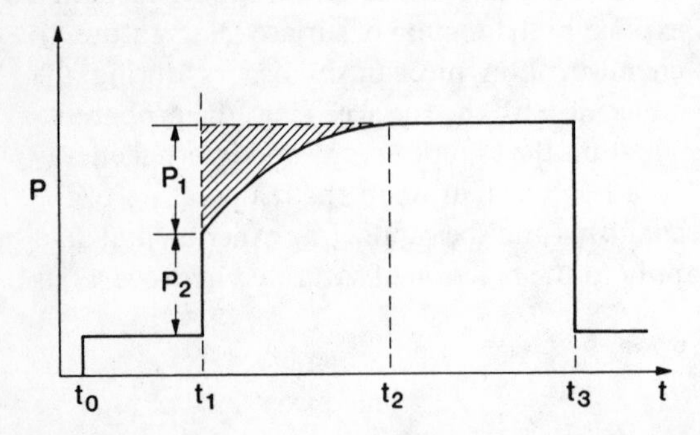

Fig. 7.1. Schematic of the partial pressure of an alkane in the scattering chamber as a function of time during a *King* and *Wells* beam reflectivity measurement. At t_0 the nozzle is filled with gas and at t_1 a flag in a differentially pumped chamber is opened so that the beam impinges on the crystal surface. The partial pressure at t_1 is P_2 greater than the baseline partial pressure. As time proceeds the partial pressure approaches a value $P_1 + P_2$ greater than the baseline and is steady after t_2. This is due to surface saturation by the alkane and thus all alkane molecules in the beam are reflected from the surface. Thus, the initial probability of dissociative chemisorption is $S_0 = P_1/(P_1 + P_2)$. Finally, at t_3 the flag is shut again

Section 7.2 will discuss in detail the mechanism of trapping-mediated dissociative chemisorption using studies of the reaction of various alkanes with the reconstructed surfaces of Ir(1 1 0) and Pt(1 1 0) as examples. First, the dynamics of trapping are presented, followed by a discussion of the surface chemical kinetics. We close this section with a short exposition on the mechanism of the surface reaction at the microscopic level. The direct dissociative chemisorption mechanism is discussed in Sect. 7.3 using the reaction of methane on various single crystalline surfaces as an example. This is followed by a brief discussion of several other studies of direct chemisorption.

It is instructive to present first, however, a brief discussion of some key studies which motivated much of the work that is discussed in later sections. Until recently it was virtually impossible to investigate the chemisorption of methane on well-characterized surfaces under UHV conditions, since the probability of adsorption of methane is very small indeed. For example, on the active early transition metal tungsten, *Winters* [7.30] observed that adsorption is activated with probabilities of adsorption ranging from 3×10^{-3} at 2600 K to below 10^{-4} at 600 K. *Yates* et al. [7.31] showed that on a Rh(1 1 1) surface, methane does not adsorb detectably (probability of adsorption below 5×10^{-5}) either in the vibrational ground state, in the first excited ν_3 vibrational (C–H stretching) mode, or in the second excited ν_4 (deformational) vibrational mode. Subsequently, these results were confirmed by *Brass* et al. [7.32]. Not surprisingly, then, *Wittrig* et al. [7.15] observed no detectable adsorption of methane on the Ir(1 1 0)-(1 × 2) surface (Fig. 7.2). What is surprising is that *Wittrig* et al. [7.15] noted the dissociative chemisorption of ethane, propane, isobutane and neopentane on the Ir(1 1 0)-(1 × 2) surface with initial probabilities near unity at temperatures below 130 K. The only observed products of desorption are hydrogen and the parent hydrocarbon adsorbed initially.

Temperature-Programmed Reaction Spectra (TPRS) of hydrogen resulting from saturation coverages on Ir(1 1 0)-(1 × 2) of the dissociatively adsorbed hydrocarbon molecules at an adsorption temperature of 100 K are shown in Fig. 7.3 with the Temperature-Programmed Desorption (TPD) spectrum of hydrogen after an exposure of 0.5 Langmuir ($L = 1 \times 10^{-6}$ Torr s) of hydrogen shown for comparison. This is a sufficient exposure of hydrogen to saturate the β_2 adstate and to begin to fill the lower temperature β_1 adstate, as may be seen in Fig. 7.3a (see Fig. 7.4 for hydrogen TPD from Ir(1 1 0)-(1 × 2)). The exposures of the hydrocarbon molecules were carried out with a directional beam doser, and the conversion factors for actual gas flux onto the crystal surface are the following: An exposure of 1 L is approximately equivalent to an exposure of 7 Torr-s for ethane, 3 Torr-s for propane, 4 Torr-s for isobutane, and 4 Torr-s for neopentane. The temperature-programmed reaction spectra of Fig. 7.2 are drawn to the same scale so that the amount of hydrogen desorbing may be compared directly. Of the four hydrocarbons compared in Fig. 7.3, ethane is qualitatively different from the others since it displays a one-peak hydrogen temperature programmed desorption spectrum, and, in particular, there is no (γ) thermal desorption state which desorbs above 500 K. The desorption spectra of

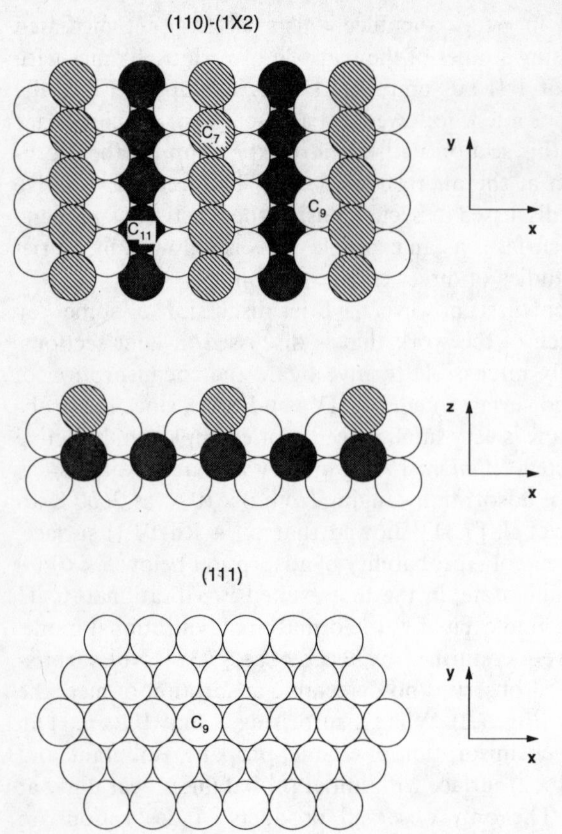

Fig. 7.2. Structural models for the (1 1 0)-(1 × 2) and (1 1 1) surfaces of platinum and iridium. The *z*-axis is perpendicular to the plane of the surface. The C_n designate the coordination numbers of the metal surface atoms

hydrogen following the adsorption of propane and isobutane show a three-peak structure, although the low-temperature peak is manifested as a shoulder in Fig. 7.3 for propane. These three peaks will be referred to as the α, the β_2', and the γ adstates proceeding from low to high temperatures, respectively. As will be shown later, the (apparent) β_2' peak of hydrogen desorbing from a surface exposed to neopentane (Fig. 7.3e) is a composite of an α and a β_2' peak. This is already intuitively apparent from Fig. 7.3 since the intensity of the apparent β_2' peak from neopentane is clearly greater than that of the β_2 peak of hydrogen following the adsorption of molecular hydrogen on the surface (Fig. 7.3a).

The α thermal desorption state, which occurs only at high surface coverages, represents desorption from a hydrocarbon fragment which is unstable on the Ir(1 1 0)-(1 × 2) surface at temperatures slightly above room temperature. The β_2' thermal desorption state represents desorption of hydrogen from β_2 adsites present on the clean surface. The γ thermal desorption states of Fig. 7.3 are indicative of the decomposition of hydrocarbon fragments which are stable on the surface to temperatures above 500 K.

Temperature-programmed reaction spectra for hydrogen evolution as a function of exposure of the surface at 100 K to ethane, propane, isobutane, and

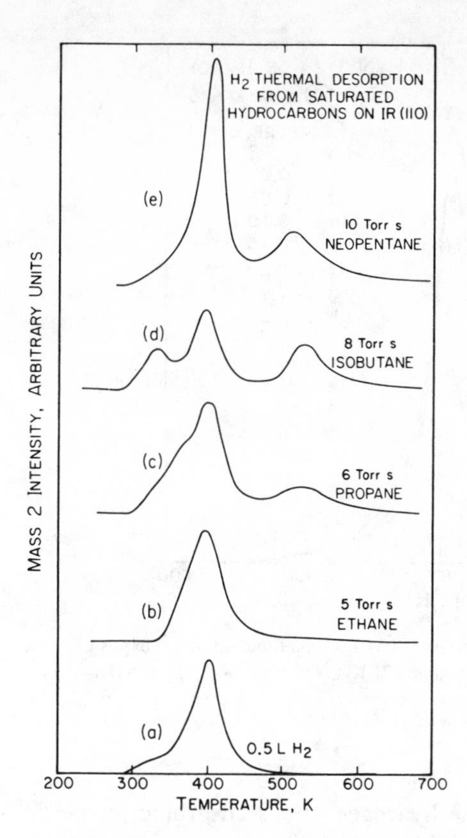

Fig. 7.3. Thermal desorption of hydrogen resulting from exposure of the Ir(1 1 0) surface to: (a) 0.5 L of H_2; (b) 5 Torrs of ethane (saturation); (c) 6 Torr s of propane (saturation); (d) 8 Torr s of isobutane (saturation); (e) 10 Torr s of neopentane (saturation) [7.15]

neopentane are shown in Fig. 7.5. Two important observations may be made from the data of Fig. 7.5. First, the α TPRS state is only populated at higher fractional coverages of the dissociatively adsorbed hydrocarbons, except for ethane, which has no γ adstate associated with it. This point is particularly clear in the hydrogen spectra from the adsorption of isobutane. Also, the ratio of the intensity of the γ peak to that of the β'_2 peak increases linearly with the surface coverage. Thus the average stoichiometry for hydrogen production from the stable hydrocarbon fragment (manifest as the γ thermal desorption state) is invariant with respect to surface coverage, whereas the stoichiometry of the unstable hydrocarbon fragment (manifest as the α thermal desorption state) is less well characterized and is a function of the surface coverage. The coverage at saturation and the number of hydrogen atoms in the stable hydrogen fragment are tabulated in Table 7.1 for ethane, propane, isobutane and neopentane. It may be seen from this table that the average compositions of the stable hydrocarbon fragments resulting from the adsorption of these four hydrocarbon molecules are C_2, C_3H_2, C_4H_4, and C_5H_4, respectively. As may be seen also in Table 7.1, the coverage at saturation of the dissociatively adsorbed hydrocarbons is constant and approximately equal to 10^{14} molecules cm^{-2}. This is consistent with the idea that the rate-limiting step in the adsorption reaction is

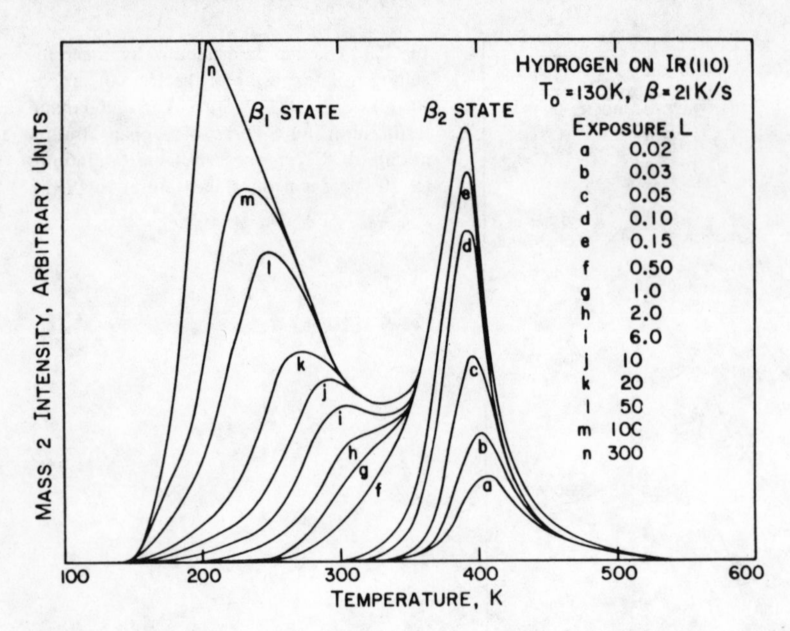

Fig. 7.4. Thermal desorption spectra of hydrogen from Ir(1 1 0) as a function of gas exposure. The adsorption temperature is 130 K, and the heating rate is 21 K/s. The ratio of the β_1 and β_2 states is 2:1 at saturation coverage

the activation of C–H bonds, with the hydrogen atoms adsorbing in the β_2' adsite. When this adsite is saturated, further dissociative chemisorption ceases. Since the number of β_2' adsites dictates the saturation coverage of the hydrocarbon, it is also to be expected that those hydrocarbons with a greater number of hydrogen atoms will give rise to a greater intensity of the unstable hydrocarbon fragment (the α desorption state). This expectation is borne out by the temperature programmed reaction spectra of Fig. 7.5.

Preadsorption of hydrogen into the β_2 adsites poisons the surface with respect to the dissociative chemisorption of these paraffinic hydrocarbons. This is illustrated by the thermal desorption spectra of Fig. 7.6 for the case of neopentane, isobutane, and propane with no preexposure to hydrogen (spectra a), a preexposure of 0.05 L of hydrogen (spectra b), and a preexposure of 0.50 L of hydrogen (spectra c). The area of the γ thermal desorption state is proportional to the number of dissociatively adsorbed hydrocarbon molecules. The linear poisoning of the reactivity of the surface with respect to the population of the β_2' adstate of hydrogen is shown clearly in Fig. 7.7, which was derived from the data of Fig. 7.6. Also, the subsequent adsorption of hydrogen onto completely dehydrogenated hydrocarbon residues results in partial population of the β_2 adsites but no population at all of the β_1 adsites. Furthermore, the γ adstate (which would correspond to the formation of C–H bonds) is not repopulated upon exposure of the dehydrogenated hydrocarbon residue to hydrogen. This

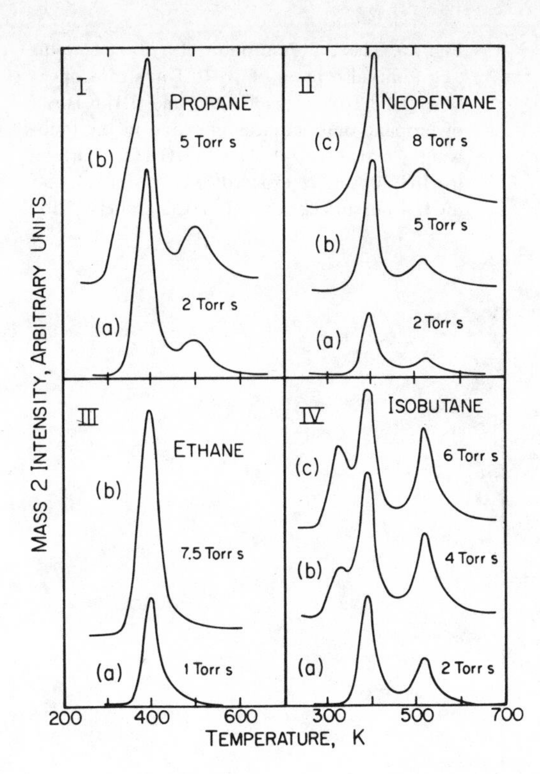

Fig. 7.5. Thermal desorption of hydrogen as a function of exposure for (I) propane, (II) neopentane, (III) ethane, and (IV) isobutane. Exposures are shown adjacent to the corresponding spectrum [7.15]

Table 7.1. Composition of paraffins adsorbed irreversibly on the Ir(1 1 0)-(1 × 2) surface

Species	Hydrocarbon coverage at saturation [10^{14} molecules cm^{-2}]	Hydrogen atoms in γ fragment (per molecule)	Hydrogen atoms desorbed below 420 K (per molecule)
Ethane	1.1 ± 0.3	0	6
Propane	1.1 ± 0.3	1.9	6.1
Isobutane	1.1 ± 0.3	3.7	6.3
Neopentane	1.5 ± 0.3	3.9	8.1

means that the dehydrogenation of the stable hydrocarbon fragments in these transient measurements is irreversible.

The effect of the postadsorption of hydrogen at 100 K after the adsorption of isobutane is revealed by the thermal desorption spectra of Fig. 7.8. The TPRS of hydrogen following the adsorption of 3.5 Torr s (approximately 0.9 L) of isobutane at 100 K is shown in Fig. 7.8a. The temperature programmed reaction spectrum of hydrogen shown in Fig. 7.8b corresponds to the postexposure of a surface prepared in the above manner to 4 L of hydrogen. There are two features

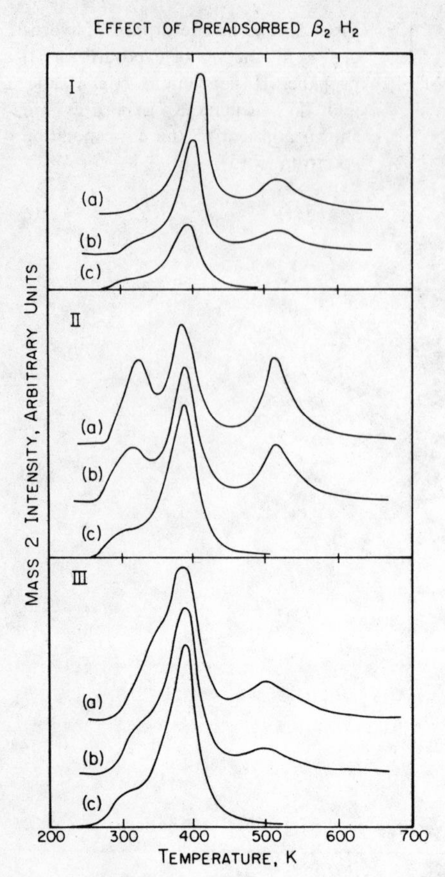

Fig. 7.6. Thermal desorption of hydrogen resulting from adsorption of: (I) 10 Torr s of neopentane, (II) 8 Torr s of isobutane, and (III) 6 Torr s of propane onto surfaces prepared in the following manner: (a) clean Ir(1 1 0) surface; (b) Ir(1 1 0) surface exposed to 0.05 L of H_2; and (c) Ir(1 1 0) surface exposed to 0.50 L of H_2 [7.15]

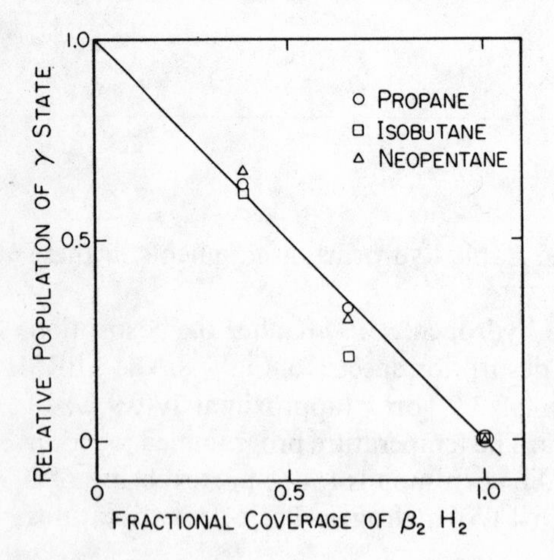

Fig. 7.7. Reactivity of surface for hydrocarbon decomposition as a function of the availability of β_2 hydrogen adsorption sites [7.15]

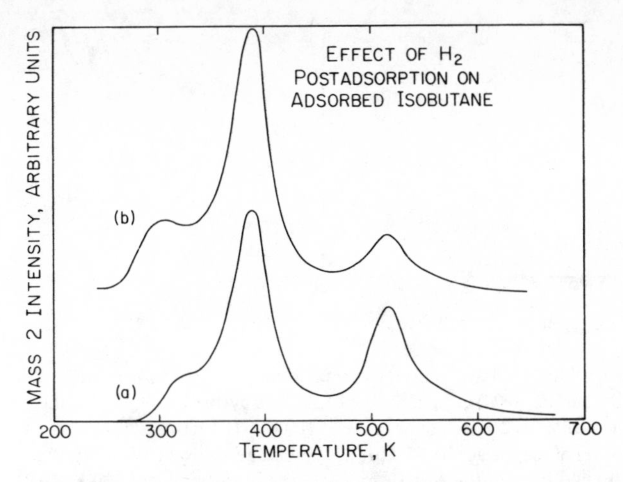

Fig. 7.8. Thermal desorption of hydrogen resulting from (a) exposure of 3.5 Torr s of isobutane to the Ir(1 1 0) surface followed by an exposure of 4 L of H_2 at 100 K; and (b) exposure of 3.5 Torr s of isobutane to the Ir(1 1 0) surface at 100 K [7.15]

of Fig. 7.8b which are noteworthy. First, since the amount of isobutane on the surface is less than saturation coverage, the hydrogen partially populates the β_1 adstate on the iridium surface, which has a temperature of desorption of 300 K. More important, however, is the decrease in intensity of the γ adstate resulting from the postexposure of hydrogen. This decrease is clear from a visual comparison of the desorption peaks above 500 K occurring in Fig. 7.8b and 7.8a. One interpretation of this result is that the postadsorbed hydrogen partially poisons the dissociative chemisorption reaction of isobutane with the Ir(1 1 0)-(1 × 2) surface. Hydrogen could intervene in this way because the dissociation of isobutane is incomplete at 100 K; it occurs at a temperature between 100 K and the temperature of desorption of molecular isobutane, which is approximately 165 K. That this is the correct interpretation, and, moreover, that the same phenomena occurs also for propane and neopentane, is demonstrated explicitly in Fig. 7.9. In these temperature-programmed reaction spectra, the evolution of the molecular hydrocarbon adsorbed initially is compared for either the postexposure of hydrogen (spectra a) or no postexposure of hydrogen (spectra b). It is clear from this figure that the postexposure of hydrogen at 100 K retards the dissociation reaction, and, furthermore, it was found that only that hydrocarbon adsorbed initially desorbs subsequent to the postexposure of hydrogen.

If deuterium is postadsorbed onto the hydrocarbon overlayers rather than hydrogen, there is no exchange of deuterium into the C–H bonds of the desorbing hydrocarbons. H–D exchange would be expected if dissociative adsorption of the parent hydrocarbon had occurred at 100 K, and the adsorption of hydrogen (or deuterium) were causing regeneration of the parent hydrocarbon. Thus, it is apparent that the adsorption of these saturated

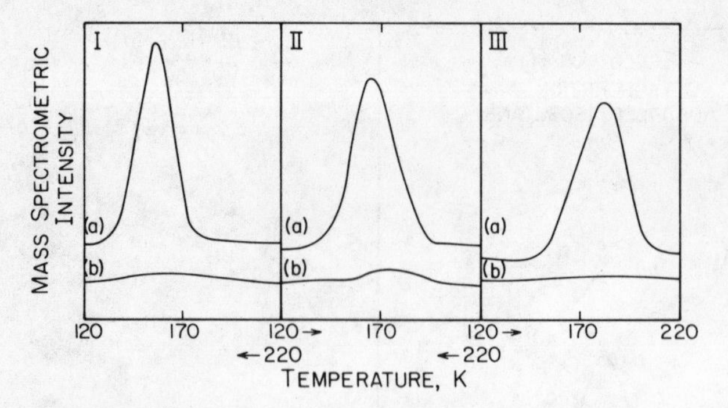

Fig. 7.9. Displacement of hydrocarbons by hydrogen. (I) Thermal desorption (monitoring mass 43) resulting from: (a) exposure of the surface to 2 Torr s of propane followed by an exposure of 2 l of H_2 at 100 K; and (b) exposure of the surface to 2 Torr s of propane at 100 K. (II) Thermal desorption (of mass 43) resulting from (a) exposure of the surface to 3.5 Torr s of isobutane followed by an exposure of 0.25 l of H_2 at 100 K; and (b) exposure of the surface to 3.5 Torr s of isobutane at 100 K. (III) Thermal desorption (of mass 41) resulting from: (a) exposure of the surface to 2 Torr s of neopentane followed by an exposure of 2 l of H_2 at 100 K; and (b) exposure of the surface to 2 Torr s of neopentane at 100 K [7.15]

hydrocarbons on the Ir(1 1 0)-(1 × 2) surface at 100 K results in an overlayer in which at least some of the molecules that will undergo dissociation (dehydrogenation) retain their molecular identity. The post-adsorbed hydrogen displaces these associatively adsorbed molecules from the vicinity of active centers (β_2 adsites) on which they would otherwise dissociate at slightly higher temperatures.

In summary, each of the paraffins adsorbs molecularly on the Ir(1 1 0)-(1 × 2) surface at 100 K or below with simultaneous dissociation and desorption occurring as the surface is heated. The initial C–H bond cleavage occurs at approximately 130 K and each alkane with the exception of ethane forms both a "stable" hydrocarbon fragment (manifest as the γ desorption state of hydrogen) and an "unstable" hydrocarbon fragment (manifest as the α desorption state of hydrogen). The more stable fragment decomposes near 525 K in all cases, whereas the temperature of decomposition of the less stable fragment varies between approximately 325 K (isobutane) and 400 K (neopentane). Both the activity and the product distribution of the hydrogenolysis of these paraffins over Ir(1 1 0)-(1 × 2) is expected to be influenced directly by the composition and the bonding of the unsaturated hydrocarbon fragments on the surface.

Interestingly, the Pt(1 1 0)-(1 × 2) surface shows remarkable similarities to the Ir(1 1 0)-(1 × 2) surface, activating all the alkanes except methane and ethane with the initial C–H bond cleaving at a surface temperature of approximately 200 K. However, neither the Ir(1 1 1) or the Pt(1 1 1) surface produce measurable dissociative adsorption under UHV conditions. Thus these reactions show a strong structural sensitivity on Ir and Pt and a significant variation in the activation barrier to surface reaction.

The results from these studies motivated, at least in part, much of the work discussed in the next section. The mechanism by which adsorbed alkanes are activated is of intense interest, as well as other means to activation. These questions, and others, motivated R. J. Madix and co-workers to begin molecular beam studies of the surface reaction dynamics and kinetics of alkanes on metals and, in particular, on Ir(1 1 0)-(1 × 2).

7.2 Trapping-Mediated Dissociative Chemisorption

As mentioned previously, the results from a large body of work, especially the molecular-beam investigations alluded to above, indicate that there are two basic mechanisms that account for most dissociative chemisorption on bare metal surfaces. There is a direct mechanism in which the molecule dissociates upon impact with the surface at relatively high incidence kinetic energy and there is also a trapping-mediated mechanism in which the molecule loses sufficient translational energy in the surface collision to trap in a molecularly adsorbed state. The probability of trapping at the surface has been found, as expected intuitively, to decrease with increasing incident kinetic energy. Once a molecule is trapped, it equilibrates with the surface and can either desorb or reorientate such that dissociation can occur. Thus, trapping-mediated chemisorption rates are typically strong functions of surface temperature. Additionally, trapping-mediated chemisorption is thought to be an important mechanism in industrial catalytic processes because it favors the relatively low kinetic energies characteristic of the gaseous reactants in such processes.

Using molecular-beam techniques to vary the incident kinetic energy and angle can provide quantitative information regarding the microscopic phenomena that are expected to be relevant in the case of trapping-mediated dissociative chemisorption. In particular, a variation of these two parameters will probe the dynamics and energy transfer processes by which the molecule traps. Intuitively one would expect that the trapping probability $S_{0,\mathrm{T}}$ of a particle impinging on a bare surface would decrease with increasing kinetic energy E_i, since a greater fraction of the incident energy must be dissipated upon collision. Indeed, several experimental studies of the trapping dynamics of rare gases and molecules have demonstrated this to be the case [7.33–42]. A molecule would be expected to trap more readily than a rare-gas atom with the same heat of physical adsorption and mass, since translational energy can be converted to rotational and vibrational energy in the molecular case, in addition to substrate excitations [7.43–45]. The dependence of the trapping probability upon the incident angle of the impinging molecule is not well understood, but measures of this are useful nonetheless. Additionally, there is frequently a strong dependence on surface temperature T_s in trapping-mediated chemisorption due to the kinetic competition between dissociative chemisorption and desorption from the trapped (physically adsorbed) state. Although experimental studies of these phenomena are of great practical and fundamental interest, only recently have relevant dynamical data appeared in the literature [7.14, 25, 46–51]. This literature can be a confusing source of information due to the jargon commonly used in the field. In particular, the phrases *precursor-mediated* and *trapping-mediated* are taken to mean the same thing in some papers and different things in other papers. Traditionally, the phrase *precursor-mediated* chemisorption has been applied to cases in which the probability of chemisorption is insensitive to adsorbate coverage [7.52]. This experimental observation is explained by assuming that an impinging molecule can trap efficiently on an *occupied* site with subsequent migration to an empty site [7.52–53]. Here, we use the phrase trapping-mediated to distinguish between, on the one hand, phenomena occurring within the traditional framework of precursor-mediated chemisorption, and, on the other hand, dissociative chemisorption in the limit of zero coverage, mediated by trapping of the molecule on the bare surface in a molecularly adsorbed state (physical adsorption for an alkane) with subsequent molecular desorption or reorientation into a configuration more favorable for chemisorption (i.e., trapping-mediated chemisorption). Indeed, we focus here on the latter subject.

In this section we will touch on several studies of trapping-mediated chemisorption dynamics. However, we will devote significant attention to investigations of the interaction of ethane [7.14, 39, 54] and higher alkanes [7.46, 47] with the Ir(1 1 0)-(1 × 2) and Pt(1.1 0)-(1 × 2) surfaces because of our own familiarity with this research and because, for the case of ethane, *independent* measurements of the trapping probability and the probability of dissociative chemisorption have been conducted. Since much of the discussion to follow is about the reconstructed (1 1 0) surfaces of Ir and Pt Fig. 7.2 has been provided to show the geometric structure of these surfaces.

7.2.1 Trapping Dynamics

Since trapping at a surface is the first step in all trapping-mediated surface reactions, it is important to have as complete a picture as is possible concerning this dynamical phenomenon. Surprisingly, there have been relatively few detailed experimental studies of the relevant dynamics inherent to trapping. While thermally averaged trapping probabilities have been extracted from accommodation coefficients (a measure of the efficiency of a gas to exchange energy with a surface under single collision conditions) from as far back as 1971 [7.33], only a small number of studies went beyond this type of measurement prior to 1989. These included, especially, the following: (1) a study in which the functional dependence of the trapping probability on incident kinetic energy was estimated for normal incidence, based on the application of detailed-balance arguments to a time-of-flight spectrum for argon desorbing from Pt(1 1 1) [7.35], and (2) a molecular beam study of the trapping of argon on Ir(1 1 0)-(1 × 2), based on the angular distribution of scattering [7.36]. Recently, however, a number of relatively detailed studies have been conducted. In particular, results regarding the trapping of argon on the hydrogen pre-covered (1 0 0) surface of *W* [7.38], xenon on Pt(1 1 1) [7.41,42] argon on Pt(1 1 1) [7.40], nitric oxide on Ag(1 1 1) [7.55], and ethane on Ir(1 1 0)-(1 × 2) [7.39] and Pt(1 1 1) [7.37] have been reported. All of these studies report the polar angle of incidence θ_i and incident kinetic energy E_i dependence of the trapping probability into a physically adsorbed or molecular state. Additionally, the study of argon trapping on Pt(1 1 1) reports on the surface temperature dependence of the trapping probability.

The phrase "dynamical phenomenon" has been adopted above because trapping involves momentum and energy exchange between the gas-phase molecule or atom and the repulsive part of the gas-surface potential. Intuitively, we might expect that a gas-phase atom or molecule would trap at a surface if it exchanged sufficient kinetic energy in the collision with the surface to have a negative total energy (kinetic plus potential), i.e., negative total energy with respect to the usual zero of energy which is the gas-phase molecule infinitely far from the surface and at rest. There are a number of dissipative (inelastic) channels available in the gas-surface encounter. If the impinging gas-phase particle is a rare gas atom then the "inelasticity" is effectively restricted to degrees of freedom associated with the solid, i.e., electronic excitations of the rare-gas atom are inaccessible for kinetic energies where trapping is likely to be important. The principal inelastic channels of the solid are phonon excitations (the more dominant channel) and electron–hole pair creation (possibly negligible for trapping). Thus, the detailed nature of the substrate, e.g., the surface Debye temperature and the local densities of phonon and electronic states, will dictate the collision dynamics. Rare-gas atoms are particularly useful for studying trapping dynamics since the "channel" for a surface chemical reaction is shut off. For impinging *molecules* additional dissipative channels are available. For alkanes, which can physically adsorb on surfaces (in the absence of dissociation) with the same sort of dispersion forces that are inherent to rare-gas atom adsorption, there can be intramolecular energy transfer that converts translational energy to either rotational or vibrational energy to facilitate trapping [7.43–45]. As an example, for ethane there are three rotational degrees of freedom and 18 vibrational degrees of freedom into which translational energy can be dissipated.

There are several methods used for determining the trapping probability experimentally. The reflectivity method of *King* and *Wells* (Fig. 7.1) requires that the surface temperature be sufficiently low that the residence time of the trapped molecule on the surface be long compared to the time scale of the experiment. For this case the trapping probability is simply the fraction of the impingement flux that adsorbs "irreversibly". The surface residence time for a physically adsorbed species, as would be the case for alkanes trapped on a metal surface, can be estimated as the reciprocal of the first-order rate coefficient of desorption which can be written as

$$k_d = k_d^{(0)} \exp\left(-E_d/k_B T_S\right), \tag{7.1}$$

where E_d is the binding energy of the physically adsorbed molecule, and $k_d^{(0)}$ is the preexponential factor of the desorption rate coefficient. For the molecular (physical) adsorption of ethane on the reconstructed Ir(1 1 0)-(1 × 2) surface the values of $k_d^{(0)}$ and E_d have been determined from thermal desorption mass spectrometry and found to be $10^{13}\ s^{-1}$ and 7.7 kcal/mol, respectively. Thus, the surface residence time may be written as

$$\tau = [k_d^{(0)}]^{-1} \exp(E_d/k_B T_S). \tag{7.2}$$

This suggests a surface lifetime of ≈ 10 s for ethane on Ir(1 1 0)-(1 × 2) at a surface temperature of 114 K. This concept will also be used in the discussion of the kinetics of dissociation. Since tens of seconds is the time scale of a typical reflectivity measurement, surface temperatures of less than 114 K must be used to measure the trapping probability of ethane on Ir(1 1 0)-(1 × 2) using this technique.

Figure 7.10 depicts the initial trapping probability $S_{0,T}$ (in the limit of zero surface coverage) of ethane on Ir(1 1 0)-(1 × 2) at three different angles of incidence as a function of $E_i \cos^{0.5} \theta_i$ [7.39]. The inset in this figure shows the same data plotted as a function of the total incident kinetic energy E_i. As seen in both plots, the initial trapping probability decreases with increasing incident kinetic energy, as discussed previously, because the impinging ethane molecule must dissipate an increasing fraction of its translational energy as this quantity becomes greater. We would say, empirically, that the data *scale* very well with the quantity $E_i \cos^{0.5} \theta_i$ since the data, when plotted as a function of $E_i \cos^{0.5} \theta_i$, nearly all fall on a single curve. The physical significance of this so-called scaling variable is not well understood at present. We originally interpretted this particular result as being due to a rather corrugated interaction potential [7.39]. A corrugated interaction potential would be consistent with the corrugated geometrical structure of this reconstructed surface [7.56, 57]. However, recent results by *Madix* and co-workers [7.37] indicate that for the trapping of ethane on the *flat* Pt(1 1 1) surface the scaling variable is $E_i \cos^{0.6} \theta_i$. These data are

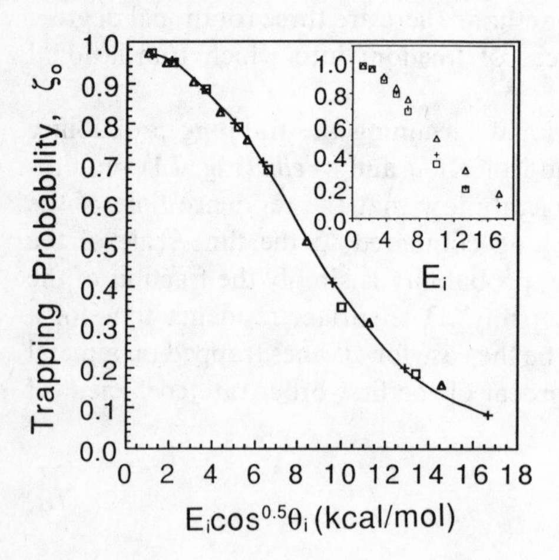

Fig. 7.10. Trapping probability of ethane on Ir(1 1 0)-(1 × 2)

exhibited in Fig. 7.11. This strongly suggests that the origin of this scaling variable is *not* due to surface corrugation. However, it is difficult to envision the corrugation not influencing the scaling variable, and more studies are needed. Thus, the physical significance of the energy scaling of the trapping of ethane on these two surfaces remains to be understood fully.

Historically, such scaling laws have evolved from fitting dynamical adsorption data to variables of the following form: $E_i \cos^n \theta_i$, varying n for the best fit [7.41]. This functional form has come about as a variation to the so-called "normal-energy scaling" variable: $E_n = E_i \cos^2 \theta_i$. The normal-energy scaling concept has as its origin some empirical evidence that the tangential component of momentum dissipated to the surface by incident particles is a small fraction of the tangential component of momentum of the incident beam [7.58]. Thus, the concept is also known as "parallel momentum conservation." The most significant loss of momentum is in the component normal to the surface, $\boldsymbol{p} \cos \theta_i$, where $\boldsymbol{p}$ is the total linear momentum. Thus, if scattering data scaled with the normal component of momentum, then these data would also scale with the normal energy, $E_n = (p^2/2m) \cos^2 \theta_i$, where p^2 is the square of the norm of $\boldsymbol{p}$.

This concept has faltered, and it appears that the dynamics are much more complicated than this simple picture allows. Even for trapping of rare-gas atoms on "smooth" surfaces, such as Pt(1 1 1), a situation where one might expect parallel-momentum conservation to hold, the trapping probability does not scale with normal energy [7.40–42]. In order to clarify these points, it is instructive to consider the trapping of argon on Pt(1 1 1) [7.40]. In these studies the trapping probability was determined in a manner complementary to the reflectivity method described above. In this case the surface temperature must be sufficiently high that the surface-residence time is short on the time scale of the

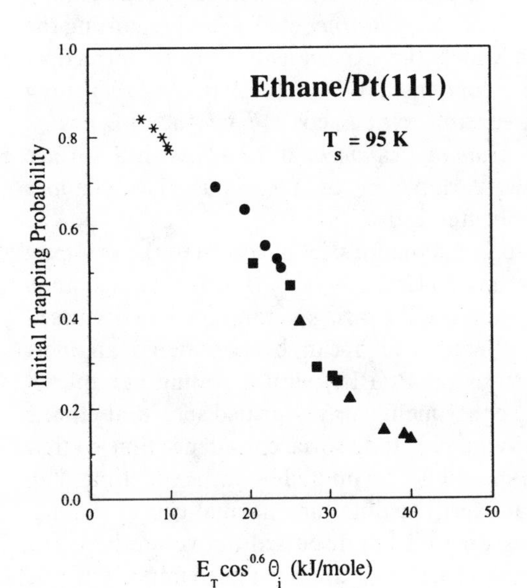

Fig. 7.11. Trapping probability of ethane on Pt(1 1 1)

measurement (in this case, a microsecond or less) which also insures that the surface coverage is negligibly small. This is because the quantity that is measured is the time-of-flight distribution of the scattered flux that has trapped and then desorbed. The relative contribution of this "trapping-desorption" flux is a measure of the trapping probability [7.38, 40]. The other component of the scattered flux, the component which does not trap, is known as the "direct-inelastic" component. Thus, the ratio of the trapping-desorption component to the total flux is the trapping probability. In these studies of the trapping of argon on Pt(1 1 1) [7.40], it was found that the scaling variable which fit the data varied with surface temperature. Indeed, at 80 K the scaling variable was $E_i \cos^{1.5} \theta_i$, while at 190 K the appropriate quantity was $E_i \cos^{1.0} \theta_i$, and at 273 K the optimized scaling variable was $E_i \cos^{0.5} \theta_i$. These results indicate that parallel-momentum dissipation is becoming increasingly more important to the trapping dynamics as the surface temperature increases. Indeed, molecular dynamics calculations by *Head-Gordon* et al. [7.59] have shown that for this system the trapping process involves very rapid equilibration of the normal component of incident momentum, but extremely slow dissipation of the tangential component. At relatively high surface temperatures, for which the residence time of argon on this surface is sufficiently short (e.g., ≈ 27 ps at 273 K), a significant fraction of the atoms which have dissipated their normal component of momentum "desorb" before their parallel component of momentum is dissipated. Thus, trapping in the usual sense of complete equilibration with the surface does not occur. These "quasi-trapped" atoms are characterized rather by full accommodation of only the normal-momentum component. This implies that at very low surface temperatures (i.e., below 80 K such that the surface residence time is extremely long) the scaling variable for trapping of argon on Pt(1 1 1) might be $E_i \cos^2 \theta_i$ so that normal energy scaling would hold. However, recent data acquired by *Rettner* et al. [7.41] regarding the trapping of xenon on Pt(1 1 1), in which the experiments were performed at approximately 85 K where xenon condenses on Pt(1 1 1) (i.e., a *very* long residence time), showed the scaling variable to be $E_i \cos^{1.6} \theta_i$ instead of $E_i \cos^2 \theta_i$ whereas recent data acquired by *Arumainayagam* et al. [7.42] with a surface temperature of 95 K scale with the variable $E_i \cos^{1.0} \theta_i$. Thus, the dynamics relevant to trapping are very complicated indeed.

Before we leave the topic of trapping dynamics, let us return to the trapping of ethane on flat (Pt(1 1 1)) and corrugated (Ir(1 1 0)-(1 × 2)) surfaces to examine the scaling variable once more. Considering the rare gas trapping data discussed above, it is very surprising that ethane, which can be considered an inert spherical particle (to first order), traps on Pt(1 1 1) with a scaling variable of $E_i \cos^{0.6} \theta_f$. For this case [7.37] the experiment was performed such that ethane condensed on the surface and thus had a very long surface residence time so that parallel momentum accomodation should not be limited by surface lifetime. Yet, the scaling was much closer to total-energy scaling than normal-energy scaling, unlike the case of rare gas trapping on Pt(1 1 1) discussed above. Perhaps the internal degrees of freedom of the ethane molecule are playing an important role in trapping, or the interaction corrugation is very strong between ethane and Pt(1 1 1). Neither of these factors is expected to be the case since to first order an ethane molecule is sphere-like with an isotropic surface potential and physically adsorbed ethane on Pt(1 1 1) has a vibrational spectrum very much like gas-phase ethane and is very unreactive toward this surface in ultrahigh vacuum [7.60]. Thus, an elucidation of these differences is crucial to achieving a complete understanding of the trapping of alkanes on transition metal surfaces.

7.2.2 Kinetics

To first order, trapping-mediated dissociative chemisorption can be separated conceptually (and mathematically) into two parts: a dynamical part, involving the trapping of the molecule on the surface with a weak dependence on surface temperature, and a kinetics part, involving the surface temperature driven competition between dissociation and desorption of the trapped molecule. Having discussed the dynamics relevant to trapping in the previous section, we will, in this subsection, discuss the kinetics of trapping-mediated chemisorption.

An important study in demonstrating that the trapping-mediated dissociative chemisorption mechanism is important for alkane activation on transition-metal surfaces was published in 1987 by *Hamza* et al. [7.47]. In this study the initial probability of the dissociative chemisorption of propane and *n*-butane on Ir(1 1 0)-(1 × 2) as a function of incident kinetic energy and parametric in surface temperature was measured, and these data are reproduced in Figs. 7.12 and 13. These measurements showed that, in *the low-kinetic energy regime*, the initial probability of dissociative chemisorption *decreased* with increasing incident

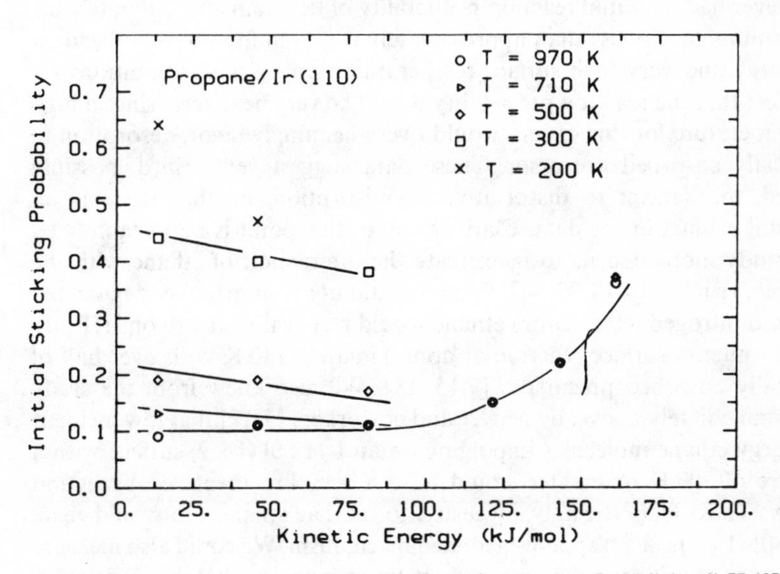

Fig. 7.12. Dissociative chemisorption probability of propane on Ir(1 1 0)-(1 × 2) [7.43]

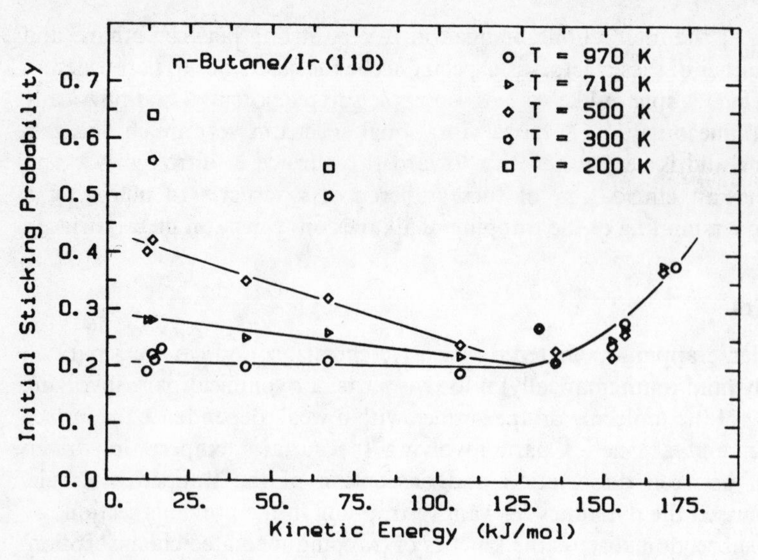

Fig. 7.13. Dissociative chemisorption probability of *n*-butane on Ir(1 1 0)-(1 × 2) [7.43]

kinetic energy, strongly suggesting a trapping-mediated chemisorption mechanism. These data also suggested a direct mechanism at higher kinetic energies. As depicted in Figs. 7.12 and 13, in the trapping-mediated regime (for both propane and *n*-butane) the reaction probability decreases with increasing surface temperature, suggesting that the barrier to reaction from the physically adsorbed state is below the vacuum zero level. Interestingly, the data for propane never had an initial reaction probability of less than approximately 0.1 and for *n*-butane never less than approximately 0.2, even for very low incident kinetic energy and very high surface temperature. Under these conditions we would expect that the reaction probability would be very near zero, since a high surface temperature for this system would overwhelmingly favor, desorption of the physically adsorbed precursor. These data suggest yet a third, possibly unactivated, mechanism to dissociative chemisorption, or that there is an experimental artifact in the data. Clarification of this point is important.

This study motivated us to investigate the interaction of ethane with the Ir(1 1 0)-(1 × 2) surface [7.14, 39, 54]. Previous studies of our group had indicated that at liquid nitrogen temperature ethane would physically adsorb on Ir(1 1 0)-(1 × 2) and undergo surface reaction at approximately 130 K with over half of the physically adsorbed precursors [7.15–18]. We also knew from the study discussed immediately above, by *Madix* and coworkers [7.47] that low incident kinetic energy ethane molecules impinging on an Ir(1 1 0)-(1 × 2) surface with a temperature of 300 K or greater would have a very low (near zero) reaction probability. These facts strongly suggested to us that ethane too would react with Ir(1 1 0)-(1 × 2) via a trapping-mediated mechanism. We could also measure the trapping probability and dissociative chemisorption probability independently and with a standard kinetic analysis of the data verify unambiguously that a trapping-mediated mechanism was indeed important for this system. The data for trapping of ethane on Ir(1 1 0)-(1 × 2) were discussed in the previous section and are shown in Fig. 7.10. Figure 7.14 displays the results of measurements for the initial dissociative chemisorption of ethane on Ir(1 1 0)-(1 × 2) as a function of incident kinetic energy and surface temperature. Importantly, the probability of dissociation decreases with increasing kinetic energy (in the low-energy regime, for a fixed surface temperature), suggesting a trapping-mediated chemisorption mechanism. The temperature dependence of the data coupled with the data for the trapping probability provide the information necessary for evaluating the kinetics. An explanation of the kinetic model is given next.

In the trapping-mediated, dissociative chemisorption of alkanes, there is a physically adsorbed intermediate on the surface which may either desorb or undergo C–H bond cleavage. The chemisorbed alkyl can then undergo further (and ultimately complete) dehydrogenation at elevated surface temperatures. This reaction sequence may be written as

$$C_nH_{2n+2}(g) \underset{k_d}{\overset{S_T F}{\rightleftarrows}} C_nH_{2n+2}(p) \xrightarrow{k_r} C_nH_{2n+1}(c) + H(c)$$
$$\downarrow$$
$$nC(c) + (n+1)H_2(g), \quad (7.3)$$

where S_T is the trapping probability of the alkane into the physically adsorbed state; F is the impingement flux (on a per surface site basis); g, p, and c denote gas phase, physically adsorbed, and chemisorbed, respectively; and k_d and k_r are the rate coefficients for desorption and reaction, both of which are of the

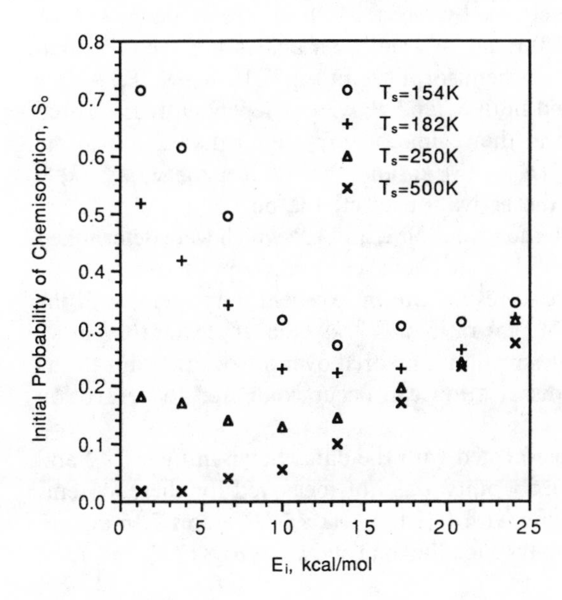

Fig. 7.14. Dissociative chemisorption probability of ethane on Ir(1 1 0)-(1 × 2) [7.43]

Polanyi–Wigner form, namely,

$$k_i = k_i^{(O)} e^{-E_i/k_B T}, \tag{7.4}$$

where the temperature here is the *surface* temperature.

If the surface temperature is sufficiently high that the fractional surface coverage of the physically adsorbed alkane, θ_p, is negligibly small, then a pseudo-steady-state analysis implies that

$$\theta_p \simeq \frac{S_T F}{k_d + k_r}, \tag{7.5}$$

and the rate of dissociation in the limit of zero conversion is given by

$$R_r = k_r \theta_p = S_T F\left(\frac{k_r}{k_r + k_d}\right). \tag{7.6}$$

The probability of the alkane activation reaction is given simply by

$$P_r = \frac{R_r}{F} = S_T\left(\frac{k_r}{k_r + k_d}\right). \tag{7.7}$$

In general, (7.7) implies that

$$\frac{S_T}{P_r} - 1 = \frac{k_d}{k_r}, \tag{7.8}$$

and for the special case of $k_r \ll k_d$, (7.7 and 8) reduce to

$$\frac{P_r}{S_T} = \frac{k_r}{k_d}. \tag{7.9}$$

In the case of activation of ethane by the Ir(1 1 0)-(1 × 2) surface, $E_r - E_d$ is negative, and, consequently, (7.8) must be used in an analysis of the measured initial probabilities of dissociative chemisorption. In Fig. 7.15, ln $((S_T/P_r) - 1)$ is plotted as a function of $1/T$, and both S_T and P_r were independently measured by molecular beam techniques in the regime of trapping-mediated activation. The slope of Fig. 7.6 gives $-(E_d - E_r)/k_B$ and the intercept gives $k_d^{(O)}/k_r^{(O)}$. These results, which quantify the activation of ethane on Ir(1 1 0)-(1 × 2), are listed in Table 7.2, together with the values of E_r and $k_r^{(O)}$ which were determined via an independent evaluation of the rate coefficient of desorption of physically adsorbed ethane [7.15–18]. These results are in excellent agreement with the earlier study by *Wittrig* et al. [7.15]. A ratio of $k_d^{(O)}/k_r^{(O)}$ that is greater than unity is expected since, entropically, desorption is favored over dissociation due to the limited phase space in which dissociation can occur compared to desorption [7.61].

A similar analysis can be performed with the data shown in Figs. 7.12 and 13 even though the trapping probability was not measured for these systems (propane and *n*-butane interacting with Ir(1 1 0) – (1 × 2)). Figures 7.16 and 17 present the results of such an analysis for the data shown in Figs. 7.12 and 13 at

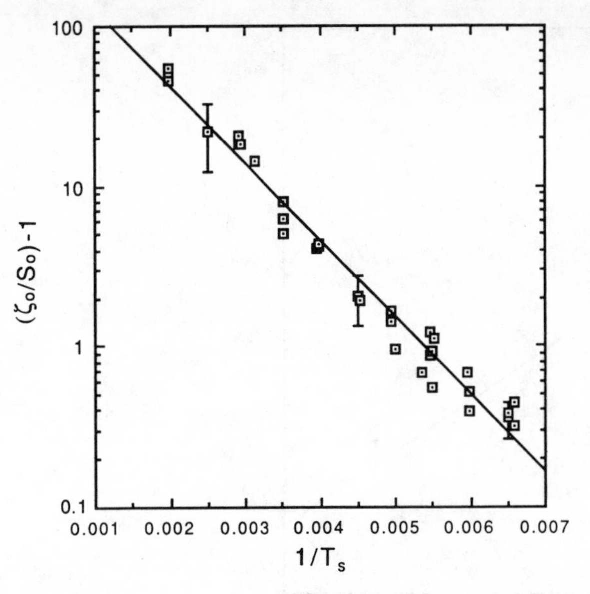

Fig. 7.15. Arrhenius construction for the trapping-mediated dissociative chemisorption of C_2H_6 on Ir(1 1 0)-(1 × 2). The slope, $-(E_d - E_r)/(k_B$, and the intercept, $k_d^{(o)}/k_d^{(o)}$, are reported in Table 7.1. The error bars represent the maximum variations in the measured rates at three different surface temperature. The abscissa is in units of inverse Kelvin (K^{-1})

Table 7.2. Rate parameters for the dissociative chemisorption of ethane on Ir(1 1 0)-(1 × 2). The separately measured values of the desorption rate coefficients were used in evaluating E_r and $k_r^{(o)}$

Reactant	E_r–E_d [cal/mol]	$k_d^{(o)}/k_r^{(o)}$	E_r [cal/mol]	$k_r^{(o)}$ [S^{-1}]
C_2H_6	− 2200	390	5500	3×10^{10}

the lowest kinetic energy for each system. We have assumed reasonably here that at the lowest kinetic energy the trapping probability is unity. This is an excellent approximation considering that ethane traps on Ir(1 1 0)-(1 × 2) at this energy with a probability of approximately 0.98, and propane and *n*-butane would be expected to trap even more efficiently since the well depth for physical adsorption of both of these molecules is greater than that of ethane. For propane the difference in activation energies is $E_d - E_r = 1.5 \pm 0.3$ kcal/mol with a preexponential factor ratio of $k_d^{(O)}/k_r^{(O)} = 19 \pm 6.6$. For *n*-butane the difference in activation energies is $E_d - E_r = 1.0 \pm 0.2$ kcal/mol with a preexponential factor ratio of $k_d^{(O)}/k_r^{(O)} = 5.4 \pm 1.0$. Note that the uncertainties presented here do not include any systematic error and are based on an assumed uncertainty in the measurements presented in Figs. 7.12 and 13 of ± 0.03. The antiintuitive trend in the data for ethane, propane, and *n*-butane is that $E_d - E_r$ decreases with increasing chain length. We would expect that the activation energy to break a C–H bond to be relatively constant across all alkanes, excluding methane, and thus we would expect that $E_d - E_r$ would increase with increasing chain length (since E_d is known to increase with increasing chain length). We

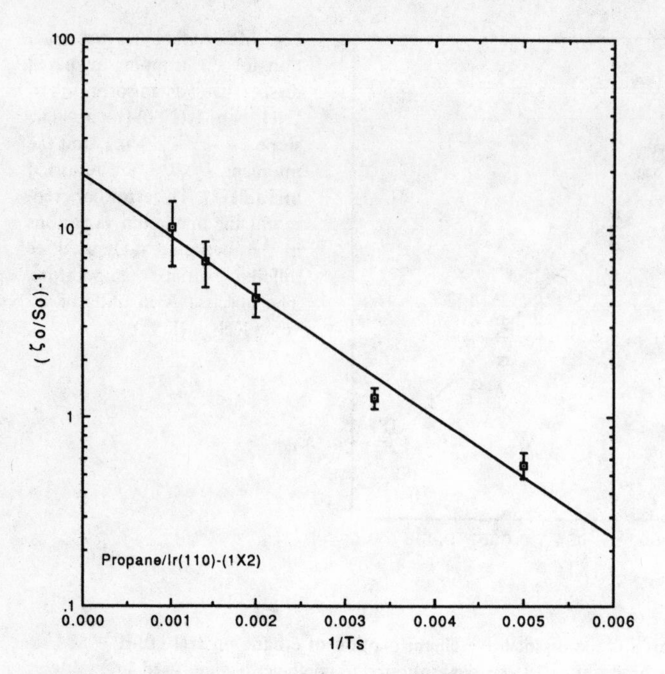

Fig. 7.16. Arrhenius construction for the trapping-mediated dissociative chemisorption of C_3H_8 on Ir(1 1 0)-(1 × 2). The abscissa is in units of inverse Kelvin (K^{-1})

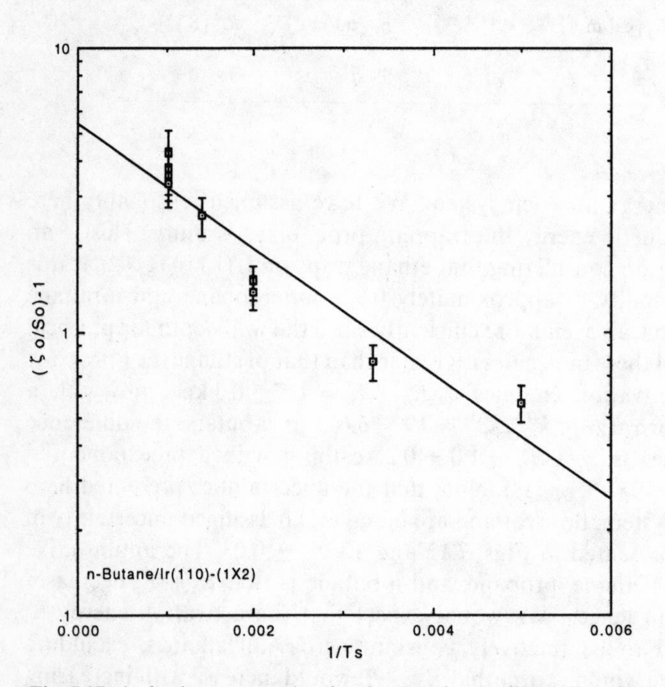

Fig. 7.17. Arrhenius construction for the trapping-mediated dissociative chemisorption of C_4H_{10} on Ir(1 1 0)-(1 × 2). The abscissa is in units of inverse Kelvin (K^{-1})

believe that it is crucial that this data be verified and examined very critically to determine the underlying cause of these apparent trends mentioned above.

Now let us consider the dissociative chemisorption of methane, ethane, and propane on the reconstructed (1 1 0) surface of platinum which *Weinberg* and *Sun* [7.24, 62] have studied in great detail. These studies are of particular interest since Pt(1 1 0)-(1 × 2) has the same geometrical structure as Ir(1 1 0)-(1 × 2) but different surface electronic properties allowing for the elucidation of this factor in catalysis. All of these measurements were performed in a bulb reactor, whereas the investigations discussed above were all conducted in a molecular beam apparatus. For methane, ethane, and propane activation on the Pt(1 1 0)-(1 × 2) surface, $k_r \ll k_d$ in all cases. Consequently, the measured values of ln P_r as a function of reciprocal surface temperature are plotted in Fig. 7.18 for CH_4 and CD_4, in Fig. 7.19 for C_2H_6 and C_2D_6, and in Fig. 7.20 for C_3H_8, $CH_3CD_2CH_3$, and C_3D_8. Both the linearity of these constructions and the observed variation of the reaction rate with *surface* temperature (at a constant gas temperature which is equal to the reactor wall temperature which is approximately equal to 300 K) confirm that the surface reaction obeys a trapping-mediated mechanism. In constructing Figs. 7.18–20, it was recognized that the trapping probability is only a weak function of surface temperature [7.63], and it was (reasonably) assumed that the trapping probability is unity in all cases [7.4].

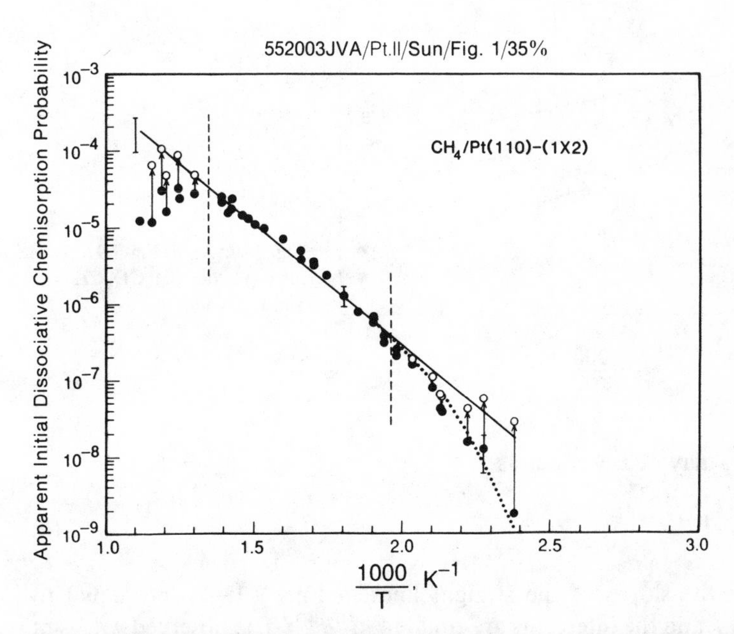

Fig. 7.18. Initial probabilities of trapping-mediated dissociative chemisorption of CH_4 (a) and CD_4 (b) on Pt(1 1 0)-(1 × 2) as a function of reciprocal surface temperature. The slopes, $-(E_d - E_r)/k_B$, and the intercept, $k_d^{(o)}/k_d^{(o)}$, are reported in Table 7.1. The error represents the maximum variation in the measured rate at each temperature

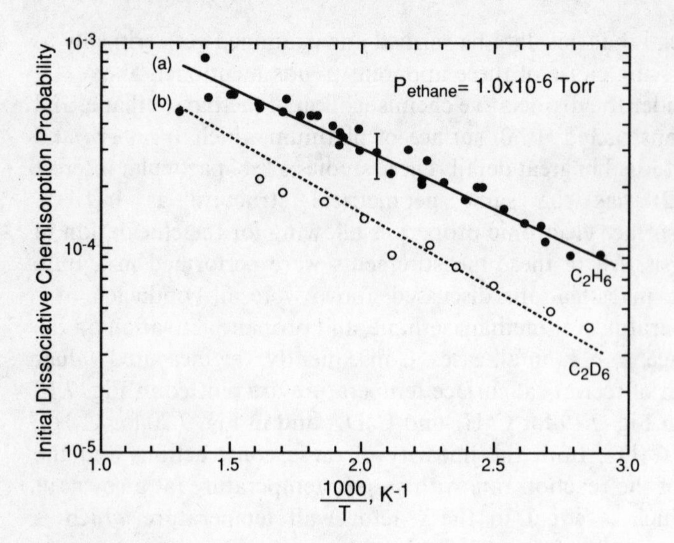

Fig. 7.19. As in Fig. 7.9, except for C_2H_6 (a) and C_2D_6 (b)

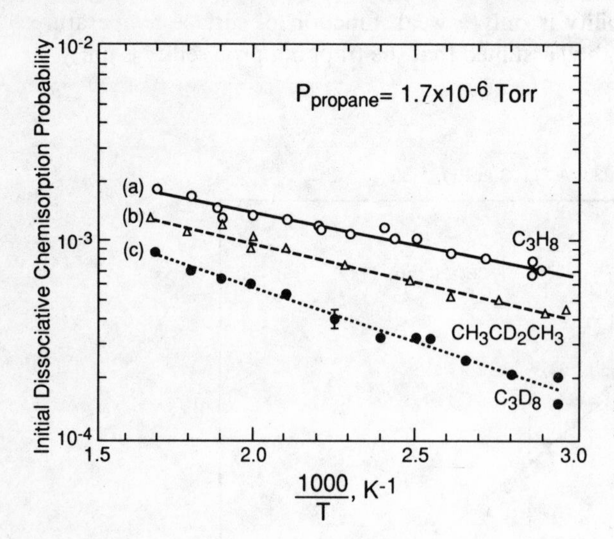

Fig. 7.20. An in Fig. 7.9, except for C_3H_8 (a), $CH_3CD_2CH_3$ (b), and C_3D_8 (c)

Since [7.9] may be rewritten as

$$\frac{P_r}{S_T} = \frac{k_r^{(O)}}{k_d^{(O)}} e^{-(E_r - E_d)/k_B T}, \tag{7.10}$$

it is clear that the slopes of the straight lines in Figs. 7.18–20 are equal to $-(E_r - E_d)/k_B$, and the intercepts are equal to $k_r^{(O)}/k_d^{(O)}$. The observed values of $E_r - E_d$ and $k_d^{(O)}/k_r^{(O)}$ are collected in Table 7.3 for activation of these alkanes on Pt(1 1 0)-(1 × 2). Notice that $E_r - E_d > 0$ in all cases, which implies that for each of these alkanes there is a real energy barrier with respect to a gas-phase energy zero, i.e., the gas-phase alkane infinitely far from the surface and at rest. Notice also that $(k_d^{(O)}/k_r^{(O)}) > 1$ in all cases, as would be expected [7.4, 64]. Since the activation energies of desorption of the physically adsorbed methane, ethane, and propane are 4500 cal/mol, 7700 cal/mol, and 9700 cal/mol, respectively [7.18], the true activation energies for dissociative chemisorption may be evaluated easily. The activation energy and preexponential factor for propane activation are actually apparent values since there are two different kinds of C–H bonds in propane that are activated (*vide infra*). These values of E_r with respect to the proper energy zero (the ground state of the physically adsorbed alkane) are also given in Table 7.3. The fact that the preexponential factors of the desorption rate coefficient, $k_d^{(O)}$, are all approximately 10^{13} s^{-1} [7.18] allows an evaluation of the preexponential factors of the reaction rate coefficients, $k_r^{(O)}$. These values of $k_r^{(O)}$ are shown in the fifth column of Table 7.3.

Notice in Fig. 7.18 there are apparent deviations from linearity for the CH_4 data both at temperatures lower than 500 K and higher than 750 K (filled circles). The nonlinearity at the low temperatures is due to poisoning of the surface by background adsorption of CO, and the nonlinearity at high temperatures is due to some dissolution of surface carbon into the bulk of the platinum. When *experimentally measured* corrections (open circles) are applied to these data [7.62], the expected linearity is achieved over the entire temperature regime from 420 to 900 K, as may be seen in Fig. 7.18. The data of Figs. 7.19 and 20 are not contaminated by poisoning by CO at the lower temperatures because its background partial pressure is much lower in these cases. This is due to the fact that both the partial pressures of the alkanes are much lower here, and the reactor is operated in a flow rather than a batch mode in these cases.

As may be seen in Table 7.3, the activation energies for C–H bond cleavage in ethane and propane are quite similar. The activation energy for C–H bond cleavage in methane is approximately 7 or 8 kcal/mol greater, however. The fact that methane is more difficult to activate than either ethane or propane is

Table 7.3. Rate parameters for the dissociative chemisorption of methane, ethane, and propane on Pt(1 1 0)-(1 × 2). The separately measured values of the desorption rate coefficients were used in evaluating E_r and $k_r^{(o)}$. As discussed in the text, $E_r - E_d$ is the activation energy (with respect to a gas-phase energy zero) of dissociative chemisorption, and these values for propane activation are apparent values; the *two* microscopic activation energies that apply to each isotope fo propane are listed in Table 7.4

Reactant	$E_r - E_d$ [cal/mol]	$k_d^{(o)}/k_r^{(o)}$	E_r [cal/mol]	$k_r^{(o)}$ [S^{-1}]
CH_4	14400	2	18900	5×10^{12}
CD_4	15600	2	20100	5×10^{12}
C_2H_6	2800	215	10500	5×10^{10}
C_2D_6	3500	215	11200	5×10^{10}
C_3H_8	1460	165	11160	6×10^{10}
$CH_3CD_2CH_3$	1800	165	11500	6×10^{10}
C_3D_8	2330	165	12030	6×10^{10}

certainly not surprising. It is quite interesting, however, that essentially all of the C–H bond-stabilization energy of the reactant methane with respect to ethane and propane [7.65] is reflected in the activation energy differences for the chemisorption reactions.

The alert reader will have noted already that there should be *two* activation energies associated with the activation of each of the propane isotopes, although this is not apparent in the measured data of Fig. 7.20. The reason for expecting two different activation energies, of course, is due to the fact that the barriers for primary (1°) and secondary (2°) C–H bond cleavage are not the same. As has been shown explicitly, however, it is possible to analyze the data of Fig. 7.20 more fully in order to clarify this important issue completely [7.62]. The analysis will be discussed briefly.

Let us define $P_{r,i}$ to be the probability of reaction of the ith isotope of propane, $\tilde{P}_p$ to be the conditional probability of formation of a 1°Pt-propyl given that the propane reacts, and $\tilde{P}_s = 1 - \tilde{P}_p$ to be the conditional probability of formation of a 2°Pt-propyl given that the propane reacts. Thus, it follows that

$$P_{r,C_3H_8}\tilde{P}_p + P_{r,C_3H_8}\tilde{P}_s = P_{r,CH_3CD_2CH_3}, \tag{7.11}$$

where $P_{r,i}$ are given, as a function of temperature, in Fig. 20. Solving (7.11) gives both $\tilde{P}_p(T)$ and $\tilde{P}_s(T) = 1 - \tilde{P}_p(T)$, where these probabilities refer to a pure isotope of propane, either C_3H_8 or C_3D_8. Each of these probabilities may be written exactly in the form $\tilde{P} = P^{(0)}e^{-\Delta E/k_BT}$. The derived slopes and intercepts of the two Arrhenius constructions implied by this relationship are the following:

$$\tilde{P}_p = 0.61e^{-(220\ \mathrm{cal/mol})/k_BT}, \tag{7.12}$$

$$\tilde{P}_s = 0.39e^{+(205\ \mathrm{cal/mol})/k_BT}. \tag{7.13}$$

Equations (7.12, 13) quantify the probabilities of 1° and 2° Pt-propyl formation for a pure isotope of propane. By taking the ratio of these probabilities, the difference in the activation energies for 1° and 2° C–H bond cleavage may be determined:

$$\frac{\tilde{P}_p}{\tilde{P}_s} = 1.56e^{-(425\ \mathrm{cal/mol})/k_BT}. \tag{7.14}$$

This equation implies that the difference in activation energy of 1° and 2° C–H bond cleavage, $E_r^{(1°)} - E_r^{(2°)}$, is 425 cal/mol, and the data of Fig. 7.20 imply directly that the difference in activation energy of C–D and C–H bond cleavage, $E_r^{(D)} - E_r^{(H)}$, is 870 cal/mol (assuming the absence of a secondary isotope effect). With these two constraints it is possible to determine unique values of the two activation energies that are contained within the single apparent value derived from Fig. 7.20 for each isotope, and reported, for example as $E_r - E_d$ in the second column of Table 7.3. Varying $E_r^{(1°)} - E_r^{(2°)}$ with the two constraints delineated above was carried out with the goal of reproducing, self-consistently, the measured data shown in Fig. 7.20 for all three isotopes of propane. The values of $E_r^{(1°)} - E_r^{(2°)}$, expressed as $E_r^{(1°)} - E_d$ and $E_r^{(2°)} - E_d$, that are reported in

Table 7.4. Activation energies for cleavage of primary C–H bond in propane (columns two and four) and for cleavage of a secondary C–H bond in propane (columns three and five) on Pt(1 1 0)-(1 × 2). The second and third columns refer to a gas-phase energy zero, whereas the fourth and fifth columns refer to the proper energy zero, which is the ground state of the physically adsorbed propane

Reactant	$E_r^{(1\,°)}-E_d$, [cal/mol]	$E_r^{(2\,°)}-E_d$, [cal/mol]	$E_r^{(1\,°)}$ [cal/mol]	$E_r^{(2\,°)}$[cal/mol]
C_3H_8	1650	1225	11 350	10 925
$CH_3CD_2CH_3$	1650	2095	11 350	11 795
C_3D_8	2520	2095	12 220	11 795

the second and third columns of Table 7.4, reproduced the experimental data of Fig. 7.20 to within 1.5% for all three isotopes over the entire temperature range. This is well within the experimental uncertainty of the data. Noticeable disagreement with the experimental data occurs if these energies are varied from their optimized values by as little as 50 cal/mol. The actual values of $E_r^{(1°)}$ and $E_r^{(2°)}$ are given in the fourth and fifth columns of Table 7.4.

It is also important to note that the preexponential factor of 7.14 gives the ratio of conditional probabilities that apply in the limit of infinite temperature. This allows an assessment to be made of "steric" or entropic effects in the reaction. If the entropic factors were identical for the two reactions, then $\lim_{T\to\infty} \tilde{P}_p/\tilde{P}_s = 3$, the ratio of the number of primary to secondary hydrogen atoms. Consequently, entropy also favors 2 °C–H bond cleavage on this surface by a factor of 1.9, i.e., 3/1.56.

In addition to providing a quantification of the activation of the specific alkanes on the particular surfaces that were studied, cf., Tables 7.2–4, some more general conclusions may be drawn for the case of ethane activation. For example, the fact that $E_r - E_d = 2800$ cal/mol on Pt(1 1 0)-(1 × 2) and $E_r - E_d = -2200$ cal/mol on Ir(1 1 0)-(1 × 2), two surfaces of identical geometric structure [7.56, 57], implies that the intrinsic *electronic difference* between platinum and iridium for ethane activation is approximately 5 kcal/mol. For ethane activation on the close-packed Pt(1 1 1) surface, *Rodriquez* and *Goodman* [7. 22] have found that $E_r - E_d \sim 8900$ cal/mol, and *Sun* and *Weinberg* [7, 24] found that $E_r - E_d = 2800$ cal/mol for ethane activation on the corrugated Pt(1 1 0)-(1 × 2) surface. This implies that the *geometrical structure* can lead to a difference in the activation energies for ethane activation of 6.1 kcal/mol, greater than the electronic difference that distinguishes platinum and iridium with the same geometrical structure! It will obviously be of interest to determine if these conclusions apply to the activation of other alkanes on platinum and iridium.

7.2.3 Microscopic Reaction Mechanism

Rather little is known about the chemical interactions of an alkane molecule and a reactive metal surface and the detailed rearrangements that occur during dissociation and surface reaction/formation of new species. This is not due to lack of interest but rather to the complexity of the problem and its inherent experimental difficulties. Indeed, this subject is of intense interest and technological importance.

The surface site for dissociation of alkanes (and other molecules) is generally unknown or the subject of speculation. For example, in the trapping-mediated reaction of ethane on Ir(1 1 0)-(1 × 2), we suspect that dissociative chemisorption occurs via an interaction with hollow sites at the bottom of rows in this corrugated surface. This proposal is supported by the following data: (1) the dissociation of ethane on Ir(1 1 1) under ultrahigh vacuum conditions in the trapping-mediated regime is not facile [7.16], implying that the close-packed (1 1 1) microfacets that compose the inclined terraces of the reconstructed (1 1 0) surface would not be expected to contain the active site, and (2) the reaction of ethane with this (1 1 0)-(1 × 2) surface of Ir is completely blocked when the β_2 sites on the surface are saturated with hydrogen [7.2, 16, 66]. The β_2 state of hydrogen on the Ir(1 1 0)-(1 × 2) surface is the state that is occupied first upon adsorption and desorbs at a temperature of ≈ 400 K, compared to temperatures of between 200 and 300 K for various coverages of the β_1 state [7.67, 68]. It is reasonable to expect β_2 hydrogen to occupy pseudo-four-fold hollow sites in the trough bottoms of this surface, since β_1 hydrogen has unambiguously been identified with desorption from the (1 1 1) microfacets. It is unlikely that hydrogen would be chemisorbed at either an on-top site or a two-fold bridging site on the apex of a row. This is all that is known about this scientific and technologically important surface reaction. In other words, very little is known about the detailed mechanisms and atomic rearrangements that occur during dissociation.

It is possible that the initial chemical interaction between a reactive metal surface and an alkane involves the formation of a three-center, two-electron bond between a filled C–H bonding orbital and an unfilled metal orbital. This postulate is motivated by results from studies in homogeneous organometallic chemistry where a large number of stable complexes containing *intramolecular* interactions of this type, the so-called agostic interaction, have been found [7.63]. This interaction has been identified and characterized by C–H bond distances that are significantly longer than expected for an sp2-hybridized carbon atom and with unusually low stretching frequencies, i.e., ν(C–H)-$\approx 2600\ \mathrm{cm}^{-1}$, or less. More recently, kinetic and mechanistic studies of several systems, in which *intermolecular* alkane activation is achieved, point clearly to the formation of a reactive intermediate prior to C–H bond cleavage, and suggest strongly that this is in fact the η^2(C–H)-alkane complex [7.70].

Interestingly, vibrational spectroscopy of cyclohexane on several transition metal surfaces has implicated "softening" of a C–H stretching mode (from $2790\ \mathrm{cm}^{-1}$ on Cu(1 0 0) to $2560\ \mathrm{cm}^{-1}$ on Ru(0 0 1) with the gas-phase value being $2890\ \mathrm{cm}^{-1}$ [7.71–73]. Studies by *Chesters* and coworkers [7.72, 73] on several transition metal surfaces have suggested that, indeed, an agostic interaction could be playing an important role in the dehydrogenation of cyclohexane to benzene. In this work it was demonstrated that the extent of ν(C–H) mode softening is directly related to the heat of adsorption and thus the strength of the cyclohexane-metal interaction (and, likely, the activation energy to dehydrogenation). *Chesters* et al. [7.60] have also studied the vibrational spectroscopy of normal alkanes on Pt(1 1 1) in ultra high vacuum and found no evidence for ν(C–H) mode softening for these systems. However, this may not be particularly surprising since the reaction of n-alkanes with this surface is not facile.

Finally, we return to the recent studies by *Weinberg* and *Sun* [7.62] concerning the activation of propane on the Pt(1 1 0)-(1 × 2) surface. Here, the quantification of primary and secondary C–H bond cleavage was determined by kinetic studies of perhydrido, perdeutero and selectively deuterated (i.e., $CH_3CD_2CH_3$) propane on this surface. *Weinberg* and *Sun* found that the activation energy for primary bond cleavage was 425 calories/mole greater than that for secondary bond cleavage, and that secondary bond breaking was also preferred on entropic grounds (preexponential factors). However, the recently discovered organometallic compound Cp*Ir(PMe_3), which so readily inserts itself into C–H bonds, generally attacks carbon–hydrogen bonds in the order primary $>$ secondary $\gg$ tertiary [7.74]. The differences between homogenous chemistry versus surface chemistry of alkane activation will require further study to be resolved but is most likely due to the steric effects of the bulky ligands on the homogeneous Ir compound.

7.3 Direct Dissociative Chemisorption

As mentioned earlier, recent studies have shown that two distinct mechanisms are responsible for dissociative chemisorption of alkanes at transition metal surfaces. In the previous section, studies regarding the trapping-mediated mechanism to dissociative chemisorption have been reviewed. In this section we discuss the direct dissociative chemisorption mechanism using several seminal investigations of the interaction of methane with different well-characterized metal surfaces [17.10, 12, 13, 75]. In particular, we will review alkane activation by translational energy, vibrational energy and by collision induced mechanisms. We can distinguish between the trapping-mediated mechanism and the direct dissociative mechanism by observing the kinetic energy dependence of the probability of dissociative chemisorption. In the case of the trapping-mediated mechanism, the probability of dissociative chemisorption decreases as the incidence kinetic energy increases because the trapping probability decreases. In the case of the direct mechanism, the probability of dissociative chemisorption *increases* as the incidence kinetic energy *increases* (and, typically, as the incidence vibrational energy increases also). The picture here is one in which the incident molecule either reacts upon collision with the surface or scatters off of the surface. For a gas-phase molecule incident on the surface, it does not trap and subsequently accommodate. However, in the case of a collision-induced reaction, the reactant (*alkane*) is physically adsorbed and certainly accommodated to the surface. Reaction occurs when a particle of sufficient kinetic energy collides with the adsorbate.

7.3.1 Activation via Translational Energy

Supersonic molecular-beam techniques provide independent control of the incident molecular kinetic energy, the surface temperature and to a lesser extent control of initial vibrational energy. Thus, beam experiments allow a more complete understanding of the effects of these variables on dissociative chemisorption.

The first supersonic molecular beam experiment to probe alkane activation was conducted by *Rettner* et al. [7.10] in a study of the dissociative chemisorption of methane and deuterated methane on the (1 1 0) surface of tungsten. As shown in Fig. 7.23, *Rettner* et al. observed a dramatic increase (approximately 10^5) in the initial probability of dissociative chemisorption S_0 as the normal component of the incident kinetic energy was increased from ≈ 0.05 eV to ≈ 1.0 eV with the $W(1\,1\,0)$ sample held at 800 K. *Rettner* et al. reported a weak effect on S_0 when the surface temperature T_s was varied; a change in S_0 by a factor of 2 or less for T_s varying from 600 to 1000 K. These researchers also reported an effect on S_0 due to variations in the supersonic nozzle temperature. Changes in nozzle temperature produce different levels of vibrational excitation in the molecular beam. We will discuss this effect in more detail in the next subsection. *Rettner* et al. proposed that the dissociative chemisorption probability was dominated by a tunneling mechanism because of the approximately

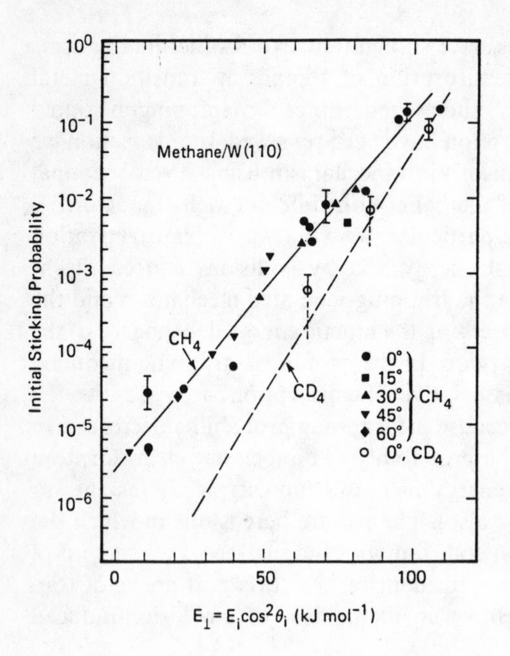

Fig. 7.21. Initial sticking probability for CH_4 (solid symbols) and CD_4 (open symbols) on $W(1\,1\,0)$ as a function of beam energy at various angles of incidence. The surface temperature for these measurements was 800 K

exponential increase in S_0 with the normal kinetic energy and the large isotope effect that was observed.

Lee et al. [7.12] conducted studies of the dissociative chemisorption of CH_4 and CD_4 on the Ni(1 1 1) surface. In their studies, *Lee* et al. observed an exponential rise in the probability of dissociative chemisorption with incident kinetic energy over the range $E_i \approx 12$–17 kcal/mol for CH_4 and $E_i \approx 16$ kcal/mol to ≈ 21 kcal/mol for CD_4. Their experiments (depicted in Fig. 7.22) were conducted with a surface temperature of $T_s = 475$ K, and *Lee* et al. state that additional experiments at $T_s = 275$ K showed S_0 to have no dependence on T_s. However, the experiments conducted at $T_s = 275$ K were performed by measuring the integrated hydrogen thermal desorption spectrum, while the data at $T_s = 475$ K were determined with Auger electron spectroscopy and it is not clear that the two sets of data can be compared unambiguously. This is an important point which will be discussed in more detail later. It is clear from Fig. 7.22 that there is a strong isotope effect, and, as with methane on *W*(1 1 0), the nozzle temperature had a strong effect on the probability of dissociative chemisorption. These data prompted *Lee* et al. to conclude that dissociative chemisorption of methane on Ni(1 1 1) was a direct process in which the initial collision of the molecule results in a deformation of the molecule from which the hydrogen or deuterium atom then tunnels in to a chemisorbed state. This prompted the researchers to explore collison-induced surface reactions of methane on Ni(1 1 1), which we will discuss later in this chapter. They also showed, for the first time, employing high-resolution electron energy loss spectroscopy that the immediate product of the surface reaction is a chemisorbed methyl and a hydrogen adatom.

Two groups have reported measurements for the dissociative chemisorption of methane on Pt(1 1 1) [7.13, 75]. *Schoofs* et al. [7.75] investigated reaction of methane with the Pt(1 1 1) surface in the high kinetic energy, high probability ($S_0 > 0.01$) of dissociative chemisorption regime. *Luntz* and *Bethune* [7.13] conducted experiments with a range of incident kinetic energies from 0.2 to

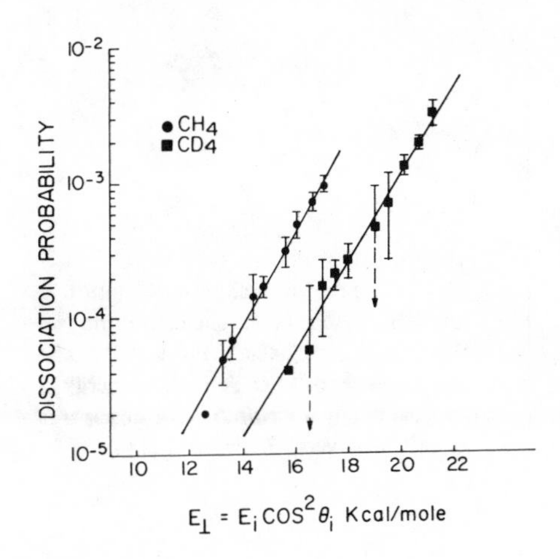

Fig. 7.22. The absolute dissociation probability of CH_4 and CD_4 on Ni(1 1 1) as a function of the normal component of translational energy [7.12]

1.3 eV with S_0 increasing from $\approx 10^{-4}$ to ≈ 0.2 over this kinetic energy range. The data of *Luntz* and *Bethune* are displayed in Fig. 7.23 for both perhydrido- and perdeutero- methane on Pt(1 1 1). Again S_0 increases dramatically with E_i and there is a strong isotopic effect. As with the two previously discussed studies regarding methane dissociative chemisorption, nozzle temperature and, hence,

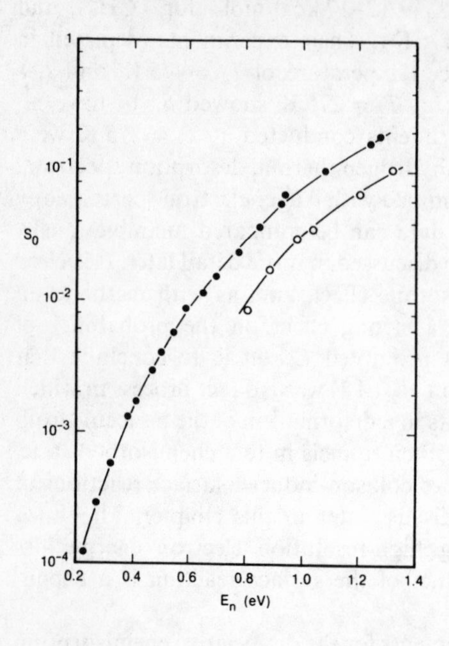

Fig. 7.23. S_0 as a function of $E_n = E_i \cos^2 \theta_i$ for $\theta_i = 0°$ and $T_s = 800$ K. The solid points are for sticking of CH_4 and the open circles are for the sticking of CD_4 on Pt(1 1 1). The lines through the points have no theoretical significance. [7.13]

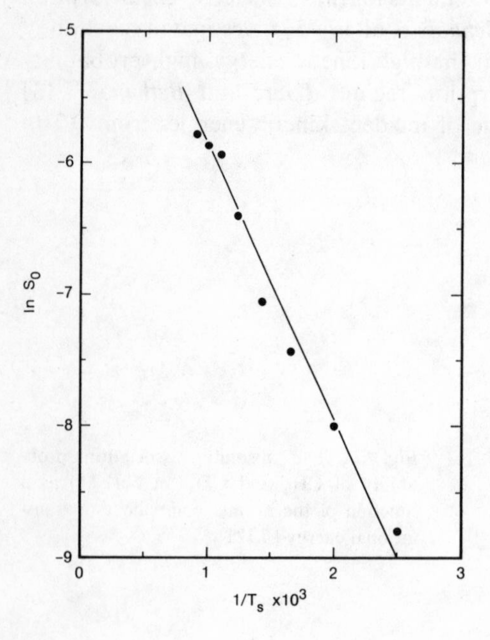

Fig. 7.24. The variance of ln S_0 with $1/T_s$ for CH_4 dissociative chemisorption on Pt(1 1 1) at $E_i = 0.42$ eV and $\theta_i = 0°$. The line corresponds to an activation energy of approximately 4 kcal/mol. The abscissa is in units of inverse Kelvin (K^{-1}) [7.13]

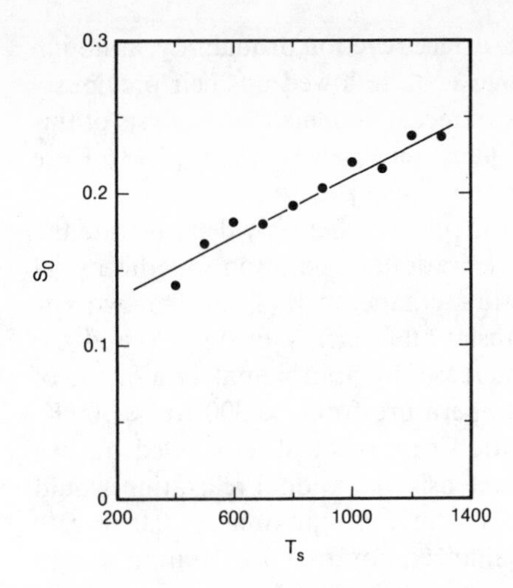

Fig. 7.25. The variance of S_0 with T_s for CH_4 dissociative chemisorption on Pt(1 1 1) at $E_i = 1.27$ eV and $\theta_i = 0°$. The abscissa is in units of Kelvin (K) [7.13]

level of vibrational excitation has a strong effect on S_0. *Luntz* and *Bethune* also reported a strong increase in S_0 with T_s for low incident kinetic energies, while the surface-temperature dependence becomes weaker with increasing normal kinetic energy (Figs. 7.24 and 25). In the other measurements of methane reacting with Pt(1 1 1) by *Schoofs* et al. a surface-temperature dependence on the probability of dissociative chemisorption was not observed. However, this study considered only a single high incident kinetic energy where the T_s dependence is expected to be smaller.

7.3.2 Activation via Vibrational Energy

As mentioned briefly above, in studies of the dissociative chemisorption of methane on *W*(1 1 0) by *Rettner* et al. [7.10] it was noted that the reaction probabilities were different for beams with the same translational energies but differing nozzle temperatures. Indeed, in Fig. 7.21, the two data points with $E_i \approx 10$ kj/mol ($\theta_i = 0°$) serve to illustrate the effect. The point with the higher reaction probability ($S_0 \approx 2.5 \times 10^{-5}$) was generated with a nozzle temperature of ≈ 750 K while the lower data point ($S_0 \approx 4 \times 10^{-6}$) was produced with a nozzle temperature of ≈ 320 K. The origin of this effect is the higher level of vibrational excitation in the beam with the higher nozzle temperature. In the supersonic expansion from the nozzle, translational energies and rotational levels (except for hydrogen) are relaxed extensively, however, due to the wide spacing of vibrational energy levels, there is very little relaxation in this degree of freedom. The extent of relaxation depends strongly on the expansion conditions, e.g., nozzle pressure and nozzle aperture diameter.

Several researchers have suspected that vibrational energy can strongly affect surface reaction probabilities. Most notably for alkane activation are the early studies by *Stewart* and *Ehrlich* [7.76], *Brass* et al. [7.32] and *Yates* et al. [7.31]. These earlier studies were unsuccessful at experimentally demonstrating that vibrational energy could enhance the surface reaction probability, although the opposite was not proven either. *Rettner* et al. followed up their previously mentioned study [7.10] with more extensive measurements and analysis of this phenomena [7.77] as have *Lee* et al. [7.12] and *Luntz* and *Bethune* [7.13]. Here we discuss the seminal studies of *Rettner* et al. [7.77].

Table 7.5 lists the results of two experiments which amply demonstrate the effect of nozzle temperature, and hence vibrational excitation, on the initial probability of dissociative chemisorption of methane on $W(1\,1\,0)$. The two sets of measurements, one at an incident translational energy of 0.11 eV and the other with $E_i \approx 0.23$ eV, both show an increase, by approximately a factor of five, in S_0 upon increasing the nozzle temperature from ≈ 300 to ≈ 700 K. Although effects due to rotational excitation cannot be strictly ruled out we believe them to be highly unlikely. First, extensive rotational relaxation would be expected since the rotational constant is 5 cm^{-1} (approximately 0.0006 eV). Second, it is difficult to imagine how enhanced rotational excitation would couple into the reaction coordinate. Rather, it is likely that the data presented in Table 7.5 be explained by the effect of vibrational excitation.

In their analysis of the data in Table 7.5, *Rettner* et al. concluded that vibrational energy is at least as effective in promoting dissociation as translational energy. In studies of the reaction of C_2H_6 with Ir(1 1 0)-(1 × 2) at relatively high kinetic energy (the direct regime) neither *Steinruck* et al. [7.78] nor *Mullins* and *Weinberg* [7.79] detected any enhancement in the dissociation probability with increased nozzle temperature for a fixed incident kinetic energy. However, these measurements were conducted in a regime where S_0 is greater than ≈ 0.02, whereas the measurements conducted by *Rettner* et al. were for $S_0 < 10^{-4}$. Recently *Verhoef* et al. [7.80] have studied vibrationally assisted direct dissociative chemisorption of deuterated methane and ethane on Ir(1 1 0)-(1 × 2) and found the enhancement to be significantly larger than in the previously mentioned studies and large enough that a *King* and *Wells* type of measurement can be employed, i.e., with $S_0 > 0.02$. Their results show that the lower-frequency modes of a perdeutero-species as compared to a perhydrido-species provides a much higher population of excited vibrational states for the same nozzle temperature and that careful analysis of these data suggest that the

Table 7.5. Effect of nozzle temperature on initial sticking probabilities of CH_4 on W(1 1 0) [7.77]

Trans. energy [eV]	Surface temp. [K]	Nozzle temp. [K]	S_0
011 ± 0.01[a]	800	770	$2.5 \pm 1.0 \times 10^{-5}$
0.11 ± 0.01	800	340	$5.0 \pm 2.0 \times 10^{-6}$
0.23 ± 0.02	800	720	$4.0 \pm 2.0 \times 10^{-5}$
0.23 ± 0.02	800	305	$7.0 \pm 5.0 \times 10^{-6}$

[a] Error bars reported here refer to approximate 95% confidence limits.

enhanced dissociation is due to a single vibrational mode. Indeed, participation of only the asymmetric stretching modes, which are 4–5% populated for deuterated methane and ethane at a gas temperature of approximately 750 K versus 1% populated for perhydrido-methane and ethane at that temperature can accurately account for their experimental observations [7.79, 80].

It is unclear at this point which vibrational modes contribute to the enhanced dissociation rate and precisely how they couple to the reaction coordinate and thus, much more work is necessary.

7.3.3 Collision-Induced Activation

In addition to the direct mechanism involving the molecule colliding with the surface, it has recently been shown that a physically adsorbed molecule can be induced to react with a transition metal surface when subjected to a collision from a relatively high kinetic energy gas-phase noble-gas atom [7.5, 81, 82]. Although this mechanism is similar to processes involving ion-assisted etching and deposition on semiconductor surfaces, these semiconductor materials modification processes typically involve much higher kinetic energy *ions* than were employed in the seminal experiments of *Beckerle* et al. They have performed the only experiment to date regarding collision-induced (i.e., collision of a rare-gas atom with physically adsorbed methane molecule) activation of an alkane on a transition-metal surface. They were motivated, in part, to conduct this experiment by their interpretation of the microscopic mechanism involved in the direct dissociation of methane in the gas-phase at high incidence kinetic energy on Ni(1 1 0), i.e., deformation of the tetrahedral molecule with subsequent tunneling of a hydrogen atom [7.5, 12]. Their results from the collision-induced studies support this idea. *Beckerle* et al. conducted the experiment by measuring the dissociation rate as an argon-atom beam bombarded a physically adsorbed monolayer of methane on a Ni(1 1 1) surface held at ≈ 47 K. The reaction cross section can be calculated from the dissociation rate, the argon atom flux and the methane coverage; and it is calculated (measured) as a function of the incident kinetic energy of the argon atoms. Figure 7.26 is a plot of the reaction cross section of physically adsorbed methane on Ni(1 1 1) versus the normal energy of the bombarding argon atoms. This figure suggests that normal-energy scaling *apparently* is not observed as with the earlier experiments regarding gas-phase methane [7.12].

In an effort to understand the data shown in Fig. 7.26 *Beckerle* et al. proposed a two step process to account for the energy dependence in a collision-induced surface reaction. The first step involves fractional energy transfer from the impinging Ar atom to the physically adsorbed methane molecule. The effectiveness of the energy transfer depends on the collisional impact parameter and can be calculated using a hard sphere collision model with impact parameter as a variable. From their earlier studies, *Beckerle* et al. [7.12] assumed that the energy crucial to surface reaction is the normal energy of the methane

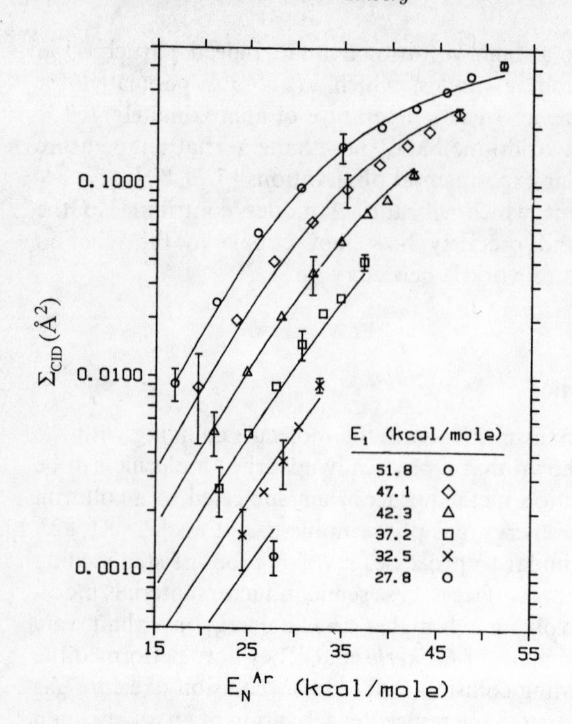

Fig. 7.26. Cross-section for dissociation of CH_4 physisorbed on Ni(1 1 1) at 47 K induced by the impact of Ar atoms vs the kinetic energy of the Ar atom in the normal direction. The methane coverage is 0.3 monolayers. Each point at the same total energy represents a different incident angle from 0° to 55°. The *solid lines* are the result of the model calculation discussed in the text [7.5 and 77]

after collision with the Ar, E_n, which is calculated as

$$E_n(E_i, \theta_i, b, \phi) = \frac{4M_{Ar}MCH_{4^4}}{(M_{Ar}+M_{CH_4})^2}E_i\left[\left(1-\frac{b^2}{d^2}\right)\cos\theta_i - \frac{b}{d}\left(1-\frac{b^2}{d^2}\right)^{1/2}\sin\theta_i\cos\phi\right]^2 \quad \text{for } b \leqslant d$$
$$= 0 \quad \text{for } b > d, \tag{7.15}$$

where E_i is the total energy, θ_i is the incident angle of the Ar atom, b is the impact parameter, ϕ is the azimuthal angle of the collision, m_{Ar} and M_{CH_4} are the masses of Ar and methane, respectively, and d is the hard sphere collision diameter.

Now, the second step in the collision-induced reaction analysis is to recall that the probability of dissociative chemisorption of methane on Ni(1 1 1) with a normal energy E_n can be obtained from the data of *Lee* [7.12], as shown in Fig. 7.22. These data can be described with the following functional form [7.5, 82]

$$P(E_n) = \frac{A}{1-e^{-\alpha}(E_n - V)}, \tag{7.16}$$

where α, V and A are simply fitting parameters. This function can be combined with the expression for methane normal energy derived from the hard sphere collision model (above) to calculate the probability of surface reaction, P_{CID}, as a function of the *argon* incidence kinetic energy and angle and the impact parameter b and azimuthal angle of collision ϕ to obtain

$$P_{\mathrm{CID}}(E_{\mathrm{i}}, \theta_{\mathrm{i}}, b, \phi) = P[E_{\mathrm{n}}(E_{\mathrm{i}}, \theta_{\mathrm{i}}, b, \phi)]. \tag{7.16}$$

Since the Ar-methane impact geometry is not fixed, it is necessary to integrate P_{CID} over all possible values of b and ϕ to calculate a value of the collision-induced reaction cross section σ that can be compared to experiment,

$$\sigma(E_{\mathrm{i}}, \theta_{\mathrm{i}}) = \frac{1}{\cos\theta_{\mathrm{i}}} \int_0^{2\pi} \mathrm{d}\phi \int_0^{\infty} P_{\mathrm{CID}}(E_{\mathrm{i}}, \theta_{\mathrm{i}}, b, \theta) b \,\mathrm{d}b. \tag{7.17}$$

Beckerle et al. [7.5, 81] have evaluated this integral and the results are represented by the solid lines in Fig. 7.26. The outstanding agreement between their model and the measured data shows that accounting for a range of impact parameters with a hard sphere collision model can accurately explain the energy transfer and the apparent breakdown of normal energy sealing.

Beckerle et al. [7.81] proposed a physical picture to describe this phenomena such that the colliding argon atom transfers some fraction of its kinetic energy to the physically adsorbed methane with the entailing methane-surface collision dynamics being the same as with translational activation of the methane [7.12]. Thus, the post-collision methane molecule with relatively high kinetic energy collides with the surface, undergoes deformation and subsequent surface reaction. *Beckerle* et al. [7.82] have also experimentally demonstrated that the surface reaction products of collision-induced dissociation are the same as for translationally activated dissociation, namely, an adsorbed methyl fragment and an adsorbed hydrogen atom. Similar experiments with impinging Kr and Ne atoms have also been conducted [7.81]. The results for Ne atoms is very similar in character to those obtained with Ar whereas experimental results with Kr suggest that the hard sphere collision model underpredicts the cross section for collision induced dissociation and this is thought to be due to the non-impulsiveness of the collision between the Kr atom and the physically adsorbed methane molecule.

Acknowledgement. The support of this work by the Department of Energy (grant no. DE-FG03-89ER14048) and the Donors of the Petroleum Research Fund, administered by the American Chemical Society (grant no. ACS-PRF-23801-AC5-C) is gratefully acknowledged. We thank S. G. Karseboom for his assistance with some of the figures.

References

7.1 J.H. Sinfelt: Prog. Solid-State Chem. **10,** 55 (1975); Catal. Rev. **3,** 175 (1969); in *Catalysis Science and Technology* ed. by J.R. Anderson, M. Boudart, (Springer, Berlin, Heidelberg 1981.) Vol. 1, p. 257

7.2 W.H. Weinberg: in *Survey of Progress in Chemistry*, ed. by A. Scott, G. Wubbel (Academic, New York 1983) Vol. 10 p. 1

7.3 R. Pitchai, K. Klier: Catal. Rev. Sci. Eng. **28,** 13 (1986)
74. W.H. Weinberg: in *Dynamics of Gas-Surface Collisions*, ed. by C.T. Rettner, M.N.R. Ashfold, (Royal Society of Chemistry, Cambridge 1991) p. 171
7.5 S.T. Ceyer: Ann. Rev. Phys. Chem. **39,** 479 (1988)
7.6 D.W. Goodman: Ann. Rev. Phys. Chem. **37,** 425 (1986)
7.7 A.G. Sault, D.W. Goodman: Adv. Chem. Phys. **76,** 153 (1989)
7.8 G. Ehrlich: In *Chemistry and Physics of Solid Surfaces VII*, R. Vanselow and R. Howe, Eds., p. 1, Springer-Verlag, Berlin, 1988
7.9 W.H. Weinberg, submitted to J. Vac. Sci. Technol. A **10,** 2271 (1992)
7.10 C.T. Rettner, H.E. Pfnur, D.J. Auerbach: Phys. Rev. Lett. **54,** 2716 (1985)
7.11 A.V. Hamza, R.J. Madix: Surf. Sci. **179,** 25 (1987)
7.12 M.B. Lee, Q.Y. Yang, S.T. Ceyer: J. Chem. Phys. **87,** 2724 (1987)
7.13 A.C. Luntz, D.S. Bethune: J. Chem. Phys. **90,** 1274 (1989)
7.14 C.B. Mullins, W.H. Weinberg: J. Chem. Phys. **92,** 4508 (1990)
7.15 T.S. Wittrig, P.D. Szuromi, W.H. Weinberg: J. Chem. Phys. **76,** 3305 (1982)
7.16 P.D. Szuromi, J.R. Engstrom, W.H. Weinberg: J. Chem. Phys. **80,** 508 (1984)
7.17 P.D. Szuromi, W.H. Weinberg: Surf. Sci. **149,** 226 (1985)
7.18 P.D. Szuromi, J.R. Engstrom, W.H. Weinberg: J. Phys. Chem. **89,** 2497 (1985)
7.19 T.P. Beebe, Jr., D.W. Goodman, B.D. Kay, J.T. Yates, Jr: J. Chem. Phys. **87,** 2305 (1987)
7.20 A.G. Sault, D.W. Goodman: J. Chem. Phys. **88,** 7232 (1988)
7.21 X. Jiang, D.W. Goodman: Appl. Phys. A**51,** 99 (1990)
7.22 J.A. Rodriquez, D.W. Goodman: J. Phys. Chem. **94,** 5342 (1990)
7.23 S.G. Brass, G. Ehrlich: J. Chem. Phys. **87,** 4285 (1987)
7.24 Y.-K. Sun, W.H. Weinberg: J. Vac. Sci. Technol, A**8,** 2445 (1990)
7.25 (a) C.T. Rettner, L.A. DeLouise, D.J. Auerbach: J. Chem. Phys. **85,** 1131 (1986)
7.26 D.A. King, M.G. Wells: Proc. R. Soc. London (Ser. A) **339,** 245 (1974)
7.27 D.W. Goodman: J. Vac. Sci. Technol. **20,** 522 (1982)
7.28 G. Ertl, J. Kuppers: *Low Energy Electrons and Surface Chemistry* (Verlag Chemie, Weinheim 1985)
7.29 G. Scoles (ed.) *Atomic and Molecular Beam Methods*, (Oxford Univ. Press, New York 1988).
7.30 H.F. Winters: J. Chem. Phys. **62,** 2454 (1975)
7.31 J.T. Yates, Jr., J.J. Zinck, S. Sheard, W.H. Weinberg: J. Chem. Phys. **70,** 2266 (1979)
7.32 S.G. Brass, D.A. Reedand, G. Ehrlich: J. Chem. Phys. **70,** (1979)
7.33 W.H. Weinberg, R.P. Merrill: J. Vac. Sci. Technol. **8,** 718 (1971); accommodation coefficients are defined in the monograph by R.P.H. Gasser, *An Introduction to Chemisorption and Catalysis by Metals* (Oxford Univ. Press, New York 1985)
7.34 W.H. Weinberg, R.P. Merrill: J. Chem. Phys. **56,** 2881 (1972)
7.35 J.E. Hurst, L. Wharton, K.C. Janda, D.J. Auerbach: J. Chem. Phys. **83,** 1376 (1985)
7.36 H.P. Steinruck, R.J. Madix: Surf. Sci. **185,** 36 (1987)
7.37 C.R. Arumainayagam, G.R. Schoofs, M.C. McMaster, R.J. Madix: J. Phys. Chem. **95,** 1041 (1991)
7.38 C.T. Rettner, E.K. Schweizer, C.B. Mullins: J. Chem. Phys. **90,** 3800 (1989)
7.39 C.B. Mullins, W.H. Weinberg: J. Chem. Phys. **92,** 3986 (1990)
7.40 C.B. Mullins, C.T. Rettner, D.J. Auerbach, W.H. Weinberg: Chem. Phys. Lett. **163,** 111 (1989)
7.41 C.T. Rettner, D.S. Bethune, D.J. Auerbach: J. Chem. Phys. **91,** 1942 (1989)
7.42 C.R. Arumainayagam, R.J. Madix, M.C. McMaster, V.M. Suzawa, J.C. Tully: Surf. Sci. **226,** 180 (1990)
7.43 J.C. Tully, M.J. Cardillo: Science **223,** 445 (1984)
7.44 J.C. Tully, C.W. Muhlhausen, L.R. Ruby: Ber. Bunsenges. Phys. Chem. **86,** 433 (1982)
7.45 C.W. Muhlhausen, L.R. Ruby, J.C. Tully: J. Chem. Phys. **83,** 2594 (1985)
7.46 A.V. Hamza, H.-P. Steinruck, R.J. Madix: J. Chem. Phys. **86,** 7494 (1986)
7.47 A.V. Hamza, H.-P. Steinruck, R.J. Madix: J. Chem. Phys. **86,** 6506 (1987)
7.48 A.C. Luntz, M.D. Williams, D.S. Bethune: J. Chem. Phys. **89,** 4381 (1988)
7.49 D.A. King: CRC Crit. Rev. Solid State Mater. Sci. **167,** 451 (1978)

7.50 C.T. Rettner, H. Stein, E.K. Schweizer: J. Chem. Phys. **89,** 3337 (1988)
7.51 C.T. Rettner, C.B. Mullins: J. Chem. Phys. **94,** 1626 (1991)
7.52 W.H. Weinberg: in *Kinetics of Interface Reactions*, ed. by M. Grunze, H.J. Kreuzer, Spinger Ser. Surf. Sci. Vol. 8 (Springer, Berlin, Heidelberg, 1987) p. 94
7.53 E.S. Hood, B.H. Toby, W.H. Weinberg: Phys. Rev. Lett. **55,** 2437 (1985)
7.54 C.B. Mullins, W.H. Weinberg: J. Vac. Sci. Technol. **A8,** 2458 (1990)
7.55 E.W. Kuipers, M.G. Tenner, M.E.M. Spruit, A.W. Kleyn: Surf. Sci. **205,** 241 (1988)
7.56 C.-M. Chan, M.A. Van Hove, W.H. Weinberg, E.D. Williams: Surf. Sci. **91,** 440 (1980)
7.57 M.A. Van Hove, W.H. Weinberg, C.-M. Chan: *Low-Energy Electron Diffraction*, Springer Ser. Surf. Sci., Vol. 6 (Spinger, Berlin, Heidelberg 1986)
7.58 F.O. Goodman, H.Y. Wachman: *Dynamics of Gas-Surface Scattering*. (Academic, New York 1976)
7.59 M. Head-Gordon, J.C. Tully, C.T. Rettner, C.B. Mullins, D.J. Auerbach: J. Chem. Phys. **94,** 1516 (1991)
7.60 M.A. Chesters, P. Gardner, E.M. McCash: Surf. Sci. **209,** 89 (1989)
7.61 A molecule can desorb from the surface in many different orientations, etc., however there are only a few configurations relative to the surface in which dissociation can occur
7.62 W.H. Weinberg, Y.-K. Sun: Science **253,** 542 (1991)
7.63 C.T. Rettner, E.K. Schweizer, H. Stein, D.J. Auerbach: Phys. Rev. Lett. **61,** 986 (1988)
7.64 C.T. Campbell, Y.-K. Sun, W.H. Weinberg: Chem. Phys. Lett. **179,** 53 (1991)
7.65 D.F. McMillen, D.M. Golden: Annu. Rev. Phys. Chem. **33,** 493 (1982)
7.66 C.B. Mullins, W.H. Weinberg: Unpublished results
7.67 D.E. Ibbotson, T.S. Wittrig, W.H. Weinberg: J. Chem. Phys. **72,** 4885 (1980)
7.68 J.R. Engstrom, W. Tsai, W.H. Weinberg: J. Chem. Phys. **87,** 3104 (1987)
7.69 M. Brookhart, M.L.H. Green: J. Organometal. Chem. **250,** 395 (1983)
7.70 R.A. Periana, R.G. Bergman: J. Am. Chem. Soc. **108,** 7332 (1986)
7.71 J.E. Demuth, H. Ibach, S. Lehwald: Phys. Rev. Lett. **40,** 1044 (1978)
F.M. Hoffmann, T.E. Felter, P.A. Thiel, W.H. Weinberg: Surf. Sci. **130,** 173 (1983)
7.72 R. Raval, M.A. Chesters: Surf. Sci. **219,** L505 (1989)
7.73 R. Raval, M.E. Pemble, M.A. Chesters: Surf. Sci. **210,** 187 (1989)
7.74 J.P. Collman, L.S. Hegedus, J.R. Norton, R.G. Finke: *Principles and Applications of Organotransition Metal Chemistry* (University Science Books, Mill Valley, CA 1987)
7.75 G.R. Schoofs, C.R. Arumainayagam, M.C. McMaster, R.J. Madix: Surf. Sci. **215,** 1 (1989)
7.76 C.N. Stewart, G. Ehrlich: J. Chem. Phys. **62,** 4672 (1975)
7.77 C.T. Rettner, H.E. Pfnur, D.J. Averbach: J. Chem. Phys. **84,** 4163 (1986)
7.78 H.P. Steinruck, A.V. Hamza, R.J. Madix: Surf. Sci. **173,** L571 (1986)
7.79 Unpublished results of C.B. Mullins and W.H. Weinberg
7.80 R.W. Verhoef, D. Kelly, C.B. Mullins and W.H. Weinberg: Surf. Sci. **287/288,** 94 (1993); Surf. Sci. **291,** L719 (1993)
7.81 J.D. Beckerle, Q.Y. Yang, A.D. Johnson, S.T. Ceyer: J. Chem. Phys. **86,** 7236 (1987)
7.82 J.D. Beckerle, A.D. Johnson, Q.Y. Yang, S.T. Ceyer: J. Chem. Phys. **91,** 5756 (1989)

Subject Index

ab initio force field calculations for vibrational spectra 65
acetaldehyde 17
acetonitrile 30
acetylene 35
acid-base theory of catalytic partial oxidation 20, 46
activated adsorption 187
  alkanes on Pt(110)(1 × 2) 261–265
  butane on Ir(110)(1 × 2) 256
  ethane on Ir(110)(1 × 2) 257–259
  Pauli repulsion 193, 201
  propane on Ir(110)(1 × 2) 255
  quantum theory for H_2 203
  role of translational energy in alkanes 268–271
  role of vibrational energy in alkanes 271–273
  rotational effects for H_2 229–232
  translational threshold 193–203
  vibrational enhancement for H_2 206, 207, 224–227
activation energy 186, 191
adsorption probability
  H_2 on Cu 193, 212
  measurement 240
agostic interactions 266
β-alkyl elimination 179
alkyl halides 136
angular distributions
  H_2 from Cu(110) and Cu(100) 195
ARUPS 90
  benzene on Pd(111) 102
  dichlorocylobutene on Pd(111) 96–100, 121
  ethene on Pd(111) 124
  ethyne on Pd(111) 116, 119
autocatalysis 177

benzene 36, 89–134
benzenethiol 59
π-bonding 32
butadiene 37, 78, 79, 94, 106, 130
t-butanethiol 59, 75
tert butanol 23
1-butene 37

carbonate 18
catalytic desulfurization 55
C_4H_4 93–100, 123, 127–131
chemical dynamics
  gas phase 185, 198
  gas-surface 185, 188
classical trajectories 199, 203
coadsorption
  C_2D_2 and C_6H_6 on Pd(111) 93, 102
  C_2D_2 and $C_4H_4Cl_2$ on Pd(111) 93
  NO and C_2H_2 on Pd(111) 101–103, 111, 113
  O and C_4H_6 on Ag(110) 130
  O and C_2H_2 on Pd(111) 94, 112
  O and $C_4H_4Cl_2$ on Pd(111) 94, 128–130
  S and C_2H_2 on Pd(111) 94, 111
  S and $C_4H_4Cl_2$ on Pd(111) 94, 111
C–S bond activation
  methanethiol 61
  thiophene 77
cyclization reactions 89–134
cycloaddition 41
cyclobutane 84
cyclohexanol 21
cyclohexanone 21
cyclopropane 169

deoxygenation
  2-propanol on Mo(110) 64
desorption dynamics of H_2
  angular distributions 214
  effect of surface motion 213–220
  internal states 224, 229–232
  velocity distributions 218
desulfurization
  effects of adsorbed sulfur 57
  general mechanism for thiolates 72–76

desulfurization (contd)
high surface area catalysts 77
Mo(110) 56–57
selectivity 62
detailed balance 194, 196, 198, 204, 209, 250
angular distributions 214–218
state resolved desorption 224, 229–232
velocity distributions 218–220
1,2 dibromopropane 142
dichlorocylobutene 93–100, 128
dihaloalkanes 160
2,5-dihydrothiophene 81
1,4-diiodobutane 163, 164
1,3-diiodopropane 165
diisobutylaluminum hydride 160
dimethylaluminum hydride 172
3,3 dimethyl butane 40
dissociative adsorption
alkanes on Ir(110)(1 × 2) 241
direct 249
methane on Ir(110)(1 × 2) 241
methane on Rh(111) 241
methane on tungsten 241
precursor mediated 249
trapping mediated 249

electronic effects 110, 125
α-elimination 180
β-elimination 21, 22, 25, 28, 44
γ-elimination 24, 73, 75
energy scaling
effect of surface temperature for Ar on Pt(111) 254
ethane trapping on Ir(110)(1 × 2) 252
ethane trapping on Pt(111) 253
etching 174
ethanol 21, 24
ethene 106, 108, 124
ethylamine 34
ethylene glycol 25
ethylene sulfide 82
ethylidyne 118, 124
ethyne 89–134
extended Huckel calculations 66, 83

formaldehyde 17
furan 37, 89, 94, 128

geometric effects 110–113, 125
Grignard reaction 138, 139, 175

H-D exchange 169
heterocylization 89, 94, 127–131

HREELS 90
CH_2CN on Ag(110) 32
dialkoxide of ethylene glycol on Ag(110) 28, 29
1,3-dibromopropane on Al(100) 161
dichlorocylobutene on Pd(111) 95
1,3-diiodopropane on Al(100) 161
ethylene glycol on Ag(110) 25
ethyne on Cu(111) and Cu(110) 125
ethyne on Pd(111) 111, 114, 118
1-iodopropane on Al(100) 141
isotopes of 2-propanol on Mo(110) 64–65
methanethiol on metals 63
methanethiol on Mo(110) 60
methanethiolate on Mo(110) 62–63, 66–67
methoxy on Mo(110) 69
H_2S
Ag(110) 59
Cu(111) 59
Mo(100) 59
Ni(110) 59
Pt(111) 59
Rh(100) 59
Ru(110) 59
β-hydride elimination 137, 138, 156, 157, 160, 167, 177
γ-hydride elimination 179
hydrogen 135
ortho-hydrogen 191
para-hydrogen 191
hydrogen cyanide 32
hydrogenolysis in desulfurization 72

impact parameter 210
infrared spectroscopy
isobutyl on Al(100) 145
methoxy on Mo(110) 69
methyl thiolate on Mo(110) 69
integrated desorption mass spectrometry 147
1,3-diiodopropane on Al(100) 167
C_6 hydrocarbons 170
3-iodo-1-propene on Al(100) 168
1-iodohexane on Al(100) 148
intermediates (also see metallacyle)
acetylide 35
alkane dissociation on Ir(110)(1 × 2) 243
alkoxides 21–29
alkyl groups 135, 152
tert butoxy 23, 45
CH_2CN 32
ethoxy 21
Fisher carbene 33
formate 17, 27

hydroxyl 20
isobutyl 136, 158, 159
metallacycle 22, 24, 43
methoxy 69
methyl thiolate 60–63, 66–69
η_2 methylene dioxy 17
phenyl thiolate 71, 74, 76
propoxide 64–65, 74
radical 172
trimethylene methane 35
internal state distributions for desorption of H_2 197, 201
1-iodohexane 149
isobutene 44
isobutyliodide 146, 155
isotopic exchange 63, 73
isotopic labelling 25, 28, 44, 64–65, 90, 152

kinetic isotope effect 209, 264
kinetics
isobutyliodide on Al(100) 156
metal film growth on Al(100) 176

LEED 90
ethyne on Pd(111) 94, 100
short chain thiolates 71

mechanistic schemes
allyl alcohol oxidation on silver 23
benzene thiol on Mo(110) 74
butadiene oxidation on Ag(110) 41
t-butoxy decomposition on Ag(110) 24
cyclohexene oxydehydrogenation on Ag(110) 37
1,4-diiodobutane on Al(100) 163
ethylamine oxidation on Ag(110) 35
formaldehyde oxidation on Ag(110) 18
i-iodopropane on Al(100) 153
methylenecyclopentane formation on Al(100) 171
propene oxidation on Rh(111) 44
thiols on Mo(110) 72
trimethylene sulfide on Mo(110) 83
metallacycle 94–100, 108, 130, 137, 160, 162, 165, 167, 170
metallorganic chemical vapor deposition 135, 175
metathesis 179
methanethiol
Cu(111), Cu(110) 59
Fe(100) 59
Mo(110) 60–63
Ni(100) 59
Ni(110) 59
Pt(111) 59
W(211) 59
β-methyl elimination 173
molecular beams 201, 220–224
seeded 205, 240
MoS_2 55

neopentyliodide 172
NEXAFS 90
benzene on Pd(111) 104
CO_3 on Ag(110) 19
dichlorocylobutene on Pd(111) 96–100
ethene on Pd(111) 124
ethyne on Pd(111) 116–118, 120
O_2 on Ag(110) 10
phenyl thiolate on Mo(110) 70–71
thiophene on Pt(111) 78
norbornene 38
normal energy scaling 194, 253, 273
H_2 on Cu(111), (110) and (100) 202, 211

olefin insertion 180
organometallic chemistry
analogies 77, 89, 96, 100, 108, 114, 115, 119
analogies to surface reactions 152, 180
homogeneous catalysis 89, 100, 108
oxidation
Au 46–47
Pd 48
oxygen addition reactions
3,3-dimethyl butene on Ag(111) 40
ethylene on silver 5, 41
isobutene on Rh(111) 45
norbornene on Ag(110) 40
norbornene on Rh(111) 45
propylene on Rh(111) 43
styrene on Ag(111) and Ag(110) 40
oxygen adsorption
Ag(110) 6, 7, 9, 12

partial oxidation
acetonitrile on Ag(110) 30
alcohols 21
amines on Ag(110) 34–35
hydrocarbon acids on Ag(110) 35–37
scavenging reactions 26
physisorbed precursor 140
poisons 110–113
potential energy surface 198–201
promoters 110–113
propadiene 169

2-propanol 64–65
n-propyl iodide 142
propylene 43
proton transfer 13–37

reactivity of O(a) on Ag(110)
  Broensted acidity 19
  nucleophilicity 17, 42
recombinative desorption 194, 195
reductive coupling 179
reductive elimination 164, 179

site blocking 158
state resolved desorption 224–227
state resolved scattering 220–224
  effect of rotation on vibrational excitation of H_2 233
  inelastic scattering 221
  time of flight of D_2 from Cu(111) 222–223, 225
  vibrational excitation 221–224
steric effects 25
STM
  O on Ag(110) 8
structure sensitive reaction 109, 122–127
styrene 40
sulfite 13
supported catalysts 89, 105–110
supported metal catalysts 55, 56
surface corrugation 193

temporal analysis of products (TAP) 41
tetrahydrothiophene 85
thiols, reactions on transition metal surfaces 58–76
thiophene
  decomposition 77
  hydrogenation 78
TPD
  hydrogen on Ir(110) 244
  neopentyliodide on Al(100) 173
TPRS 90, 137
  acetonitrile oxidation on Ag(110) 31–34
  alkanes on Ir(110)(1 × 2) 241, 243, 245
  benzene on Pd(111) 103
  coadsorbed C_2D_2 and $C_4H_4Cl_2$ on Pd(111) 92
  coadsorbed C_2D_2 and C_6H_6 on Pd(111) 93
  coadsorbed hydrogen and alkanes on Ir(110) 246, 247
  coadsorbed NO and C_2H_2 on Pd(111) 101–103
  coadsorbed O_2 and C_2H_2 on Pd(111) 94, 112
  computer multiplexing 39
  1,4-diiodobutane on Al(100) 163
  ethene on Pd(111) 124
  ethylene glycol oxidation on Ag(110) 26, 28
  ethyne on Cu(110) 125
  ethyne on Pd(110) and Pd(100) 122
  ethyne on Pd(111) 91, 109, 114, 122
  ethyne on supported Pd catalysts 106
  1-iodo-2-methylpropane on Al(100) 156
  1-iodopropane on Al(100) 153
  mass spectrometry 75
  methanethiol on Mo(110) 73
  norbornene oxidation on Ag(110) 39
  propyliodide on Al(100) 144
  triisobutylaluminum on Al(100) 156
transition state 158, 159
trapping dynamics 250, 251
  argon on Pt(111) 251
  ethane on Ir(110)(1 × 2) 251, 252
  ethane on Pt(111) 251
  NO on Ag(111) 251
  xenon on Pt(111) 251
triethylamine alane 176
triisobutylaluminum 145
trimethylaluminum 172
trimethylene sulfide 82
trineopentylaluminum 180
tunneling 194, 210

velocity distributions 196
  H_2 from Cu(110) 196
vinylidene 118–121

water 20

XPS 90
  benzene on Pd(111) 104
  dichlorocylobutene on Pd(111) 96–98
  2,5-dihydrothiophene 61
  ethene on Pd(111) 124
  ethyne on Pd(111) 120
  methanethiol on Mo(110) 60–61

Springer-Verlag and the Environment

We at Springer-Verlag firmly believe that an international science publisher has a special obligation to the environment, and our corporate policies consistently reflect this conviction.

We also expect our business partners – paper mills, printers, packaging manufacturers, etc. – to commit themselves to using environmentally friendly materials and production processes.

The paper in this book is made from low- or no-chlorine pulp and is acid free, in conformance with international standards for paper permanency.

Printing: Mercedesdruck, Berlin
Binding: Buchbinderei Lüderitz & Bauer, Berlin